Integrated Drainage Systems Planning and Design for Municipal Engineers

Urban water management has to take an integrated approach that prioritizes sustainable drainage systems (SuDS) over gray infrastructure. This book elaborates on the planning and evaluation of pipework drainage systems with a focus on modern-day constraints to deliver a solution that favors sustainability as the overarching goal. The book includes a technical section on design of gray and green infrastructure, considering the total lifecycle costs of drainage systems. Advanced computer simulation techniques are discussed after covering the derivation of both standard and empirical equations for appropriate hydrology and hydraulics. The book provides an incorporation of reliability analyses for both green and gray infrastructure starting with techniques for forecasting flows, hydraulic performance, and lifecycle costs. The work also involves 3-D modeling, geospatial and big data analysis, and how these techniques are applied into city management—particularly beneficial to municipal engineers who are increasingly becoming involved in mapping the underground. Soil mechanics and subsurface drainage systems are analyzed and structural aspects of sewers are included. Finally, soil behavior in shear, retaining wall structures, and tunneling is briefly featured in the book.

This book will be of interest to (under)graduate and postgraduate engineering students, drainage engineers, urban planners, architects, water engineers, developers, construction contractors, and municipal engineers.

Patrick Ssempeera possesses a master's degree in Engineering (Civil) and BSc Civil Engineering, both from Makerere University, Kampala UG. Ever since Patrick Ssempeera graduated from the College of Engineering, Design, Art, and Technology (CEDAT), he has worked and trained on various civil engineering projects as a project manager, project supervisor, and design engineer. He has gained 11 years' practical experience in the fields of project management, project appraisal, project planning, project design, tender preparation, tender evaluations, project costing, construction procurement, and contract management of various projects such as roads and highways, drainage systems, water supply systems, and buildings in line with various standards such as British Standards (BS), Eurocodes, The International Federation of Consulting Engineers (FIDIC) Conditions of Contract—The Red Book, and The World Bank Conditions of Contract. Patrick Ssempeera has had great opportunities to work with both contractors and consultants. He is a Chartered Civil Engineer and a Member of the Institution of Civil Engineers, UK.

Integrated Drainage Systems Planning and Design for Municipal Engineers

Patrick Ssempeera

CRC Press
Taylor & Francis Group
Boca Raton   London   New York

CRC Press is an imprint of the
Taylor & Francis Group, an **informa** business

A BALKEMA BOOK

Designed cover image: © Shutterstock

First published 2023
by CRC Press/Balkema
Schipholweg 107C, 2316 XC Leiden, The Netherlands
e-mail: enquiries@taylorandfrancis.com
www.routledge.com – www.taylorandfrancis.com

CRC Press/Balkema is an imprint of the Taylor & Francis Group, an informa business

© 2023 Patrick Ssempeera

The right of Patrick Ssempeera to be identified as author of this work has been asserted in accordance with sections 77 and 78 of the Copyright, Designs and Patents Act 1988.

All rights reserved. No part of this book may be reprinted or reproduced or utilised in any form or by any electronic, mechanical, or other means, now known or hereafter invented, including photocopying and recording, or in any information storage or retrieval system, without permission in writing from the publishers.

Although all care is taken to ensure integrity and the quality of this publication and the information herein, no responsibility is assumed by the publishers nor the author for any damage to the property or persons as a result of operation or use of this publication and/or the information contained herein.

Library of Congress Cataloging-in-Publication Data
Names: Ssempeera, P., author.
Title: Integrated drainage systems planning and design for municipal engineers/P. Ssempeera.
Description: Boca Raton : CRC Press 2023. | Includes bibliographical references and index.
Identifiers: LCCN 2022052381 (print) | LCCN 2022052382 (ebook) |
ISBN 9781032186290 (hbk) | ISBN 9781032186559 (pbk) | ISBN 9781003255550 (ebk)
Subjects: LCSH: Drainage. | Sewerage. | Urban hydrology. | Municipal engineering.
Classification: LCC TC970 .S66 2023 (print) | LCC TC970 (ebook) | DDC 628/.2—dc23/eng/20230103
LC record available at https://lccn.loc.gov/2022052381
LC ebook record available at https://lccn.loc.gov/2022052382

ISBN: 9781032186290 (hbk)
ISBN: 9781032186559 (pbk)
ISBN: 9781003255550 (ebk)

DOI: 10.1201/9781003255550

Typeset in Times New Roman
by Newgen Publishing UK

To Joan, Peace, Alex, Patricia, Mark, Martin, Oprah, and Charlotte.

To Joan Grace Nansubuga Naloongo for her understanding.

Contents

List of figures   xxv
List of tables   xxix
Preface   xxxi
Acknowledgments   xxxv

1 Introduction   1

1.1 Origins of present-day practice 1
1.2 Why are drainage systems so important? 2
1.3 Storm sewer system 2
1.4 History of sustainable drainage systems 2
1.5 Role of government through public authorities 3
    1.5.1 The effect of urbanization 4
1.6 General features of standard pipework drainage solutions 4
1.7 Outline of several types of SuDS 5
    1.7.1 Tree planting 5
    1.7.2 Tree trenches 5
    1.7.3 Bioretention gardens 5
    1.7.4 Infiltration systems 6
    1.7.5 Porous pavers 6
    1.7.6 Swales 6
    1.7.7 Green spaces 6
    1.7.8 Green streets 6
    1.7.9 Downspout disconnection 7
    1.7.10 Vegetated kerb extension 7
    1.7.11 Rainwater harvesting systems 7
    1.7.12 Green roofs (vegetated rooftops) 7
    1.7.13 Detention ponds 7
    1.7.14 Retention ponds 8
    1.7.15 Soakaways 8

　　　　　1.7.16　Planter boxes 8
　　　　　1.7.17　Artificial rills 8
　　1.8　Combined and uncombined systems 8
　　1.9　Modern drainage master planning projects 9
　　1.10　Conclusion 11
　　References 11

2　Drainage planning and evaluation process　　　　　　　　　　　　　13

　　2.1　Introduction 13
　　2.2　Importance of drainage systems 15
　　2.3　Sources of funding 15
　　　　2.3.1　*Direct funding through loans and tax revenues 16*
　　　　2.3.2　*Fee-based stormwater utility 16*
　　　　2.3.3　*Collaborating with nongovernmental organizations 16*
　　2.4　Constraints and data collection 16
　　　　2.4.1　*Public participation 16*
　　　　2.4.2　*Required data and documents review 17*
　　　　　　2.4.2.1　*Meteorological data 17*
　　　　　　2.4.2.2　*Geotechnical and site investigations 17*
　　　　　　2.4.2.3　*Soils and geological data 18*
　　　　　　2.4.2.4　*Socioeconomic data (demographics) 18*
　　　　　　2.4.2.5　*Topographical surveys and information systems 18*
　　　　　　2.4.2.6　*Physical plans/land use plans 19*
　　　　　　2.4.2.7　*Past study reports 19*
　　　　　　2.4.2.8　*Maps 19*
　　　　　　2.4.2.9　*Utility service corridors 19*
　　　　　　2.4.2.10　*Site visits to take area photographs 20*
　　　　2.4.3　*Selection of appropriate routes 20*
　　　　2.4.4　*Site survey and evaluation 21*
　　　　2.4.5　*Defining the catchment characteristics 22*
　　　　2.4.6　*Environmental and social issues 23*
　　2.5　Drainage systems planning 26
　　　　2.5.1　*Setting goals 26*
　　　　　　2.5.1.1　*Stormwater quantity reduction 26*
　　　　　　2.5.1.2　*Pollutant removal 26*
　　　　　　2.5.1.3　*Peak flow reduction 27*
　　　　　　2.5.1.4　*Reduced costs 28*
　　　　　　2.5.1.5　*Low-carbon design 28*
　　　　　　2.5.1.6　*Construction materials 30*
　　　　　　2.5.1.7　*Land stabilization and embankments 30*
　　　　　　2.5.1.8　*Hydrologic and hydraulics methods 30*
　　　　　　2.5.1.9　*Other constraints 31*
　　　　2.5.2　*Defining drainage networks 31*

		2.5.3	Cost of SuDS and gray infrastructure 34
2.6	Software applications 35		
2.7	The decision-making process 35		
	2.7.1	Reliability and sustainability 39	
	2.7.2	Hydraulics and flow routing 40	
2.8	Case studies 45		
	2.8.1	Case study 2.1: Kabale municipality drainage master planning, Uganda, 2016 45	
2.9	Conclusion 49		

References 49

Further reading 50

3 Planning for sustainability   51

- 3.1 Introduction 51
- 3.2 Greenfield and retrofitting 54
- 3.3 What are sustainable drainage systems and how do they work? 54
- 3.4 Approaches to develop SuDS 55
  - 3.4.1 Review standards and codes of practice 55
  - 3.4.2 Identify funding opportunities to install and maintain SuDS 56
  - 3.4.3 Plan for maintenance 57
  - 3.4.4 Identify high-visibility projects 57
- 3.5 The role of sustainable drainage systems 58
- 3.6 Achievable SuDS (site-specific SuDS) 58
  - 3.6.1 Topography/slope 59
  - 3.6.2 Depth to water table and to bedrock 59
  - 3.6.3 Soils 60
  - 3.6.4 Potential hotspots 61
  - 3.6.5 Maximum drainage area 61
  - 3.6.6 Climate of the area 61
  - 3.6.7 Native species/vegetation 61
  - 3.6.8 Space 61
  - 3.6.9 Sacred sites and culture 62
  - 3.6.10 Integrated factors 62
- 3.7 SuDS application and feasibility potential going by area physical plan 62
- 3.8 Planning, design, and development of SuDS 63
  - 3.8.1 Infiltration systems 65
    - 3.8.1.1 Description and purpose 65
    - 3.8.1.2 Potential limitations for infiltration systems 67
    - 3.8.1.3 How does an infiltration system work? 67
    - 3.8.1.4 Infiltration systems' design criteria 67
    - 3.8.1.5 Probable cost of infiltration systems 74

	3.8.2	Bioretention gardens 74
		3.8.2.1 Description and purpose 74
		3.8.2.2 How does a bioretention work? 75
		3.8.2.3 Bioretention design criteria 76
		3.8.2.3.1 Bioretention design goals and objectives 78
		3.8.2.3.2 General filtration design process for sizing the bioretention garden 78
		3.8.2.3.3 Sizing the bioretention, North American standard 79
		3.8.2.3.4 Summary of the bioretention design criteria, North American standard 80
		3.8.2.3.5 Probable cost 81
	3.8.3	Ponds/basins 81
	3.8.4	Detention ponds/basins (dry ponds/basins) 82
		3.8.4.1 Description and purpose of detention ponds/basins 82
		3.8.4.2 Features and benefits of surface detention system 83
		3.8.4.3 Features and benefits of below surface (subsurface) detention system 84
		3.8.4.4 Potential limitations for detention systems 84
		3.8.4.5 Environmental benefits of detention ponds 84
		3.8.4.6 How does a detention pond work? 85
		3.8.4.7 Detention ponds' design criteria 85
		3.8.4.8 Probable detention pond cost 85
	3.8.5	Retention ponds/basins 86
		3.8.5.1 Description and purpose of retention ponds 86
		3.8.5.2 Integrated on surface detention and retention system 86
	3.8.6	Vegetated/bio-swales 86
		3.8.6.1 Description and purpose of swales 86
		3.8.6.2 How does a vegetated swale work? 87
		3.8.6.3 Potential limitations of swales 88
		3.8.6.4 Variations in bio-swale designs 88
		3.8.6.5 Probable cost 89
	3.8.7	Reforestation and afforestation 89
		3.8.7.1 Description and purpose 89
		3.8.7.2 How does reforestation and afforestation work? 90
		3.8.7.3 Potential afforestation and reforestation limitations 90
		3.8.7.4 Afforestation and reforestation design criteria 90
	3.8.8	Planter boxes 90
		3.8.8.1 Description and purpose 90
		3.8.8.2 Planter box probable cost 91
	3.8.9	Tree trenches 91
		3.8.9.1 Description and purpose 91
		3.8.9.2 Tree trench design criteria 93
		3.8.9.3 Potential limitations of tree trenches 93
		3.8.9.4 Probable cost for tree trenches 93

   3.8.10 Rainwater harvesting systems 94
     3.8.10.1 Description and purpose 94
     3.8.10.2 Benefits of rainwater harvesting 95
     3.8.10.3 Limitations of potential rainwater harvesting systems 96
     3.8.10.4 How does a rainwater harvesting system work? 96
     3.8.10.5 Rainwater harvesting systems design criteria 96
     3.8.10.6 Probable cost 97
   3.8.11 Open spaces 98
   3.8.12 Porous/permeable pavers 98
     3.8.12.1 Description and purpose 98
     3.8.12.2 Probable cost 99
   3.8.13 Downspout disconnection 100
     3.8.13.1 Description and purpose 100
     3.8.13.2 Potential limitations of downspout disconnection 101
     3.8.13.3 How does the downspout disconnection work? 101
     3.8.13.4 Downspout disconnection design criteria: What are the factors to consider? 101
     3.8.13.5 Probable cost 101
   3.8.14 Vegetated kerb extensions 101
   3.8.15 Green roofs 102
     3.8.15.1 Description and purpose 102
     3.8.15.2 Benefits and effectiveness of green roofs 103
     3.8.15.3 Potential limitations of green roofs 104
     3.8.15.4 Key design features for green roofs 104
     3.8.15.5 Probable cost of green roofs 105
3.9 Stormwater quality improvement and quantity reduction potential for different SuDS types 105
3.10 River restoration programs 105
3.11 Flood control structures 110
3.12 Conclusion 111
References 112
Further reading 114

4 Useful hydrology 115

4.1 Introduction 115
4.2 Hydrologic cycle 116
4.3 Hydrologic models, tools, and techniques 116
4.4 Rainfall analysis 118
   4.4.1 Average recurrence interval 118
   4.4.2 What does a design year flood such as a 100-year flood mean? 119
   4.4.3 Runoff estimation 119
     4.4.3.1 Hydrologic methods 120

xii Contents

                        4.4.3.1.1   EPA SWMM 120
                        4.4.3.1.2   Rational method 120
                        4.4.3.1.3   UK modified rational method 120
                        4.4.3.1.4   The soil conservation service curve number method 122
              4.4.3.2   Nonconventional design methods 129
    4.5   Statistical models and probability distribution functions 130
        4.5.1   Applications of continuous probability distribution functions 131
              4.5.1.1   Frequency and return period 131
              4.5.1.2   Forecasting behaviors 133
              4.5.1.3   Extreme value analysis 134
    4.6   Case studies and worked examples 136
        4.6.1   Case study 4.1: Generating IDFs for the rational method, Kabale Municipal Council, Western Uganda (2016) 136
        4.6.2   Case study 4.2: Kabale municipality rainfall forecasting with use of two-parameter Weibull distribution, Western Uganda (2016) 141
              4.6.2.1   Findings 144
              4.6.2.2   Analysis implications—decisions that followed 144
        4.6.3   Case study 4.3: Extreme value analysis for river Kiruruma in Kabale, Uganda (2016) 145
              4.6.3.1   Conclusion 148
              4.6.3.2   Recommendation from the analysis 148
    4.7   How to deal with runoff quality 149
    4.8   SuDS hydrology 150
    4.9   Conclusion 151
References 151
Further reading 152

5 Hydraulics design principles   153

    5.1   Introduction 153
    5.2   Hydraulic design—overview 154
    5.3   Energy equations 156
        5.3.1   Relevance of energy equations 157
        5.3.2   Derivation of key energy equations 158
              5.3.2.1   Specific energy 160
        5.3.3   Normal depth used for design and analysis 163
    5.4   Flow profiles/regimes for a sustainable design 163
    5.5   The momentum principle 170
        5.5.1   Application of momentum equations 174
    5.6   Case studies with worked examples 179
        5.6.1   Case study (5.1): Typical design project, proposed warehouse park on plot 31, Namanve, Kampala, Uganda (2021) 179

    5.6.1.1 Overview 179
    5.6.1.2 Evaluation 179
    5.6.1.3 General assumptions, conditions, and guiding design criteria 181
    5.6.1.4 Option A 182
    5.6.1.5 Option B 183
    5.6.1.6 Design 183
        5.6.1.6.1 Option A—Part A: Storm sewers 184
    5.6.1.7 Analysis 190
    5.6.1.8 Conclusion and recommendations 191
  5.6.2 Case study (5.2): Typical audit and design project for collapsed retaining wall following flash floods, Kitende, Kampala, Uganda (August 14, 2021) 191
    5.6.2.1 Overview 191
    5.6.2.2 Problem evaluation 191
    5.6.2.3 Justification of the probable cause 194
    5.6.2.4 Drainage design 194
    5.6.2.5 Remodeling the U-drain 194
    5.6.2.6 Analysis 196
    5.6.2.7 Conclusion and recommendations 197
    5.6.2.8 Final drainage scheme route and retaining wall 198
  5.6.3 Worked example 5.1: Application of Equation (5.2) 198
5.7 Conclusion 200
Reference 201
Further reading 201

6 Flow routing techniques   203

 6.1 Introduction 203
 6.2 Hydraulic routing techniques 204
 6.3 Applications of hydraulic flow routing techniques 205
   6.3.1 Evaluating the impact of flash floods 205
   6.3.2 Predicting the arrival time of floods 206
   6.3.3 Urban drainage design 206
   6.3.4 Analyzing flood control system's safety 206
 6.4 Derivation of hydraulic flow routing techniques 206
   6.4.1 Deriving the continuity equation 209
   6.4.2 Deriving the momentum equation 210
      6.4.2.1 The grid 212
 6.5 Steady flow 212
 6.6 Kinematic wave routing technique 213
   6.6.1 Worked example—Kinematic wave routing technique 215
 6.7 Dynamic wave routing technique 219

- 6.7.1 *How to solve the full 1-D Saint Venant equations using the method of characteristics (finite-difference scheme)* 221
- 6.7.2 *Analysis of dynamic wave routing technique* 230
  - 6.7.2.1 *Understanding energy and momentum equations as we route flows in the design* 230
- 6.7.3 *MATLAB simulation* 232
- 6.8 Comparison of hydraulic routing techniques 233
  - 6.8.1 *Kinematic wave routing* 233
  - 6.8.2 *Dynamic wave routing* 234
- 6.9 Stormwater runoff hydraulic modeling and simulation 234
  - 6.9.1 *Available software used to develop stormwater models* 236
- 6.10 Conclusion 237
- References 237
- Further reading 237

7   Useful topics in soil mechanics   239

- 7.1 Introduction 239
- 7.2 Subsurface drainage 240
  - 7.2.1 *Introduction* 240
  - 7.2.2 *Common subsurface drainage terminologies* 241
  - 7.2.3 *Primary materials used in construction of subsurface drainage systems* 242
    - 7.2.3.1 *Nonwoven geotextile fabrics* 242
    - 7.2.3.2 *Perforated pipes* 242
    - 7.2.3.3 *Filter materials* 242
  - 7.2.4 *Subsurface drain types* 242
    - 7.2.4.1 *French drains* 243
    - 7.2.4.2 *Stabilization trenches* 244
  - 7.2.5 *Road construction subsurface drain categories* 244
    - 7.2.5.1 *Subsoil drains* 244
    - 7.2.5.2 *Subpavement drains* 244
    - 7.2.5.3 *Foundation drains* 244
  - 7.2.6 *Drainage mats in road construction* 245
  - 7.2.7 *Filter material differences* 245
- 7.3 Subsurface drainage design criteria 246
  - 7.3.1 *Subsurface drainage special design guidelines or considerations* 246
  - 7.3.2 *Maintenance design considerations for subsurface drainage* 246
- 7.4 Drainage and groundwater recharge 247
- 7.5 Micro-tunneling, pipe jacking, and underground drainage systems 247
- 7.6 Conventional tunneling 248
- 7.7 Soil behavior in shear 249

- 7.7.1 Horizontal component of stress 249
- 7.7.2 Mohr–Coulomb model 250
- 7.8 Flood defense structures 253
  - 7.8.1 Introduction 253
  - 7.8.2 On design of flood control structures 254
    - 7.8.2.1 Seepage control in flood control earthen embankment structures 255
    - 7.8.2.2 Control of water movement in basements 258
    - 7.8.2.3 Concrete flood defense structures 258
    - 7.8.2.4 Coastal flood defenses 260
- 7.9 Retaining walls 260
  - 7.9.1 Goal setting for retaining wall design 260
  - 7.9.2 Factors influencing the choice of retaining wall 261
    - 7.9.2.1 Soil 261
    - 7.9.2.2 Location 261
    - 7.9.2.3 Drainage 262
    - 7.9.2.4 Design 262
  - 7.9.3 Types of retaining walls 263
    - 7.9.3.1 Gravity retaining wall 263
    - 7.9.3.2 Cantilever retaining wall 264
    - 7.9.3.3 Segmental retaining walls 264
    - 7.9.3.4 Counterfort retaining walls 265
    - 7.9.3.5 Sheet or bored pile walls 265
    - 7.9.3.6 Gabion mesh walls 266
- 7.10 Case study 7.1: Collapsed retaining wall following flash floods—continuation of case study (5.2): Typical audit and design project (Section 5.7.2) (2021) 267
- 7.11 Conclusion 268
- References 268
- Further reading 268

8 Structural aspects of storm sewers   269

- 8.1 Introduction 269
  - 8.1.1 Differential settlement 270
  - 8.1.2 Excessive pressure 270
  - 8.1.3 The expansion of the joints 271
  - 8.1.4 Loading conditions 271
  - 8.1.5 Restraining thermal expansion 271
  - 8.1.6 Concrete shrinkage 272
  - 8.1.7 Chemical attacks due to corrosion of storm sewers 272
- 8.2 Pipe loading conditions/patterns 272
  - 8.2.1 Types of loads that act on buried pipes 273
  - 8.2.2 Theories applicable in pipe load calculations 273

- 8.3 Matching imposed loads to pipe strength 273
- 8.4 Loads due to earth overburden 274
  - 8.4.1 Derivation of Marston's equation 275
  - 8.4.2 Variation in the use of Marston's equation 277
    - 8.4.2.1 The complete embankment case or wide-trench case (positive projection) 278
    - 8.4.2.2 The incomplete wide-trench case (positive projection) 279
      - 8.4.2.2.1 The settlement-deflection ratio r_{sd} and projection ratio p 280
    - 8.4.2.3 Negative projection cases 285
      - 8.4.2.3.1 Complete case negative projection 285
      - 8.4.2.3.2 Incomplete case negative projection 286
      - 8.4.2.3.3 Settlement-deflection ratio for negative projection 286
    - 8.4.2.4 Zero projection case 288
- 8.5 Loads superimposed by vehicle wheels 288
  - 8.5.1 AASHTO LRFD design method 289
- 8.6 Internal water load 289
- 8.7 Pipe strength and bedding classes 290
  - 8.7.1 Concrete pipe design—conclusion 291
  - 8.7.2 Worked example 8.1 for illustrative purposes (loading case: Ordinary [normal] trenches) 291
    - 8.7.2.1 Solution 291
- 8.8 BS 9295:2020 Guide to the structural design of buried pipes 293
- 8.9 Storm sewer construction materials 294
  - 8.9.1 Reinforced concrete pipes 294
  - 8.9.2 Corrugated steel pipes 294
  - 8.9.3 Unplasticized polyvinyl chloride 295
  - 8.9.4 Vitrified clay drainage (clayware) pipes 295
  - 8.9.5 Brickwork 295
  - 8.9.6 High-density polyethylene pipes 295
  - 8.9.7 Ductile iron pipes 296
  - 8.9.8 Cast iron pipes 296
- 8.10 Conclusion 296
- References 297
- Further reading 297

9 Evaluation of integrated drainage system designs — 299

- 9.1 Introduction 299
- 9.2 Advantages and potential disadvantages of SuDS 300
  - 9.2.1 Advantages of SuDS 300
    - 9.2.1.1 Reduction in stormwater runoff quantities 300

                9.2.1.2   Reduction in runoff velocities 301
                9.2.1.3   Improve aesthetics 301
                9.2.1.4   Recreational activities 301
                9.2.1.5   Reduce greenhouse gases 301
                9.2.1.6   Provide shades for the city dwellers 302
                9.2.1.7   Regulate temperatures 302
                9.2.1.8   Enormously contribute to climate action policy 302
                9.2.1.9   Educational opportunities 302
                9.2.1.10   Recharges groundwater 302
                9.2.1.11   Harvesting rainwater for domestic/commercial use 303
                9.2.1.12   Biodiversity and resilience 303
                9.2.1.13   Preserving ecological sites 303
                9.2.1.14   Reduce pollutants 303
                9.2.1.15   Enhancing conservation practices for sustainable development 304
                9.2.1.16   Ease of working with underground utility service 304
                9.2.1.17   SuDS foster collaboration with NGOs 304
                9.2.1.18   Ease of funding compared with traditional gray infrastructure 304
                9.2.1.19   Economic benefits 305
        9.2.2   Potential disadvantages of SuDS 305
                9.2.2.1   Tropical climates 305
                9.2.2.2   Engineering, art, and continuing education 305
                9.2.2.3   Heavy initial cost 305
                9.2.2.4   Large space requirements 306
                9.2.2.5   Compliance with regulatory frameworks 306
9.3   Integrated Drainage system reliability analysis 306
        9.3.1   Introduction 306
        9.3.2   Drainage system model reliability 307
        9.3.3   Sensitivity analysis 308
                9.3.3.1   Model inputs variables for uncertainty checks 308
                9.3.3.2   Model output variables 308
                9.3.3.3   Using what-if analysis to model system reliability 309
        9.3.4   Environmental lifecycle assessment 309
                9.3.4.1   Introduction 309
                9.3.4.2   Factors considered in LCA for drainage schemes 311
                           9.3.4.2.1   Capital cost 311
                           9.3.4.2.2   Project design life 311
                           9.3.4.2.3   Material service life 312
                           9.3.4.2.4   Inflation/interest factor 312
                           9.3.4.2.5   Maintenance cost 313
                           9.3.4.2.6   Rehabilitation cost 313
                           9.3.4.2.7   Replacement cost 313
                           9.3.4.2.8   Residual value 313
                9.3.4.3   The ASTM procedure 314
                           9.3.4.3.1   Identify objectives, alternatives, and constraints 314
                           9.3.4.3.2   Establish basic criteria 314
                           9.3.4.3.3   Compile data 314

		9.3.4.3.4	Compute LCA for each material, system, or structure *314*
		9.3.4.3.5	Evaluate results *315*
	9.3.4.4	ASTMC-1131 formula *315*	
		9.3.4.4.1	Residual value *315*
		9.3.4.4.2	Inflation factor *315*
		9.3.4.4.3	Maintenance cost *316*
	9.3.4.5	Worked example—for illustrative purposes *316*	
	9.3.4.6	Analyzing reliability goals—typical example for illustrative purposes *318*	
	9.3.4.7	Design life and service life used as reliability indicators *322*	

9.4 Achieving the safe net-zero goal *323*
    9.4.1 Estimating carbon offsets for integrated drainage systems *324*
        9.4.1.1 Embodied carbon *326*
            9.4.1.1.1 Introduction *326*
            9.4.1.1.2 Embedded carbon from concrete constituents *326*
                9.4.1.1.2.1 Portland cement *326*
                9.4.1.1.2.2 Aggregates *327*
                9.4.1.1.2.3 Other major constituents *327*
            9.4.1.1.3 Embedded carbon from concrete constituents, C32/40 mix *327*
            9.4.1.1.4 Variations of eCO2 in concrete strength and mix design *327*
            9.4.1.1.5 Embedded carbon in reinforcement *327*
        9.4.1.2 What to do in drainage schemes to reduce carbon emissions? *329*
        9.4.1.3 Operational carbon in drainage systems *330*

9.5 Embodied carbon worked example for illustrative purposes *330*
    9.5.1 Analysis of example *331*

9.6 Conclusion *332*

References *332*

Further reading *333*

10 Construction and management   **335**

10.1 Introduction *335*

10.2 Procedural management in drainage construction *335*

10.3 Excavation of trenches for storm sewers *336*
    10.3.1 Dewatering of excavations *338*
    10.3.2 Excavation techniques *339*
        10.3.2.1 Drill and blast application *339*
        10.3.2.2 Conventional methods: Excavators +/− rock breakers application *340*
        10.3.2.3 Trenching application (hand excavation) *340*
    10.3.3 Excavation limitations *340*

10.4	Excavation supports 341		
10.5	Pipe laying and jointing 343		
10.6	Construction and environmental management 345		

    10.6.1 Environment and social impact assessment for drainage systems 345
        10.6.1.1 Introduction 345
        10.6.1.2 Potential project benefits of drainage schemes 345
        10.6.1.3 Potential project negative impacts 346
            10.6.1.3.1 Potential negative impacts of mobilization phase 346
                10.6.1.3.1.1 Loss of vegetation 346
                10.6.1.3.1.2 Nuisance from dust and noise 346
                10.6.1.3.1.3 Increased incidence of diseases 346
            10.6.1.3.2 Potential negative impacts of construction phase 346
                10.6.1.3.2.1 Traffic flow interruption 346
                10.6.1.3.2.2 Air and noise pollution 346
                10.6.1.3.2.3 Erosion 347
                10.6.1.3.2.4 Contamination of water sources (water pollution) 347
                10.6.1.3.2.5 Increased disease prevalence in quarry areas 347
                10.6.1.3.2.6 HIV/AIDS and other sexually transmitted diseases 347
                10.6.1.3.2.7 Increased competition for water sources 348
                10.6.1.3.2.8 Flooding potential 348
            10.6.1.3.3 Presence of environmentally sensitive areas 348
                10.6.1.3.3.1 Impacts due to waste 348
                10.6.1.3.3.2 Loss of flora and fauna 348
            10.6.1.3.4 Resettlement of people 348
        10.6.1.4 Environmental management system 348
        10.6.1.5 Environment and social impact assessment 350
            10.6.1.5.1 Scope of environmental activities 351
        10.6.1.6 Environmental screening 351
            10.6.1.6.1 Overall impacts assessment 351
    10.6.2 Environmental and social management plan 355
        10.6.2.1 Construction phase 355
        10.6.2.2 Operation, repair and maintenance phase 355
        10.6.2.3 Decommissioning phase 363
    10.6.3 Stakeholder's responsibilities 364
        10.6.3.1 Contractor's project manager 364
        10.6.3.2 Contractor's environmental and social manager 364
        10.6.3.3 Contractor's site supervisor 364
        10.6.3.4 Contractor's staff 364
    10.6.4 ESIA conclusion and recommendations 365
        10.6.4.1 Conclusion 365
        10.6.4.2 General recommendations 365

- 10.7 Safety in drainage construction 366
  - *10.7.1 Introduction 366*
  - *10.7.2 Hierarchy of preventive and protective measures in drainage system planning, design, and construction 366*
    - *10.7.2.1 Elimination 367*
    - *10.7.2.2 Substitution 367*
    - *10.7.2.3 Engineering controls 367*
    - *10.7.2.4 Administrative controls 369*
    - *10.7.2.5 Personal protective equipment 369*
  - *10.7.3 Excavations 369*
    - *10.7.3.1 Hazards and risk assessment for excavation work 369*
    - *10.7.3.2 Control measures for excavation work 371*
      - *10.7.3.2.1 Deploy safe digging elements and methods 371*
      - *10.7.3.2.2 Supporting excavations 372*
      - *10.7.3.2.3 Safe means of access 372*
      - *10.7.3.2.4 Installing barriers at the site 372*
      - *10.7.3.2.5 Installing lighting and warning and prohibitory signs 373*
      - *10.7.3.2.6 Controlling, positioning, and routing the movement of vehicles, plants, and equipment 373*
      - *10.7.3.2.7 Providing personal protective gear for workers 374*
      - *10.7.3.2.8 Contaminated ground 374*
      - *10.7.3.2.9 Dewatering excavations 374*
      - *10.7.3.2.10 Detection, identification, and marking of buried services 374*
- 10.8 Quality control and management in drainage construction 374
  - *10.8.1 Introduction 374*
  - *10.8.2 Quality of storm sewer pipe 375*
  - *10.8.3 Handling of storm sewer pipes 375*
  - *10.8.4 Installation of storm sewer pipes 376*
- 10.9 Equipment and machinery used in construction and sustainable drainage structures maintenance 376
  - *10.9.1 Excavators 376*
  - *10.9.2 Backhoes 376*
  - *10.9.3 Draglines 377*
  - *10.9.4 Compactors 377*
  - *10.9.5 Tunnel boring machine 377*
  - *10.9.6 Sewer cleaning truck 377*
  - *10.9.7 Tree spades 377*
- 10.10 Conclusion 378
- References 378

11 Operation and maintenance   381

- 11.1   Introduction   381
- 11.2   Storm sewer appurtenances   382
  - *11.2.1   Inlets   382*
  - *11.2.2   Types of inlets   384*
    - *11.2.2.1   Kerb inlets   384*
    - *11.2.2.2   Gutter inlets   385*
    - *11.2.2.3   Combination inlets   385*
  - *11.2.3   Catch basins or catch pits   385*
  - *11.2.4   Clean-outs   385*
  - *11.2.5   Manholes   386*
    - *11.2.5.1   Precast concrete manholes   386*
      - *11.2.5.1.1   Advantages of precast manholes   388*
    - *11.2.5.2   Brick manholes   389*
    - *11.2.5.3   Cast-in-place concrete manholes   389*
  - *11.2.6   Mini-manholes   390*
  - *11.2.7   Deep manholes   391*
    - *11.2.7.1   Drop manholes   392*
    - *11.2.7.2   Junction boxes   394*
  - *11.2.8   Outfalls and outlets   394*
  - *11.2.9   Lamp-holes   394*
  - *11.2.10   Flushing devices   395*
  - *11.2.11   Grease and oil traps   396*
  - *11.2.12   Stormwater regulators   397*
- 11.3   Legal and policy frameworks   398
- 11.4   Stormwater drainage network policy framework   398
  - *11.4.1   How does a standard pipework drainage system work?   399*
  - *11.4.2   Who's responsible for what?   399*
- 11.5   How do SuDS work?   400
- 11.6   A guide to drainage design   400
- 11.7   Legal point of stormwater discharge   400
- 11.8   Drainage information, reporting, and record keeping   401
- 11.9   Enforcing laws, policies, rules, and regulations for effective drainage asset management   402
- 11.10   Maintenance of SuDS   403
  - *11.10.1   Procure equipment   403*
  - *11.10.2   Maintenance budgets   404*
  - *11.10.3   Working with partners   404*
  - *11.10.4   Inspection and maintenance personnel   404*
  - *11.10.5   Identify SuDS maintenance triggers   404*

- 11.10.6 Update the infrastructure asset register and standard operating procedures 404
- 11.11 Specific SuDS maintenance practices 405
  - 11.11.1 Postconstruction 405
    - 11.11.1.1 Tree trenches 405
    - 11.11.1.2 Bioretention gardens 405
    - 11.11.1.3 Infiltration systems 406
    - 11.11.1.4 Porous pavers 406
    - 11.11.1.5 Vegetated swales 407
    - 11.11.1.6 Open/green spaces 407
    - 11.11.1.7 Downspout disconnection 408
    - 11.11.1.8 Vegetated kerb extension 408
    - 11.11.1.9 Rainwater harvesting systems 408
    - 11.11.1.10 Green roofs (vegetated rooftops) 408
    - 11.11.1.11 Detention and retention ponds 409
    - 11.11.1.12 Soakaways 410
    - 11.11.1.13 Planter boxes 410
- 11.12 Maintenance of storm sewers and drains 410
  - 11.12.1 Excessive surface loads 410
  - 11.12.2 Corrosion of sewers 411
  - 11.12.3 Root intrusion 411
  - 11.12.4 Sediment and grit 411
  - 11.12.5 Trash and debris 411
  - 11.12.6 Hazardous materials 412
- 11.13 Storm sewer repair and maintenance practices 412
  - 11.13.1 Roots invasion in sewers 413
  - 11.13.2 Corrosion control measures 413
  - 11.13.3 Grease and oils 414
  - 11.13.4 Sediment and grit 414
  - 11.13.5 Trash and debris 414
- 11.14 Commonly recommended operational and routine maintenance management practices for standard pipework drainage solutions 414
- 11.15 Routine oversight, inspection, monitoring, and maintenance plan 416
  - 11.15.1 What constitutes an inspection and maintenance checklist for a drainage asset? 416
    - 11.15.1.1 Site conditions 416
    - 11.15.1.2 Vegetation 417
    - 11.15.1.3 Structural conditions 417
    - 11.15.1.4 Earthworks 417
    - 11.15.1.5 Spills/releases 417
- 11.16 Conclusion 417
- References 418
- Further reading 418

12 The future of urban drainage — 419

- 12.1 Likely future developments in the drainage field 419
  - 12.1.1 The concept of a sponge city 420
  - 12.1.2 Flood warning systems 421
    - 12.1.2.1 Remarkable flood impacts 422
  - 12.1.3 Drainage easements 422
  - 12.1.4 Vertical forests and forest cities 423
  - 12.1.5 Verticulture 424
  - 12.1.6 The future of trenchless technology 424
  - 12.1.7 Polyhedral pipes 424
    - 12.1.7.1 Advantages of polyhedral composite pipes 425
  - 12.1.8 Digital monitoring of sewers 425
  - 12.1.9 Work planning and budgeting 426
  - 12.1.10 Nonconventional drainage design methods 426
- 12.2 Funding opportunities for drainage schemes 426
- 12.3 Cost implications of drainage schemes 427
- 12.4 Flash floods 428
  - 12.4.1 Case study 12.1: Flash floods that occurred at the construction of US$350 m Lubowa international specialized hospital, Kampala, Uganda (2019) 428
    - 12.4.1.1 Introduction 428
    - 12.4.1.2 Lessons from the case 430
  - 12.4.2 Case study 12.2: The November 5, 2021, flash floods at seroma workshop premises: Seroma workshop flood damage evaluation and analysis, Kampala, Uganda 432
    - 12.4.2.1 Background to the problem 432
    - 12.4.2.2 Interpretation 436
    - 12.4.2.3 Rainfall analysis 440
    - 12.4.2.4 Recommendations to the contractor 441
    - 12.4.2.5 Lessons from the case for a drainage engineer or planner 442
- 12.5 Embedding UN'S SDGs in public authorities' mainstream drainage activities 442
- 12.6 Conclusion 443
- References 443
- Further reading 444

Glossary 445
Index 457

Figures

1.1	Typical conventional design for a residential area drained on a separate system	9
1.2	Typical modern SuDS site design for a residential area	10
2.1	Cantilever retaining wall holding back soil and an infiltration system	21
2.2	Drain with controls	27
2.3	Backdrop manholes on a steep terrain	28
2.4	Steep slope negotiated stepwise to reduce flow velocity	28
2.5	Grass swale along a highway	29
2.6	Conceptual diagram for a drainage network	33
2.7	Site planned to have little impact on site hydrology in a suburban environment	39
2.8	U-drain	41
2.9	Two links feed into a node and one link routes the flow away from the node	42
2.10	Backwater effects and pressurized flow	44
2.11	Pipe is reduced in link CD	44
2.12	A section of Kabale municpality topographical map	46
2.13	Digital elevation model for Kabale central business district	47
2.14	Kabale municipality soil types	48
3.1	Sustainable development pillars	52
3.2	A summary of steps to consider while planning for SuDS	55
3.3	Simple site illustration to locate the optimal solution between built and unbuilt areas	64
3.4	Typical infiltration system	65
3.5	A conceptual bioretention garden	75
3.6	Schematic of a bioretention cell with underdrains (filter design)	78
3.7	Schematic diagram of a typical dry detention pond	83
3.8	Typical subsurface detention system	84
3.9	Illustration of interconnected SuDS (detention pond) and gray infrastructure	85
3.10	Typical retention system	86
3.11	Integrated surface detention and retention system	87
3.12	Vegetated swale	88
3.13	Planter box	91
3.14	Example of a street tree trench with structural soil and adjacent infiltration trench	92
3.15	Tree trench with porous pavers and subsurface infiltration bed	92
3.16	HDPE rainwater harvesting system and stainless steel tanks	94
3.17	Cross-section of a ferro-cement rainwater harvesting tank	97

3.18	Permeable pavers in parking lot	99
3.19	Downspout disconnection	100
3.20	Typical green roof	102
4.1	The hydrologic cycle	117
4.2	A flowchart for the SWMM model process	121
4.3	Modified rational method hydrograph	122
4.4	Construction of the unit hydrograph	124
4.5	Graph of maximum daily rainfall (mm) against return period, t	140
4.6	Intensity–duration–frequency curves for Kabale	142
4.7	ln(ln(1/(1-median rank))) versus ln(maximum daily rainfall (mm))	144
4.8	River Kiruruma in Kabale municipality, Western Uganda	145
4.9	Graph of forecasted peak discharge against return period.	148
5.1	A conceptual diagram illustrating mass conservation	154
5.2	Demonstrating U-channel hydraulic radius	155
5.3	Trapezoidal section	156
5.4	Drain passing through terrain of slopes with miniature changes	158
5.5	Derivation of energy in open channel flow	159
5.6	Slope of the energy grade line and channel bed	159
5.7	Discharge versus depth for constant specific energy	162
5.8	Normograms for solution of Manning's equation for circular pipes flowing full (n = 0.013)	164
5.9	Variation of flow and velocity with depth in circular pipes	167
5.10	Depth versus specific energy at constant discharge	167
5.11	Drain with controls	169
5.12	Pipe flowing partially full	170
5.13	Variation of R and A with depth in circular pipes	171
5.14	Forces on fluid in a control volume for open channel flow	172
5.15	Relationship of depth of flow with specific force	174
5.16	Channel with spatially increasing flow	175
5.17	Demonstrating a hydraulic jump	177
5.18	Momentum function versus depth at constant discharge	178
5.19	Topographical survey with some existing features	180
5.20	Intensity-duration-frequency curve for Kampala city	181
5.21	Option A—proposed location of drainage sewers	182
5.22	Option B—proposed location of drainage sewers	183
5.23	Subcatchment delineations	183
5.24	Isometric view of drainage sewers	184
5.25	Profile of the drainage sewer	185
5.26	Isometric view of U-drain	186
5.27	Option B—subcatchments delineations	187
5.28	Isometric view of U-drain, precast unit	188
5.29	U-drain located on the longest wing	189
5.30	Site layout (drainage and retaining wall)	192
5.31	Built apartment and collapsed wall	192
5.32	Collapsed wall	193
5.33	Entrance and exit pipes	196

5.34	3-D representation of drainage and retaining wall	197
5.35	Modified estate's drainage system	197
5.36	Water profile in flat-bottomed channel	200
6.1	Flow moves from point 1 to point 2 in the channel	207
6.2	Hydrographs for the flow at point 1 and 2	207
6.3	Control volume defining continuity and momentum equations (profile view)	208
6.4	Control volume defining continuity and momentum equations (plan view)	208
6.5	3-D representation of the control volume	209
6.6	Mesh points for a numerical solution of the wave equation	213
6.7	Hydrograph routing for the linear kinematic wave routing	218
6.8	Typical conduit/node sectional illustration	219
6.9	Characteristic grid for open channel flow	224
6.10	Updating internal points	225
6.11	Boundary conditions in the characteristic equations solving strategy with a regular grid	229
6.12	Typical hydraulic model sketch (1-D analysis of storm sewers)	234
7.1	Typical French drain	243
7.2	Mohr–Coulomb failure criterion	252
7.3	Mohr–Coulomb criterion (undrained conditions)	252
7.4	(a) Earthen embankment flood control structure (low permeability central core). (b) Earthen embankment flood control structure (grout curtain). (c) Earthen embankment flood control structure (impermeable blanket)	257
7.5	Impermeable membranes used to control seepage together with a geotextile filter and drain tube	258
7.6	Artistic impression of how an impermeable membrane controls water movement in building basement.	259
7.7	Concrete interlocking (lego) block flood defense wall	259
7.8	Typical coastal concrete flood defense wall	260
7.9	Gravity retaining wall	263
7.10	Cantilever retaining wall	264
7.11	Segmental retaining walls	265
7.12	Counterfort retaining walls	266
7.13	Gabion retaining walls	266
7.14	Collapased retaining wall due to excessive pore water pressure. Interim response to the flash floods; temporal retaining wall	267
8.1	Storm sewer culvert with manholes	270
8.2	Socket and spigot pipes	271
8.3	Springing line on a pipe	272
8.4	Derivation of Marston's equation	276
8.5	Derivation of wide-trench coefficient (complete embankment)	279
8.6	Derivation of wide-trench coefficient or embankment case (incomplete case)	280
8.7	Settlement-deflection ratio for positive projection	281
8.8	Complete negative projection case	285
8.9	Incomplete negative projection case	286
8.10	Settlement-deflection ratio for negative projection	287
8.11	Bedding classes	292

9.1	Lifecycle costs	310
9.2	ln(ln(1/(1-median rank))) versus ln(maintenance cost), division A	319
9.3	ln(ln(1/(1-median rank))) versus ln(maintenance cost), division B	321
9.4	Reduce and specify green	324
9.5	Embedded carbon (eCO_2) is contributed by each of the concrete constituents	328
10.1	Establishment of line and grade of a sewer	337
10.2	Shoring the sides of an excavation by 'close vertical sheeting'	343
10.3	Footbridge segregates pedestrians from moving vehicles	368
10.4	Designated crossing point	373
11.1	Manhole inlet located at the junction receiving flows from minor sewers that feed the main sewer	383
11.2	Typical precast manhole	387
11.3	Brick manhole	390
11.4	Standard deep manhole up to 5-m depth to invert	392
11.5	Drop-junction manhole	393
11.6	Lamp-hole	395
11.7	Grease and oil trap	397
11.8	Combined sand, grease, and oil trap	397
11.9	Ideal stormwater movement pattern for residential areas	399
11.10	Moving from 2-D to 3-D planning	403
12.1	Extent of flooding on one of the fateful days. The flooding from the cleared site blocked roads for many hours and severely damaged the drainage system	429
12.2	Aerial view of Seroma workshop area	432
12.3	Kampala flyover construction and road upgrading project (the contractor built two drainage ditches in blue dots)	433
12.4	Flash floods reached a height of 1.25 m in one of the store rooms	434
12.5	Topographical map	435
12.6	Clock Tower area	436
12.7	Kibuye roundabout	436
12.8	Runoff flow movement pattern	437
12.9	Collapsed eastern wall fence	438
12.10	Fallen southern fence due to flash floods	439
12.11	10-Year rainfall pattern for Kampala city	440

Tables

3.1	Areas suitable for particular SuDS	63
3.2	Different types of infiltration systems	66
3.3	Stormwater quality indicators and lifecycle considerations for SuDS	106
4.1	TRRL classification of soil characteristics	127
4.2	Standard contributing area coefficient (wet zone catchment, short grass cover)	127
4.3	Catchment wetness factor	128
4.4	Land use factor (base assumes short grass cover)	128
4.5	Catchment lag time	128
4.6	Rainfall time (T_P) for East African 10-year storm	129
4.7	Common continuous probability distributions used in integrated drainage systems analysis and design	132
4.8	Maximum rainfall received by Kabale per year	137
4.9	Ranking maximum rainfall received by Kabale for 10 years	137
4.10	Plotting position formulas	139
4.11	Estimates of return period following Weibull plotting position formula (Kabale)	139
4.12	Calculating a^T values for return periods	141
4.13	Intensity-duration-frequency table generated using a rectangular hyperbola, Equation (4.29)	141
4.14	Maximum annual rainfall ranked	143
4.15	Median rank table	143
4.16	Data for Gage Station No. 81249 (R. Kiruruma North at Kabale Kisoro Road)	145
4.17	Peak discharge per year for R. Kiruruma ranked from highest to lowest	146
4.18	Peak flow for River Kiruruma sorted from 1955 to 1997	147
4.19	Four statistical distributions forecasting ARI peak discharges	147
5.1	Manning's roughness coefficient, n, for pipe flows	169
6.1	Integrating downstream	216
6.2	Open channel flow left boundary condition	228
6.3	Open channel flow right boundary condition	229
7.1	Properties of drained granular and cohesive soils	253
8.1	Value of the product $\mu'K$ and μK	277
8.2	Unit weight of backfill material	277
8.3	Design value for settlement-deflection ratio	282
8.4	Error table for Equation (8.17)	284

9.1	ISO 14040:2006 environmental management—lifecycle assessment—principles and framework	310
9.2	SuDS maintenance expenditure in divisions A and B	318
9.3	Weibull analysis for SuDS maintenance expenditure in division A	319
9.4	Weibull analysis for SuDS maintenance expenditure in division B	320
9.5	Calculation of CO_2 emissions	325
9.6	Transport conversion table	325
9.7	UK Environmental agency (transport emissions (gCO_2/t km)	326
9.8	Embedded carbon (eCO_2) figure presented is based on fair estimate in UK industry	327
9.9	Embedded carbon by cement product for UK, USA, and Finland	327
9.10	Embedded (eCO_2) varies by strength class	328
9.11	How embodied carbon varies with steel reinforcement	328
9.12	Dimensions of reinforced concrete pipe	329
10.1	Potential environmental and social impacts due to project location, design process, equipment, during construction, operation and maintenance in reference to drainage and stormwater works	352
10.2	Detailed environment management plan for the construction phase of a drainage system	356
10.3	Environmental management/monitoring plan for the decommissioning phase	363
12.1	Kampala rainfall data (2012–2021)	441

Preface

This account of a particular branch of civil engineering practice lies somewhere between the extremes of a practitioner and an academic text. The publication has come at a time when Sustainable Drainage Systems (SuDS) are now recognized as a vital part of modern water infrastructure systems. Thinking SuDS increases the value of development by significantly reducing the cost spent in mitigating floods and responding to the aftermath of floods.

This book is intended primarily to supplement the already available books on drainage planning and design by promoting a new wave of holistic thinking. The text extends beyond the immediate needs of the undergraduate. However, undergraduate students carrying out design projects as a means of consolidating and synthesizing their knowledge of urban drainage master planning and design will find the book very helpful. Therefore, much of the material in the text will already be familiar to the graduate student. The book is primarily concerned with stormwater management and stormwater runoff control and management.

The hot topic of the 21st Century is sustainable development, and therefore, it calls for planning cities, municipals, and towns that are more resilient to the effects of climate change thus making SuDS an integral part of drainage system planning and design. This book presents the opportunities and benefits that well-planned SuDS, which are achievable in given circumstances, can bring to society when properly integrated with gray infrastructure.

There has been a rapid change in this field in the last decade than in the previous two or three decades. Cities are now planning for a more robust and resilient built environment that incorporates greening activities to supplement the standard pipework drainage solutions. The reader will find this book useful specifically on the planning of achievable SuDS, the design, evaluation, and analysis of traditional standard pipework stormwater drainage solutions. Site-specific SuDS are described together with appropriate gray infrastructure to guide the drainage planning for a given area.

The book's main focus is on stormwater and subsurface sustainable drainage solutions. It primarily deals with the design and implementation of stormwater collection and conveyance systems as separate systems. It further deals with the design and implementation of SuDS, the environmental concerns, and climate action. Please note that planning and design of wastewater collection, conveyance, and treatment are out of the scope of this book.

The major aim of the text is to promote holistic thinking as a new wave of thinking to solve drainage problems in a sustainable approach by viewing things as system of interrelated elements, the elements themselves also being systems interacting with one another. Therefore, any intervention made to solve drainage problems may trigger ripple effects on other systems so that localized constraints can be adequately addressed relying on the knowledge of the wider environment.

The text, therefore, sets out an evaluation criterion to decide on alternative schemes for standard pipework solutions being informed by greening activities or SuDS in a back-and-forth approach to achieve sustainability. The basic theory and practice is provided in sufficient depth to promote basic understanding, while also ensuring extensive coverage of all topics deemed essential to students and practitioners. The text seeks to place the drainage topic in the 'pillars of sustainable development', that is, economic, social, and environment. In this way, it becomes more relevant and interesting to sustainable development enthusiasts!

Chapter 1 (Introduction) addresses the origins of present-day practice, history of SuDS, the role of government through public authorities, general features of standard pipework drainage solutions, an outline of several types of SuDS, combined and uncombined systems, and an example of modern drainage master planning projects.

Chapter 2 (Drainage planning and evaluation process) defines drainage planning and details the decision-making process in line with public participation. The evaluation process considers a number of localized and global constraints to derive a proper solution for a drainage scheme emphasizing holistic approaches for modern drainage schemes and/or drainage master planning projects. The chapter provides a feasibility study process outlining the importance of public participation on selection of appropriate drainage routes, SuDS, defining drainage networks, drainage prioritization, and environmental/social issues.

Chapter 3 (Planning for sustainability) delves more into the role of SuDS in building more resilient communities by fostering a sustainable built environment. The chapter provides a number of reasons to decide on site-specific SuDS for a given area from both local and global perspectives bringing sustainable development at the heart of society. Chapters 2 and 3 communicate with each other to bring out holistic interventions that are most appropriate to solve drainage problems in a sustainable approach.

Chapter 4 (Useful hydrology) defines the hydrologic cycle, the techniques and tools used to produce hydrologic models and to forecast future climatic conditions of a specific area. It briefly involves the estimation of flow quantity and how to deal with quality of runoff economically.

Chapter 5 (Hydraulics design principles) outlines how energy and momentum equations are applied to solve drainage problems.

Chapter 6 (Flow routing techniques) outlines the routing techniques and provides in detail the commonest hydraulic routing methods used in drainage evaluation, design, and analysis. It specifically outlines the dynamic wave routing technique and kinematic wave routing technique. The chapter also briefly tackles the development of hydraulic models and provides reasons to choose from the dynamic and kinematic wave techniques to route stormwater runoff or a flood wave.

Chapter 7 (Useful topics in soil mechanics) outlines how to deal with subsurface drainage, drainage and groundwater recharge, tunneling and underground drainage systems, soil behavior in shear, seepage control in flood defense structures, and retaining structures.

Chapter 8 (Structural aspects of drainage sewers) briefly outlines bedding classes of drainage pipework drainage solutions, loads due to earth overburden, construction materials, loads superimposed by vehicle wheels, internal water load, and pipe strength to match imposed loads.

Chapter 9 (Evaluation of integrated drainage system designs) provides in more detail the advantages and potential disadvantages of SuDS over standard pipework drainage systems. It also provides a procedure to conduct integrated drainage system reliability analysis focusing on the stormwater models. Finally, the chapter provides a roadmap for estimating the amount of carbon emissions stopped from being pumped into the atmosphere as a result of sustainable drainage practices. This is more beneficial in determining the long-term benefits of SuDS over pipework solutions as far as sustainability is concerned.

Chapter 10 (Construction and management) outlines the procedures applicable in drainage construction, which includes safety issues, environmental management, excavation of trenches for sewers, and quality processes. It also involves several machinery/equipment used in drainage construction and maintenance works.

Chapter 11 (Operation and maintenance of sewers) provides a legal and policy framework to guide management of drainage infrastructure. It also involves sewer appurtenances. The chapter also provides ideas 'on polluter must pay sustainable development principle' and how public authorities can fix a surcharge and a legal point of stormwater discharge fees.

Chapter 12 (The future of urban drainage) outlines current drainage issues and the likely future developments in the field, the funding opportunities for drainage schemes, more on sustainability issues, embedding UN SDGs in the mainstream drainage activities of public authorities to contribute to net-zero targets and cost implications of various drainage schemes.

In overall terms, the text sets out procedures and techniques needed for the planning, design, and construction of drainage systems which are sustainable. There has been significant effort made to ensure that up-to-date sources are referenced particularly on applicable standards and codes of practice. My particular gratitude is expressed to the late Prof. Albert Rugumayo, a professional civil engineer, for his support, advice, and guidance. Without him, this book would never come to exist.

Acknowledgments

This book would not have been possible without the extraordinary support of the late Prof. Albert Rugumayo. Without his support, advice, and guidance, this book would never come to exist. I extend my sincere gratitude to the following for their support:

- Michael Daka, Dickson Berabose, Benjamin Olobo, Charles Kizito, Peter O. Lating (PhD), Richard Li Chengzhe, Prof. Jackson Mwakali, and Andrew Grace Naimanye (PhD).
- The Institution of Civil Engineers (ICE), UK.
- Associated Design and Build Engineers Limited, Kampala UG.

Chapter 1

Introduction

1.1 ORIGINS OF PRESENT-DAY PRACTICE

A drainage system is a system that collects, stores, and conveys wastewater and stormwater to a safe discharge point. Sometimes, a drainage system may be designed to collect and convey both wastewater and stormwater—in this case, referred to as a combined system. However, current modern drainage systems are designed and constructed as uncombined systems—specifically stormwater collection and conveyance is separated from wastewater. Usually, traditional drainage systems are constructed with concrete, stones, metallic pipes, cast iron, uPVC, HDPE, etc., collectively referred to as gray infrastructure (standard pipework design solutions or sewer systems). The traditional system has a point that collects the flows, that is, entrance, a section that conveys the flow, and exist, that is, outlet or outfall.

Several communities globally have been greatly adopting standard pipework drainage sewer solutions for a long period. However, with the world increasingly anxious about the sustainability of life as we know it on Earth, it's a great time to consider designing and developing integrated sustainable drainage practices, both sewers and green infrastructure. The topic of sustainable development aimed at significantly reducing greenhouse gas effects has had a worldwide impact on the way urban development is planned in recent years specifically on stormwater management solutions.

This book primarily deals with sustainable stormwater control and management, surface water drainage, and subsurface drainage. Stormwater is the runoff from rainfall, melting snow, or ice. If stormwater is left uncontrolled, it can contribute to costly flooding and pollution issues that wreak havoc on communities. This book treats sanitary and storm sewers as separate systems, corresponding to modern practice. In some cases where combined sewers are required, household drains are connected to storm sewer lines. Usually, the sanitary sewage flow is negligible with respect to stormwater flow.

The terms adopted for conveyance system are stormwater sewer pipelines, stormwater drainage system (pipes and drains), standard pipework solutions, surface water sewers, stormwater sewer networks subsurface pipe networks, and stormwater conveyance systems. The terms constantly adopted worldwide for green techniques aimed at boosting environmental conservation, ecological sustainability, and reducing the greenhouse gas effects are sustainable drainage systems (SuDS), green infrastructure (GI) techniques, low-impact development (LID), and sustainable urban drainage systems (SUDS). However, for the purposes of this book, SuDS terminology is used.

1.2 WHY ARE DRAINAGE SYSTEMS SO IMPORTANT?

Drainage schemes are essential for a country's economic growth and development. The primary goal of drainage systems is to collect and convey wastewater and stormwater from buildings and built-up areas to a safe discharge point in a systematic manner to maintain healthy conditions. Adequate stormwater drainage systems safely route stormwater runoff from the impervious surfaces to discharge it into natural watercourses thereby solving flooding problems, contributing to smooth running of economic activities, saving the communities from waterborne diseases, reducing downtime-related costs due to flooding, reducing traffic delays resulting from flooding of roads, and overall improvement in the standard of living.

Integrated drainage systems planning incorporates greening activities, which makes the cities and municipalities drainage networks aesthetically appealing. Thus, it contributes to net-zero carbon targets because greening activities extensively sequester carbon contributing to a carbon sink while building resilient communities. Drainage systems convey sewage, rainwater, and other foul water from buildings, industries, roads, and other built environments to a safe disposal point.

Overland surface drainage systems remove excess water through drains and ditches, while the subsurface drainage systems typically remove excess water through pipe networks. Both overland and subsurface drainage systems serve the purpose of safely dealing with excess flows for correct disposal and management. A cast-in-place channel is an excellent example of overland surface drainage system, and a perforated HDPE pipe is an example of a subsurface drainage controlling device.

1.3 STORM SEWER SYSTEM

The storm sewer system is a structure or 'asset drainage infrastructure' collecting, conveying, and providing water quality improvement and treatment for stormwater or snow melt runoff from roads, highways, streets, and built-up areas. Storm sewers are commonly constructed using concrete, clay pipes, cast iron, uPVC, and HDPE. The storm sewer systems sometimes transport pollutants as runoff flows across pervious and impervious areas.

The entire storm sewer system includes sewer appurtenances such as manholes, inlets, outlets, and kerbs. The system is always interconnected with gutters, ditches, pipes or culverts, basins, and drainage channels. In modern storm sewers, sustainable drainage systems form an integral part of the drainage system.

1.4 HISTORY OF SUSTAINABLE DRAINAGE SYSTEMS

The increasing pollution of receiving water bodies globally gave rise to think new ways to control the pollution. In North America and Canada, best management practices (BMPs) were thought to control pollution in early 1990s. The U.S. Environmental Protection Agency (EPA) issued a guidance manual for developing best management practices in 1993. The guidance manual set out the procedure to prevent the release of toxic and hazardous chemicals. The guidance manual was of a wide range encompassing many BMPs of different types of facilities. This controlled water pollution safeguards the receiving waters from excessive pollution. The BMPs approach demonstrated flexibility in controlling releases of pollutants to the receiving water bodies.

Categorically, stormwater best management practices were identified as effective approaches to managing stormwater at the source. These were classified in two categories: (a) structural BMPs and (b) nonstructural BMPs. Stormwater BMPs demonstrated flexibility in controlling stormwater and stormwater runoff and improving the stormwater runoff quality, safeguarding the receiving waters from excessive pollution. Since then, the entire concept of stormwater management has shifted from overreliance on pipework standard solutions to an integrated approach that mimics the natural environment, to safeguard receiving water bodies from pollution and addressing the call for climate action.

The concept of SuDS developed from BMPs to the traditional way of collecting and safely conveying stormwater runoffs. These best management practices included the structural and nonstructural approaches to managing stormwater. As the world moved to appreciate sustainable development, SuDS gained much momentum using nature as a model to manage rainfall at the source. Therefore, the overarching goal of SuDS, from the beginning until now, has been to manage and control stormwater and stormwater runoff, in a way that mimics the natural environment. In this way, many benefits may be achieved.

First and foremost, the conservation of the ecological system is promoted by the way of promoting SuDS as the overarching goal. The regional flora and fauna is preserved in a way of incorporating SuDS in the planning and construction of drainage systems or the built environment. This is of course, a very promising technique in the 21st century where sustainability has become a hot topic of discussion globally. Many strategies are used which include runoff mitigation and treatment controls to remove pollutants.

The terminologies used for stormwater control and management vary widely. In England, SuDS is used. In North America and Canada, stormwater BMPs, that is, structural and nonstructural, and LID, and GI techniques are the terminologies adopted. SUDS is the term adopted in Scotland. LID is a process for land development in the USA that attempts to minimize impacts on water quality and the similar concept of SuDS in the UK.[1] These terms are adopted interchangeably with slight differences globally. Throughout this book, the standard terminology used is SuDS. However, in a few sections other terminologies are used.

In the USA, BMPs are recognized as an important part of the National Pollutant Discharge Elimination System (NPDES) permitting process to prevent the release of toxic and hazardous chemicals.

In the USA, the NPDES, state and local water quality management regulations increasingly limit impacts to receiving water. Therefore, stormwater management and the use of BMPs have become a primary design consideration for most civil engineering projects.

1.5 ROLE OF GOVERNMENT THROUGH PUBLIC AUTHORITIES

As a result of climate change, cities around the world are fast-tracking measures to develop robust and resilient infrastructure. The variances in climatic conditions, topography, and constraints across the globe necessitate planning needs that address these unique factors from one location to another. That means, adaptation planning is vital as opposed to reactive emergency response planning. This is because climate change mitigation strategies alone could not prevent the climate change effects completely.[2] Therefore, constructing large flood defense barriers and implementing sustainable drainage systems only do not warranty reliable schemes that would contribute to a resilient community. Implementing policies on greening, effective physical planning, climate change adaptation strategies, and effective water management solutions must be integrated as a daily routine measure in order to achieve a robust and resilient built environment.

Sustainable development is a topic that requires sound government policy. The need to carbon offset all activities in the country to attain the net-zero target requires the government to enact statutes and policies that consider sustainability as the overarching goal. In some instances, the feasibility studies and lifecycle cost analysis conducted may show the initial investment cost of a SuDS project as quite big at a micro level. However, with government support, aggregating SuDS projects at a macro level is the only way the net-zero target would be realized. This therefore, requires government to subsidize some activities that contribute to the overarching goal of attaining a carbon neutral economy.

Some SuDS are classified as building drainage and not as public drainage assets. These may include permeable paving, rainwater harvesting systems, planter boxes, green roofs, rainsave planters, water barrels, cisterns, and water-butts. These, when holistically integrated with public SuDS through appropriate government policies, support the main goal of reducing the risks associated with stormwater.

Therefore, sound government policy is important to support projects aimed at offsetting carbon emissions. Without government support, some SuDS may prove not cost-effective in some instances. However, holistic planning interventions that consider the project lifecycle costs and benefits, the wider environment, and service delivery would find an optimal solution that compensates for the high SuDS initial investment costs.

1.5.1 The effect of urbanization

Urbanization increases impervious surfaces and reduces the rate at which rainfall infiltrates into the ground. Urbanization modifies the hydrologic characteristics of a region. The big percentage of the rain that falls converts into runoff. Naturally occurring watercourses are distorted when buildings and other structures are constructed. This becomes more detrimental in urban areas where buildings and other infrastructure are not well-planned. When naturally occurring drainage networks are distorted as a result of urban development, it becomes pertinent to plan for new reliable and sustainable drainage infrastructure.

The movement of a flood wave always follows a natural path until when it is interrupted by a structure developed in the natural course. In poorly planned urban areas, the risk of flooding is greatest. This is because naturally occurring drainage easements and natural watercourses are encroached on by the developers. It is the role of the government through environmental management and protection agencies to foster organized development that will leave the drainage easements and natural watercourses intact. Therefore, government always enforces sustainable urban planning.

1.6 GENERAL FEATURES OF STANDARD PIPEWORK DRAINAGE SOLUTIONS

Typical standard pipework drainage solutions have the following features: junctions, manholes, inlets, outlets, pipelines, and drains. They are commonly laid in straight lines. However, curved storm sewers are not uncommon. Intermediate junctions and manholes are provided in the storm sewer networks to control stormwater runoff. In several cases, storm sewers are designed to follow road networks underneath the roads avoiding interference with private property especially during maintenance and repair activities. Installing storm sewers underneath roads usually saves costs in land compensation and avoids delays.

Manholes are provided at junctions, change of direction, pipe size, and slope. On steep terrains backdrop manholes drastically drop flow energy. Backdrop manholes connect pipes of differing invert levels and allow for the pipe at a higher level to connect to the lower-level pipe with minimum fall. Manholes support cleaning activities acting as points detaining sediment load, garbage, and trash, and they account for a small proportion of a large stormwater sewer project unless they are too deep.

Adequate ventilation is vitally important in deep manholes and concrete surrounds are recommended for storm sewer pipes giving the pipeline additional strength. Usually, about 100-to-300-m distance is the range in which spacing of manholes is commonly placed. However, the spacing of manholes on storm sewer lines in which humans can work through while standing may be larger than this.

1.7 OUTLINE OF SEVERAL TYPES OF SUDS

1.7.1 Tree planting

Tree planting is the process of transplanting tree seedlings, generally for forestry, land reclamation, or landscaping purpose. However, when properly done, tree planting can effectively work as a SuDS practice. Tree planting is an excellent practice for habitat conservation. For trees to serve effectively as a SuDS does not come without challenges. For example, large tree planting initiatives are faced with many problems, including losses from climate change, tree pests and invasive species, and urban development. When trees are planted on a large scale to create a forest, it can either be afforestation or reforestation.

Afforestation involves planting trees in areas that had no tree cover, in order to create a forest. The type of land planted could include areas that have turned into desert (through desertification), places that have long been used for grazing, disused agricultural fields, or industrial areas.

Reforestation mainly refers to the replanting of plants on those land which has been destroyed for the benefits of humankind.

1.7.2 Tree trenches

A tree trench is a system of trees that are connected by an underground infiltration structure. On the surface, it looks like a row of ordinary street tree pits, but underground there is a system to manage incoming runoff. It is categorized as an infiltration system practice.

1.7.3 Bioretention gardens

It is a SuDS practice, usually developed in depressions, that captures and treats the first runoff flush from nonporous surfaces. The primary purpose of the bioretention garden is to capture pollutants form the first flush of the runoff. The first flush usually contains a large portion of the pollutants that leave a nonporous area. The bioretention, also referred to as the rain garden, infiltrates the first flush runoff into the soil profile, where it is treated and released to the ground or surface water. The infiltration rates of local soils vary, and where they are far greater than 5 mm/h, the treated waters can be easily infiltrated into the ground. However, where infiltration rates are far below 5 mm/h, the treated runoff returns to the surface waters via an underdrain. These are primary design considerations.

1.7.4 Infiltration systems

Infiltration system are devices, structures, or practice designed specifically to encourage the entry and movement of surface water into or through underlying soil. An infiltration system captures stormwater runoff and allows it to infiltrate into the soil. Infiltration systems include infiltration basins, infiltration trenches, bioretention gardens, and porous pavers.

1.7.5 Porous pavers

Unlike nonporous pavers, which do not allow runoff to infiltrate into the subsoil layers, porous pavers are designed with holes that enable runoff to infiltrate into the subsoils. Porous pavers have a cellular grid system filled with sand or gravel. This system provides grass reinforcement, ground stabilization, and gravel retention. The grid structure reinforces infill and transfers vertical loads from the surface, distributing them over a wider area. They are generally designed to allow water to pass through and infiltrate in a well-designed soil layer.

1.7.6 Swales

These are termed either bio-swales, vegetated swales, or eco-swales. A swale may be either natural or human-made and they are beneficial in recharging groundwater. Artificial swales are often infiltration basins, designed to manage runoff, filter pollutants, and increase rainwater infiltration into the subsoil. Bio-swales are commonly applied along streets and around parking lots, where substantial automotive pollution settles on the pavement and is flushed by the first instance of rain, known as the first flush. Bio-swales, or other types of biofilters, can be created around the edges of parking lots to capture and treat stormwater runoff before releasing it to the public storm sewer. Bio-swales involve the inclusion of plants or vegetation in their construction. They are channels designed to concentrate and convey stormwater runoff while removing debris and pollution.

1.7.7 Green spaces

Green spaces are kind of open spaces provided in built-up areas for aesthetic purposes, leisure, proving a cool temperature range, natural habitat development, air circulation, and a relaxed environment for dwellers. It is typically an area of grass, trees, or other vegetation set apart for recreational or aesthetic purposes in an urban environment. In land-use planning, urban green space is open-space areas reserved for parks including plant life, water features (also referred to as blue spaces), and other kinds of the natural environment. Most urban open spaces are green spaces, but occasionally include other types of open areas.

1.7.8 Green streets

A green street is a SuDS practice that incorporates vegetation (perennials, shrubs, or trees), soil, and engineered systems (e.g., permeable pavements) to slow, filter, and cleanse stormwater runoff from impervious surfaces (e.g., streets, sidewalks). It uses natural processes to manage stormwater runoff at its source.

1.7.9 Downspout disconnection

This is the process of disconnecting the roof downspout from the sewer system and instead redirecting the roof runoff toward pervious surfaces, most commonly a lawn. The primary goal of disconnecting the downspout is to reduce the size of the connected impervious area that contribute to the gray drainage system. Depending on the size of the downspout installed on the building, heavy rains usually can significantly increase the risk of basement backups and sewer overflows. Downspout disconnection from the sewers and storm drains is usually the first step to reduce such overflows. This SuDS practice is mandatory in some cities, and homeowners that fail to act could be subject to a fine.

1.7.10 Vegetated kerb extension

Vegetated kerb extensions are landscaped areas within the parking zone of a street that capture stormwater runoff in a depressed planting bed. The landscaped area can be designed similar to a rain garden or bio-swale, utilizing infiltration and evapotranspiration for stormwater control and management.

1.7.11 Rainwater harvesting systems

Rainwater harvesting (RWH) is a process or technology used to conserve rainwater by collecting, storing, conveying, and purifying it for future use. Rainwater is harvested from rooftops, parks, roads, and open spaces. RWH systems include underground and above-the-ground systems. They include cisterns and rain barrels. They are designed to harvest stormwater runoff from building commonly for domestic use, irrigation of crops, feeding animals, and washing, among others.

1.7.12 Green roofs (vegetated rooftops)

A green roof or vegetated roof is a roof of a building partially or completely covered with vegetation and a growing lightweight engineered soil medium, planted over a waterproofing membrane. It is a low-maintenance system that stores rainwater in a soil medium, where the water is taken up by plants and transpired into the air. Green roofs may also include additional layers such as a root barrier and drainage and irrigation systems.

Other terminologies used for a green roof are 'living roof' and 'eco-roof'. Much less water runs off the green roof, as compared to conventional rooftops. Green roofs provide an extra layer of insulation reducing heating and cooling costs, and they are likely to last much longer than conventional roofs, since the roofing material itself is shielded from ultraviolet light and thermal stress. Green roofs are easy to incorporate into new construction and can even be used on many existing buildings. The vegetation on green roofs improves air quality, enhances the appearance of the building, and reduces the urban 'heat island' effect.

1.7.13 Detention ponds

These are ponds designed for detaining stormwater runoff for a certain period and later release it. They are also known as 'dry ponds' because they only hold water when it rains. A detention or dry pond has an orifice level at the bottom of the basin and does not have a permanent pool of water. All the water runs out between storms and it usually remains dry.

1.7.14 Retention ponds

Unlike detention ponds outlined in Section 1.7.13, retention ponds have a permanent pool of water. A retention basin, sometimes called a wet pond, wet detention basin, or stormwater management pond, is an artificial pond with vegetation around the perimeter and includes a permanent pool of water in its design. Retention ponds can provide both stormwater attenuation and treatment. They are designed to support emergent and submerged aquatic vegetation along their shoreline.

1.7.15 Soakaways

This is a pit, typically filled with hard core, into which stormwater is piped so that it drains slowly out into the surrounding soil. Soakaways are provided on relatively very small watersheds where the runoff volume can be contained in the soakaway. Soakaways are primarily provided in instances where combined or partially combined systems are required. For example, one design that is capable of treating effluent water is a 'soakaway drainage field', in which liquid waste or sewage is dispersed into a soil field, where it is cleaned (to an extent) by aerobic soil bacteria, which occurs naturally in the soil. Note that soakaways are not well-equipped to treat effluent water emerging from a sewage system.

1.7.16 Planter boxes

Planter boxes are normal boxes usually planted with vegetation, flowers, and native vegetation. They are usually enclosed features either below or above-the-ground boxes planted with vegetation that capture stormwater within the box. They are in many cases located close to downspouts to receive stormwater runoff from the rooftops.

1.7.17 Artificial rills

Artificial rills are channels constructed to carry a water supply from a distant water source. In landscape architecture and garden design, constructed rills are primarily an aesthetic water feature. Together with other water features such as fountains, pools, ponds, streams, or artificial waterfalls, rills can be aesthetically appealing and can attract additional greening practices that can collectively contribute to low-impact site development when properly integrated in modern architecture. Rills are not only used for aesthetics but also connect multiple ponds which prove very helpful in fish farming.

1.8 COMBINED AND UNCOMBINED SYSTEMS

Modern-day drainage system practice separates urban stormwater runoff from foul sewage. Current practices exclude surface runoff by constructing sanitary sewers. It becomes more understandable when sustainable development is the overarching goal in delivering projects. Sustainable management of natural resources like lakes and rivers greatly requires reducing pollution. In many instances foul sewage would need treatment before discharging to the streams, rivers, and lakes.

Urban runoff is always so great that treatment works are so costly. Previously, combined sewers assumed that urban stormwater runoff would heavily dilute the foul sewage. However, the increasing pollution of cities globally calls for a robust separate system to protect water bodies from excessive pollution.

During combined sewer overflow events, combined sewers can excessively pollute the receiving waters especially when flows exceed the capacity of the sewage treatment plant. Therefore, modern practices separate surface runoff from foul sewage.

In some cases, a partially separate system is warranted especially in industrial and commercial areas where oil spillage from tank vehicles and manufacturing plants has to be drained. In that case designing for a partially separate system might be necessary.

SuDS such as retention basins, infiltration systems, bioretention gardens, and swales are helpful in mitigating the effects of combined sewer overflows. Some older cities and towns used combined systems and still continue using the same. However, the increasing urbanization globally greatly requires separate systems in the interest of sustainable development.

1.9 MODERN DRAINAGE MASTER PLANNING PROJECTS

Modern drainage systems are designed as **separate systems**. This means they deal with stormwater and wastewater as separate systems. Modern drainage systems incorporate **SuDS** mimicking the natural environment to reduce stormwater runoffs as much as possible through infiltration practices and storage facilities.

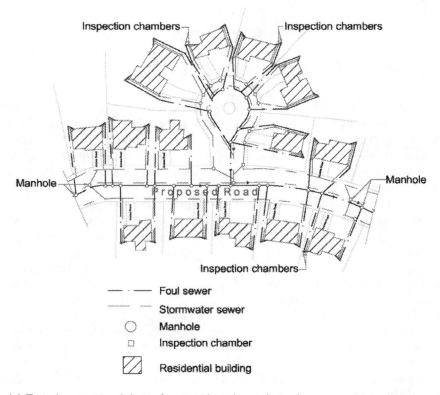

Figure 1.1 Typical conventional design for a residential area drained on a separate system.

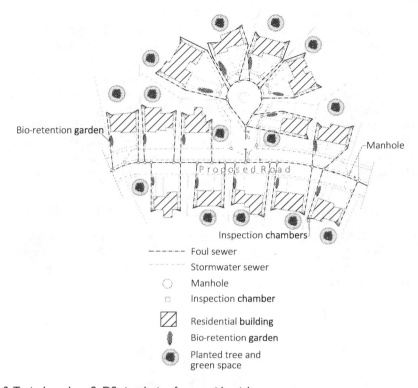

Figure 1.2 Typical modern SuDS site design for a residential area.

Conventional stormwater management techniques direct all of the stormwater to storm sewers to remove it from the site as quickly as possible. Figure 1.1 gives a conventional design of a typical layout for sewers in residential areas. Conventional design largely utilizes end-of-pipe facilities that are typically designed to store and detain runoff to reduce peak flows for infrequent storm events, such as the 10-year, 24-hour storm. However, controls are often not in place to reduce flows for smaller, more frequently occurring events. Controls also are not structured to address nonpoint source **pollution problems** or to **recharge the groundwater.** Since runoff needs to be managed on the site to meet sustainability goals, large ponds, or a series of ponds, are required. However, these controls take up a significant portion of land.

SuDS, specifically infiltration practices, are an alternative site design strategy that uses natural and engineered infiltration and storage techniques to control stormwater where it is generated. These SuDS combine conservation practices with distributed stormwater source controls and pollution prevention to maintain or restore watershed functions. The objective is to disperse SuDS devices uniformly across a site to minimize runoff. These SuDS reintroduce the hydrologic and environmental functions that are altered with conventional stormwater management.

SuDS as bioretention systems help to maintain the water balance on a site and reduce the detrimental effects that traditional end-of-pipe systems have on waterways and the groundwater supply. SuDS devices provide temporary retention areas, increase infiltration, allow for nutrient (pollutant) removal, and control the release of stormwater into adjacent waterways. Figure 1.2 gives a modern master plan layout where standard pipework stormwater and foul sewers are integrated with SuDS, that is, bioretention gardens, green spaces, and trees.

1.10 CONCLUSION

Today's modern drainage practices are separate systems and incorporate sustainable drainage systems on a large scale to achieve sustainability. If combined system is required, sanitary sewers are simply connected to storm sewers. The assumption held is that sanitary sewer flows are usually negligible compared to stormwater flows. In the current practice, much emphasis is placed on a low-carbon drainage assets development. Achieving this goal requires the planner to evaluate the potential of integrating SuDS as outlined in Section 1.7 with standard gray drainage infrastructure assets.

References

1. Fletchera, T. D, Shusterb, W.,. Huntc, W. F., Ashleyd, R., Butlere, D., Arthurf, S., Trowsdaleg, S., (2014). SUDS, LID, BMPs, WSUD and more—The evolution and application of terminology surrounding urban drainage. *Urban Water Journal*, *12*(7), 525–542.
2. Harvey, D., Gregory, J., Hoffert, M., Jain, A., Lal, M., Leemans, R., Raper, S., Wigley, T., & De Wolde, J. (1997). *An introduction to simple climate models used in the IPCC Second Assessment Report.* IPCC Technical Paper II. Intergovernmental Panel on Climate Change, Geneva, Switzerland.

Chapter 2

Drainage planning and evaluation process

2.1 INTRODUCTION

Drainage system planning and evaluation encompasses three primary items: measurement, assessment, and prediction. The evaluation process is an essential part of planning because the planner or designer must be able to see ahead. Unlike in the recent past, today's planning and design roadmap must be clearly known and defined for sustainable drainage projects. Inadequacies in the evaluation process can affect the functional and operational performance of the drainage system. Therefore, building reliability in the drainage system must be done at the earliest start while planning different facilities to collect, convey, store, and treat stormwater and stormwater runoff in a sustainable approach. That means, where possible, operation and maintenance teams are highly recommended to be part of the planning and design of the drainage infrastructure deemed sustainable. This is because sustainable drainage systems (SuDS) require operation and management plans and agreements that are in perpetuity. Sound planning would support a reduction in retrofitting and remodeling after construction which is sometimes costly.

It is essential that the responsible SuDS' operation and management teams have substantial funding and commitment to undertake the projects, and they must be identified at the onset of the project and integrated in the planning activities.

The modern integrated drainage systems' evaluation, planning, and design is multidisciplinary and transdisciplinary. It requires drawing knowledge from a multitude of fields to appraise lively projects in the built environment. It is an approach to solving a problem that involves several professional specializations. While tackling drainage problems, people's psychosocial well-being, environmental management, sustainable management of natural resources, and climate action require careful planning within the regional legal and institutional frameworks to promote efficiency and sustainability.

Today's modern drainage systems, like any other project, emphasize low-carbon planning to tackle climate change globally. This means, interventions to reduce the drainage activities' impacts on the environment are fostered worldwide. Holistic planning, therefore, involves several stakeholders at all stages of the planning process promoting diversity and inclusion.

Holistic drainage planning considers all areas, such as community goals, objectives, values, mission, vision, and ambitions, and uses that information to build a customized and comprehensive drainage systems plan. Holistic planning requires all interest groups and stakeholders to contribute to designing and developing drainage systems that are sustainable. Stakeholder inclusion and participation at all stages of project development form the strongest pillar in designing and developing drainage systems that are sustainable.

The overarching goal is to offset carbon emissions as much as possible, mitigate the risk of flooding, find optimal lifecycle costs, and build reliability throughout the project lifecycle phases. Offsetting carbon emissions takes into consideration the capital (embodied) carbon and operational carbon. The designer weighs several options to significantly reduce the lifecycle project emissions. As a designer, the decision taken must be corroborated with substantial evidence to defend it, in the interest of sustainable development.

In terms of construction, embodied carbon (also known as capital carbon or embedded carbon or embodied energy) means the total greenhouse gases emitted in delivering a construction project. For a drainage scheme, the embodied carbon would involve emissions from plants, vehicles, equipment, materials, etc. All carbon emissions that associate with the construction process is termed 'embodied carbon'. Therefore, by using low-carbon embodied construction materials, the project embodied carbon is significantly reduced. For example, drainage sewers and channels may be required to use locally available construction materials within the region to reduce emissions associated with long-distance transportation of materials. This affects the planning since the critical transport routes would influence the optimal solution.

As a designer, scaling down the use of concrete, steel, and other materials contributing to the standard pipework gray infrastructure in drainage systems construction significantly reduces the embodied carbon. The production of steel and cement comes with enormous production of carbon globally. In some instances, recycling concrete is a technique adopted to promote renewability in developing gray infrastructure.

The cobblestones paving technology is also used as a local material to develop pervious pavements for low-volume roads. Also, sewers and drains that are ecofriendly can use cobbles. Stone-pitched drains provide economical solutions because cobbles are readily available in some small-to-medium cities/towns and would produce low-carbon emissions to build the drainage schemes. However, as the designer contributes to shaping net-zero, the decision is always taken considering drainage system reliability, operability, and maintainability. Supplementing gray infrastructure with SuDS would significantly reduce the runoff volumes, consequently influencing the level of hydraulic efficiency needed of the drainage sewers and channels. When the hydraulic efficiency required is lowered, materials with a high frictional resistance like cobblestones available locally in the region especially in financially constrained areas would be adopted and would perform satisfactorily and reliably, thus significantly contributing to net-zero in the long run.

SuDS, such as infiltration systems especially bioretention systems, would absorb greenhouse gases, collect pollutants from the runoff, and reduce runoff velocities. These, when designed and developed effectively, are excellent approaches to scaling down the size of gray infrastructure. In that case, the size of standard pipework solutions connected to the bioretention systems would significantly reduce. However, checking the carbon emitted during the maintenance of the infiltration systems and the pipework solutions must be conducted in this case, weighing the available options before zeroing to what is termed the optimal solution.

The hierarchy of planning for sustainable drainage practices includes first, the energy in use to operate and maintain the drainage systems and second, the embodied carbon when developing the drainage systems. The predicted carbon is tabulated and graphed to locate optimality.

Therefore, bringing sustainable drainage practices at the heart of society necessities planning that considers a spectrum of factors including rampant urbanization that quickly modifies the hydrologic conditions of the area. And looking at several constraints such as limited funding for drainage schemes, topography, soils, the built environment, and other environmental conditions, a wide knowledge base is required to find the most appropriate field applications to solve today's drainage problems.

As a drainage planner, you need to juggle a number of local and global constraints focusing on the availability and accessibility to the required resources, that is, time, cost, equipment, machinery, etc. The accessibility to data and essential information influences the quality of the drainage system. The type of contractor required to build the drainage system is assessed basing on the complexity of the works. Taking into consideration both the local and global constraints while planning for sustainable drainage schemes implies tackling the extent of ripple effects in the system, that is, what is done upstream would influence what is done downstream and vice versa.

2.2 IMPORTANCE OF DRAINAGE SYSTEMS

Standard pipework drainage systems serve many purposes. They collect and convey stormwater runoff to a safe discharge point, which is usually a water body such as oceans, lakes, rivers, or streams. They also collect and convey foul water from areas such as residential areas, commercial, and industrial areas.

Standard pipework stormwater drainage systems are usually huge because they collect and convey large volumes of runoff. Some drainage systems function as fish passage conduits, serving two purposes.

On the other hand, SuDS provide additional benefits. They promote greening, which is a foundation for sustainability. As they mimic the natural environment, many SuDS collect stormwater and infiltrate it into the ground. In this case, they collect pollutant loads and significantly reduce the risk of excessively polluting the receiving waters. This is more beneficial in protecting the ecosystem, flora and fauna. Properly landscaped areas that integrate SuDS present a cool temperature range suitable for people to live, aesthetically appealing, and act as water detention and retention systems, among many other benefits.

2.3 SOURCES OF FUNDING

In many cases, stormwater projects do not come with direct revenue streams. Projects will direct revenue streams utilize traditional public sector (TPS) funding and private finance initiative (PFI) frameworks. As a drainage systems planner, looking at a bigger/holistic picture which is sustainable for the community is a must—engaging locals to own the stormwater management program and adopting a more decentralized approach, than relying on the central government/authorities to fund stormwater programs. The future promises regional decentralized approaches to solve proximate drainage problems. Therefore, both SuDS and gray infrastructure opportunities will increasingly become integrated, shifting and distributing stormwater burdens between the individual developers and the public. In this way, financial modeling of stormwater programs will become more promising, and funding of drainage schemes will become more feasible and meaningful.

A majority of stormwater management alternatives especially the gray infrastructure projects present one of the biggest challenges to the public authorities since they have no immediate direct payments from end users. Contrary to government sourcing for a loan to invest in electricity generation, hospital or school where direct payments from end users may be achieved, drainage infrastructure, especially gray infrastructure, is all different. It entails the public authorities' efforts to create a shift in the community thinking toward stormwater in order to view stormwater more or less like any other public utility say portable water or power. A number of funding alternatives can be realized for stormwater management and drainage programs. Some of these sources are listed below:

2.3.1 Direct funding through loans and tax revenues

Local funding sources such as tax revenue or loans can provide a stable source of funding for construction and maintenance of stormwater drainage infrastructure. Without adequate funding of drainage assets, infrastructure assets' sustainability is risked because maintenance of the systems is compromised. Loans and tax revenues provide the commonest method of funding drainage infrastructure.

2.3.2 Fee-based stormwater utility

A stormwater utility fee charges based on impervious surfaces directly related to the amount of stormwater generated from the property. This creates a more service-based allocation of cost. The creation of a stormwater utility fee creates a reliable funding source to pay for the municipal's stormwater services. Stormwater utility fees (SUFs) are an alternative dedicated revenue source to fund stormwater management. When complemented with stormwater utility credits or discounts, SUFs provide greater flexibility to adopting best management practices and reducing stormwater runoff at a lower overall cost to the community.[1] The inability of some communities to establish and maintain sustainable revenue sources (such as user fees) limits their access to low-cost debt financing to help them meet the challenge of paying for stormwater infrastructure and management.

2.3.3 Collaborating with nongovernmental organizations

Nongovernmental organizations (NGOs) provide alternative means for financing drainage systems. Municipalities may engage NGOs as partners to finance the stormwater management programs through grants and donations creating long-term performance partnerships between public and private (or NGO).

2.4 CONSTRAINTS AND DATA COLLECTION

2.4.1 Public participation

Community and stakeholder engagement is an essential component when planning a drainage scheme. The stakeholders always know the extent of the drainage problem and will quickly cite the impact floods cause to the community. The information gathered from discussions with the communities is relevant and would later be used to prioritize the drainage schemes, also, defining appropriate routes for the drainage networks. Therefore, as a designer, doing things **with** the community is the best approach but not doing things **to** the community.

Stormwater discharge points, available easements, diversity and inclusion issues, and fostering resilient communities by incorporating greening activities, drainage prioritization, and public health are pertinent topics for discussion with the stakeholders. This is conducted through public consultations. Collaborating with stakeholders is crucial to determine the drainage system layouts and alignment. This is important to evaluate the available easements to avoid utility services conflicts and clashes.

Sensitizing the community about sustainable drainage practices at the planning stage would be beneficial in the long run. Sensitization and dissemination workshops on sustainable drainage practices would make communities and stakeholders more engaged. These and other software activities would form part of municipal annual budgets and work plans promoting sustainable drainage practices. This improves better understanding of stormwater management programs making communities feel ownership of the stormwater program projects. Enforcing sustainable drainage management polices is done by community leadership. Therefore, a drainage planner ought to shift from planner-centered approach to community-centered approach, so that the community appreciates controlling stormwater at the source as the essential component of sustainable drainage practices.

2.4.2 Required data and documents review

The data collection and survey involve site visits to the project area and gathering data and information from the stakeholders, for example, city or municipal planning units, engineering department, environment and natural resources management office, relevant government agencies, and gleaning information from documentations such as the development plans and available copies of other previous works done on stormwater drainage and management in the town. Different materials and existing data relevant to the survey are always reviewed as part of desk reviews. Information, data, and documents that are mostly required from the city or municipal council and associated government agencies include the following:

2.4.2.1 Meteorological data

Depending on the scope of the project, meteorological data may include rainfall, humidity and temperature, sunshine, and river flows. These can be used to create hydrologic or precipitation or runoff models. Hydrologic data are necessary for hydrological analysis, forecasting/prediction of future events. This is vitally important in studying the hydrology of the place. Climatological data of the project area, for example, rainfall, temperature, etc., can be obtained from the meteorological stations.

2.4.2.2 Geotechnical and site investigations

Nature of soils, rocks, landslides, and potential of earthquakes would require an investigation before taking decisions on drainage infrastructure development in an area. Ground investigations are part of the evaluation process to ascertain the water table, nature of soils to use as backfill, and bedding materials. In some cases, excessively waterlogged areas might increase the cost of construction through pumping and ground freezing. Where such conditions are expected, a thorough site investigation is paramount, and the design must be done accordingly. This influences the scheme route since the choice would greatly be affected by cost to dewater the working area. The geotechnical report would be produced describing the on-site soils, expansive soils, poor chemical properties, groundwater conditions, and more.

Through site investigations, the applicability of various types of SuDS can be evaluated based on the nature of soils, the water table, and the extent of land contamination. Land contamination influences the runoff pollution levels. Specifically, afforestation and reforestation may require the assessment of trees water demand and the plasticity of soils. This is because the location of building structures is affected by the proximity of trees. Similarly, infiltration systems require infiltration tests conducted on native soils.

In waterlogged areas, the ground would need to be dewatered by a well-pointing system alongside the trench and freezing has to be resorted to infrequently. Such methods considerably increase the construction cost, and in areas where there is a possibility that they would be required, an extensive site investigation is warranted at the design stage to evaluate whether a cheaper alternative scheme can be located elsewhere. Therefore, in several cases, site investigations should be carried out and the results supplied to bidders to make well-informed offers that would result in minimal variations at contract execution.

2.4.2.3 Soils and geological data

Information on soil and geology is used to come up with soil moisture model and permeability. These parameters are helpful in deciding on site-specific SuDS. The geology of the area is crucial to understand how drainage assets infrastructure will respond when constructed in the soil strata. Locating rocks, important metals, and ground water is essential while planning for drainage systems in line with statutory laws.

2.4.2.4 Socioeconomic data (demographics)

Planning for sustainable stormwater and drainage assets management requires the social and economic aspects incorporated in the decision-making process. To properly inform the social and economic pillars of sustainable development, area demographics must be collected, for example, community member's education levels, number of households, ages, population, and economic activities within the community. This would support land acquisition and the preparation of resettlement action plans for project affected persons. It also helps to conduct economic analyses to forecast how the drainage infrastructure would benefit the community. On source revenue collection would be evaluated following the socioeconomic data collected and projections made based on how drainage improvement would support the growth of economic activities in the area.

The linkage between stormwater and drainage assets management and other sectors would be analyzed following the community's development plan and physical development plan. Collecting sufficient amounts of socioeconomic data would definitely provide an understanding of the priority areas that need immediate attention, the challenges communities face, and help delineate flood zones. This would support sustainability goals going forward.

Stakeholder engagement when collecting socioeconomic data makes the community feel involved in the stormwater planning activities. In that way, community sensitization is enhanced in a consultative approach, doing things with the stakeholders rather than having to do things to them. Encouraging participation of community members furnishing as much information as possible supports comprehensive analysis producing projections and the overall community growth in welfare that the drainage infrastructure improvement and proper stormwater management would present.

2.4.2.5 Topographical surveys and information systems

Topographical maps from which digital elevation maps are derived are produced. The digital maps are zoomed in and out (enlarged) to the appropriate scales to give a clear view of topography and features of the project area. These will be obtained from the Survey and Mapping Department, Entebbe. Topographical maps are used to delineate catchment areas in relation to the altitude of the project area and the length of drainage areas during diagnosis of existing situation. Information is processed from topographical surveys, which include digital elevation models (DEM), spatial models, geo-referenced maps, and satellite imagery.

A geographic information system (GIS) is used to create, manage, analyze, and map all types of data. It is used to process spatial data, geo-referencing, data collection and storage, and analyze and output deliverables. Existing situational analysis can be interesting, for example, picking drainage assets data to generate a corresponding map. Therefore, GIS connects data to a map, integrating location data (where things are) with all types of descriptive information (what things are like there).

Invert levels are obtained from the topographical surveys. During evaluation, the stormwater sewers need not be buried so deep below the ground surface. Therefore, the initial slope tried should be that providing minimum cover. In storm sewers, it is desirable to prevent the flow line from falling too far below the ground surface, hence the initial slope should be one which provides minimum cover. All these are made possible through generating a detailed topographical survey.

2.4.2.6 Physical plans/land use plans

Approved physical, structure, and 3-D master plans are essential in drainage planning. The plans are used to ascertain the activities carried out in the region under study. Ultra-urban, urban, commercial, residential, industrial downtown, and other areas have a certain level of imperviousness. This is represented by runoff coefficients which is taken as a standard in drainage planning and design. The importance of this coefficient is to determine the extent of water draining on the surface and ends in the sewers. Different impervious areas will provide variations in the quantity of runoff. Suppose the area under study is covered with bituminous roofs, concrete pavers, and asphalt roads. In that case, the weighted coefficient helps to determine the extent of infiltration and runoff generated in the area.

2.4.2.7 Past study reports

These reports include existing drainage structure drawings, operation and maintenance reports, as-built drawings, data on existing drainage infrastructure, and flooding hotspots. Studies by other agencies are reviewed and possibly referenced.

2.4.2.8 Maps

Different kinds of maps may be collected including geographical maps, hydrogeological maps, cadastral maps, aerial maps, topographical maps, etc. These can be used for several purposes, which may include locating existing drainage systems, infrastructure and hotspots with frequent flooding occurrence will be marked on these maps. Geological maps provide information on the soils and basement rock types and location and type of geological hazards such as landslides and faults. Drainage area characteristics, namely, size, topography, land use, soils, and other developments may affect the site drainage patterns.

2.4.2.9 Utility service corridors

It is well-known that congestion of underground utilities is one of the items that most likely causes variations during project implementation. The location of underground utilities such as portable water service lines and sewerage lines is essential. Therefore, utility service corridors must be delineated on maps and specifically utility maps obtained.

A utility map shows the positioning and identification of buried pipes and cables beneath the ground. The procedure involves detecting things like sewers, electric cables, telecoms cables, gas, and water mains. When the mapping process is conducted together with a topographical survey exercise, the results give a comprehensive detailed map of anything that is hidden underground or directly related to any above-ground features.

Utility maps are important when breaking ground because they show accurate positions of the buried utilities. It also helps to prevent digging into or damaging any utilities that may cause harm to the public or workforce. For example, health and safety guidance published in the UK recommends checking for underground services prior to undertaking any ground works to meet the company's obligations under CDM regulations (2015) and the Health and Safety at work Act (1974) to ensure the safety of workforce and general public.

Utility maps enable avoidance and re-rerouting of utilities to be considered at the planning stage of a contract hence limiting unexpected project costs during construction. Since, urban drainage systems largely utilize road reserves in some countries, underground utilities and overhead utilities utilize the road reserve corridors. In many cases, the sewer inlets are usually located by the side of the roads. In some cases, relocation of service utilities is provided as a provisional sum in the bill of quantities. However, where utility maps are accessed and used, proper engineering judgment is used, which is usually founded on the adequacy and completeness of the design. That provides a reliable provisional sum. In Uganda and other emerging countries, there is a problem of inaccessibility to utility maps at the planning stage that risks wrong prediction on utility relocation costs, and this always brings about unexpected project costs during construction. In several emerging countries, utility surveys are poorly conducted, and in some cases, inadequate, which poses significant risks to workers and estimated project budgets.

2.4.2.10 Site visits to take area photographs

Area photographs are essential in the planning and design of drainage schemes. The planner or designer would need to evaluate the existing condition of the area and this necessitates to have adequate information which makes a collection of photographs essential. Site visit notes would indicate sediment or debris problems and the effect of stormwater on nearby structures. Also, environmental risk assessments would show properties and buildings near floodplain to clearly delineate sensitive floodplain values and convenient detours.

2.4.3 Selection of appropriate routes

Selection of appropriate routes necessitates juggling several constraints, for example, nature of available easements, underground utilities, earthworks involved, construction time, and environmental limitations. Where interruption of traffic needs to be avoided, micro-tunneling by jacking and boring is adopted.

Micro-tunneling is warranted where interruption to traffic would cause delays or where a scheme is more feasible when direct. Where sewers need to be laid under buildings, railroads, airfields, highways, crossroads, breakthrough rocks, tunneling and jacking pipes are inevitable. Thrust boring allows pipes to be laid easily with finite accuracy compared to other methods while evading disruption to traffic and the public.

Once underground construction by tunneling becomes the most appropriate option, a sustainable plan must be produced, clearly outlining the equipment, machinery, time, and skills required and associated carbon emissions. A best combination of open-cut methods and underground construction must be evaluated. When drainage sewers are raised, they may be anchored or inserted in retained soil masses. In most cases, stormwater sewers are not raised on anchors unlike sewage (foul) sewers.

2.4.4 Site survey and evaluation

Site survey in regard to integrated drainage systems planning and design looks at the big picture juggling between the applicability of SuDS and storm sewers for a given area going by site-specific constraints. In some areas, the topography and nature of soils may pose a constraint that necessitates the planner or designer to think about designing retaining walls, subsurface drainage structures, and flood defense systems. Several homes have backyard features with 0.6-1-m-tall terraced retaining walls that create a unique look and provide usable outdoor space. Other areas are prone to flash floods.

Is a retaining wall needed? A retaining wall is intended to hold back soil when there is a drastic change in elevation. Often retaining walls are used to terrace yards that originally had a steep slope. In some cases, SuDS may call for use of retaining walls to hold back soil, raise levels, and make the site aesthetically appealing. The retaining walls can also be used where the pipe is to be laid on the original ground with an embankment tipped on top (see Section 8.4.2.2). The drainage engineer working with other professionals would require information and data about the soils to be retained and the type of retaining wall most suitable for the site (see Section 7.9). Based on the site-specific constraints, both localized and global, the drainage engineer works through the optimal sustainable solution. Low retaining walls are often used as planting beds and can add interest to an otherwise flat yard. Figure 2.1 shows a cross-section of a cantilever retaining wall holding back the backfilled soil in which an infiltration system is developed. A combination of such structure would be applicable while terracing yards that originally had a steep slope and their evaluation together with backfill evaluation becomes pertinent. The design may opt for weep holes near the base of the retaining wall (see

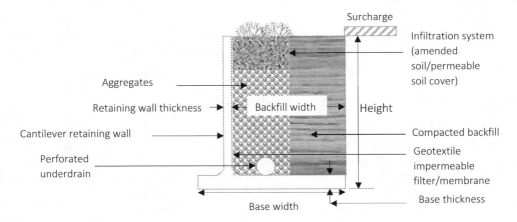

Figure 2.1 Cantilever retaining wall holding back soil and an infiltration system.

Figure 7.10) or an impermeable membrane depending on the purpose of the retaining wall. Perforated underdrains may also pass through the retaining wall to connect to the drainage system. The choice is left to the designer. Whichever wall system is found suitable for the project, the location, soil, and drainage requirements must be thoroughly evaluated.

Are flood defense or control structures required? Flood defense structures are primarily designed to protect coastal and riverbank areas, including urban and agricultural communities, households, and other economically valuable areas, and the people located within these areas. Nowadays, flash floods are also increasingly becoming common in recent years due to climate change. Municipal engineers and planners are increasingly tasked to plan ahead by providing flood emergency mechanisms, warning systems, and flood defense systems. Among the many systems that would be in place include flood defense walls well-built to mitigate seepage. The rising sea levels globally due to climate change is another issue that requires mitigation measures, adaptation strategies, and engineering controls aimed at avoiding devastating incidences when there are heavy rainfalls.

Planning for flash floods is gaining priority and some areas require planning for such floods; there would be need to control the impact due to flash floods by designing flood defense structures, such as flood control channels, seawalls, and other barriers. Such defense structures are designed to stop rising sea levels from flooding nearby communities. Evaluation of such flood defense structures primarily lies on the size and scope of the project governed by the overarching goal of sustainability.

Also, interlocking (lego) blocks are used to stop flash floods generated outside the area from entering the estate. Retention and detention ponds can be used as flood control structures to scale down the volume and velocity of the runoff generated upstream clearly withstanding seepage and well-engineered to control sediment load. For example, the disturbance of new areas for construction purposes, especially in hilly areas, would trigger voluminous runoffs. This may require the designer or engineer to plan ahead and have flood defense control structures in place to save the communities in lower areas from the devastating incidences due to flash floods. In order to reduce climate change impacts on flood control structures and the resulting damage and destruction to coastal and low-lying communities, designers and planners must adapt flood control structures to future climate stressors.

Are restoration programs necessary? The impact of improved drainage systems can lead to huge amounts of runoff collected from impervious surfaces which could potentially harm the receiving water bodies. The capacity of receiving water bodies may be overwhelmed by the runoff quantity and therefore planning for restoration programs may be vitally important. See Section 3.10.

2.4.5 Defining the catchment characteristics

Defining catchment characteristics refers to taking an inventory of features within a watershed under study usually taken out of the physical plans, site layouts, and available maps. These maps would include geological maps, topographical sheets, cadastral maps, and base maps. The land area from which all runoff flows to form a waterway is a watershed, which has several natural and artificial features. The boundary of a watershed is usually made up of natural features such as hills and mountains. A watershed is theoretically defined as the dividing ridge between two catchments with water flowing down each side.

Delineating catchments would identify several natural and artificial features. Artificial features include roads, buildings, townscapes, cultural heritage sites, ecological sites, utilities, dams, and other types of the built environment. Natural features include natural landscapes, hills and mountains, receiving water bodies, plantations, forests, and grasslands. Water bodies (oceans, lakes, rivers, and streams) are usually the receiving bodies; these may warranty gauging, flood protection structures such as river training, river restoration programs, and creating buffer zones. However, in some locations, rivers may not be gauged providing scanty data for use in the assessment. On the other hand, mountainous and hills, sometimes risk landslides—and therefore, geological surveys need to be properly evaluated in such areas.

By identifying the catchment characteristics, the project scope is defined aiming to produce a sustainable drainage design output. Modern sustainable drainage plans require holistic approaches by listing multidisciplinary teams and using advancing technologies that fit within available funding and regional legal frameworks. Several features within a watershed would influence the drainage scheme routes and type of SuDS. The applicability of SuDS depends on the physical plans characteristics of the watershed, that is, commercial, residential, institutional, ultra-urban, industrial, and recreational areas call for different kinds of SuDS.

The project scope would delineate the extent of hydrologic and hydraulic models to develop for the catchment. Catchments adaptable to a variety of SuDS can be different from those that call for only specific and limited SuDS. In many cases, a study on the receiving water bodies necessitates teams with appropriate skills and knowledge as part of the drainage planning teams—these might call for designing a package of restoration programs. River and stream restoration programs such as greening, in-stream enhancement, floodplain reconnection and wetland creation, managing catchments through a community-led based approach, removing or passing barriers (i.e., dams or weirs), restoring a more natural river course, sediment transport control, erosion risk mitigation, riverbank/shoreline erosion stabilization, and environmental remediation—contaminated riverine sites, erosion and sediment control permitting, habitat/ecological restoration, and creating buffer zones may be essential in order to reinstate natural processes for receiving water bodies to restore biodiversity and provide benefits to both people and wildlife hence meeting sustainability goals. River restoration techniques to adopt depend on river type, modification extent, and adjacent restrictions, for example, infrastructure such as roads or buildings.

The variance in catchment characteristics, globally, necessitates unique approaches to drainage planning assessing the potential for retrofitting, advanced water treatment methods for water with high ecological value, and multidisciplinary teams' expert levels. Ultra-urban areas usually limit SuDS practices, and sometimes greatly focus on stormwater quantity reduction and quality improvement through developing large infiltration and bioretention systems. Sometimes peak flow rates are lowered by integrating detention and retention systems in the planning. On the other hand, areas with cultural heritage sites greatly call for SuDS practices. Agricultural lands, residential, recreational areas, industrial, and commercial areas have close to unlimited potential for developing SuDS.

2.4.6 Environmental and social issues

The construction of drainage schemes comes with environmental and social considerations to avoid negative impacts severely damaging the ecological system. The receiving waters, that is, rivers, lakes, streams, ponds, swamps, and oceans constantly get polluted after rain events. In the pursuit of reduction in excessive pollution of the receiving waters, governments worldwide constantly require quality control measures adhered to ensuring that stormwater runoff quality is improved as one of the primary goals of a sustainable drainage system. In the United States of America, the National Pollutant Discharge Elimination System (NPDES), state and local water quality management regulations progressively limit impacts to receiving water.

Today's integrated drainage systems planning prepares drainage plans focusing on economic, social, and environmental issues. In the past, about 70 decades ago, economic assessments were the basis of undertaking a project. However, the 21st century emphasizes striking a balance between economic, social, and environmental factors. With origins in the USA, in the 1960s, the environmental and social impact assessments (ESIA) are usually conducted for infrastructure projects globally. In that case, while defining drainage schemes, planners and engineers focus on the impact of drainage systems' development on air quality, water quality, and cultural heritage. They also consider potential noise generation during construction, disturbance of ecological system, disruption to service delivery during construction and how communities are affected, that is, resettlement planning.

Drainage scheme planning and development requires sound environmental and social impact assessments because developed areas always greatly increase volumes of runoffs. Therefore, drainage systems improvement must conform to standards that control stormwater runoff volumes and quality. Also, drainage comes with voluminous excavations to lay sewers and construct open drains and modifies the hydrology of the area especially where SuDS are an integral part. Drainage schemes require sediment and garbage control mechanisms because stormwater runoff always collects sediment load and garbage as it routes off impermeable surfaces. Evaluation of potential sediment or debris problems in the area planned for drainage systems upgrade is essential.

In industrial and commercial areas, runoffs always mix with pollutants such as oil spills and would require oil interceptors and sedimentation tanks as appropriate. In some instances, sediment basins and grit traps are installed to control sediment loads carried in the runoffs. During impact assessments, it is imperative to evaluate the probable sediment loads, oil pollutants, and other chemicals of great impact on local water quality. Receiving waters of high ecological value may require advanced treatment techniques to maintain the quality of waters.

Resettlement action plans are drawn where people need to be relocated to provide for drainage easements. Rapid urbanization has led to haphazard construction in many developing cities globally where physical planning policies are poorly enforced which causes adverse effects from extreme flood events because developers always encroach on drainage easements. When natural watercourses are interrupted, and no new provisions for drainage easements are created, flooding becomes inevitable. Where routes seem feasible when constructed more direct, compensation for land to evict siting occupants is warranted. Adequate compensation costs for new drainage easements always influence a drainage scheme route choice. In that case, deciding on underground tunneling, for example, under buildings or developing open drains removing siting occupants would be guided by the cost of current buildings sited on the land, complexity of work, time, and interruption in service delivery such as causing traffic delays, and peak flows.

Environmental risk assessments consider buildings near floodplain, sensitive floodplain values, and existence of convenient detours. Buildings in floodplain areas require evaluation of underground construction tunneling and detouring around the buildings. Therefore, consequences of drainage schemes on nearby structures are evaluated.

Drainage schemes come with immediate economic benefits like employment to locals during construction and drainage systems maintenance which is counted as a social benefit because it creates an immediate social transformation in the affected communities especially developing communities worldwide. This socioeconomic benefit improves the welfare of the involved people.

The integrated drainage systems planning would holistically consider the environment alongside economic and social benefits. In terms of the environment, climate action for sustainable development is the overarching goal which emphasizes low-impact development by minimizing impervious surfaces whilst promoting use of low-carbon construction materials and techniques, and greatly mimicking the natural environment to create a constancy in the hydrological conditions of the area. In that perspective, SuDS are greatly chosen going by site-specific conditions to control any adverse impacts on native species and regional soils, while standard pipework solutions are dictated by soil shear strength parameters, groundwater conditions, and construction materials.

In the sustainable development frameworks, preserving cultural heritage is a social factor where their disturbance is greatly assessed. In that case, demolition of ancient monuments, buildings, and archeological remains is prohibited. Similarly, landscapes are maintained to the extent possible.

Excessive excavations pose safety risks during construction and an effective safety management plan must be provided during construction. Accidents on sites can result in prolonged and costly civil cases that can delay drainage projects for a long time hence influencing total investment costs. Therefore, assessing the geology and soils of the area is essential because drainage construction practices greatly destabilize the soil structure and also expose rock formations. Measures to minimize adverse effects of excavations must be assessed during drainage planning and construction. The geology and soils of an area are evaluated to predict potential landslides.

Excessive excavations for drainage structures may lead to collapse of adjacent structures. Therefore, digging very close to, or under, the foundations that support nearby buildings or structures may need to be evaluated because it can undermine the support of the building and causes collapse of the building or structure into the excavation. This would even be more dangerous in case the excavation itself were also to collapse. Therefore, excavation techniques as outlined in Section 10.3.2 must be evaluated.

Areas with landfills influence drainage systems practices and scheme routes direction. The need to always keep landfills dry necessitates planning drainage schemes giving way to trash not combining with groundwater. In the same way, stormwater sewers and drains will always be diverted not to cross through landfills. Planning for detours is essential in areas with landfills, contaminated land, and environmentally sensitive areas.

In the same way, drainage construction materials, that is, concrete pipes, kerbs, precast manholes, box-culverts, and blocks are heavy, and their transportation needs sound planning during loading, offloading, and installation. Selecting appropriate routes to transport materials would significantly influence embodied carbon as a result of vehicle emissions. For safety purposes that significantly form part of social issues, the involved stakeholders are required to minimize manual handling, and materials wastage as a result of transportation structural failures.

2.5 DRAINAGE SYSTEMS PLANNING

2.5.1 Setting goals

As a drainage planner, goals setting is a vital step in the integrated drainage systems planning and design to effectively manage and control stormwater and stormwater runoff. The goals differ from one project site to another, based on the vision and mission and community policy frameworks. However, effective control of stormwater at the source remains the overarching goal of constantly evolving drainage planners and designers worldwide. The decision-maker outlines the desired result or outcome of the process. The range of primary goals includes the following:

2.5.1.1 Stormwater quantity reduction

The designer might require to reduce stormwater quantity entering the sewers and drains. This can be accomplished through implementing SuDS such as infiltration systems, bioretention system, permeable pavers, and rainwater harvesting techniques upstream. These must be integrated in the drainage network and the quantity of flow would be reduced.

2.5.1.2 Pollutant removal

As a primary goal, the designer might find it relevant to remove pollutants from stormwater. Several factors influence the decision to design a particular SuDS or integrated SuDS to remove pollutant loads, which may include the site location, amount of pollutant loads, activities conducted on the site, extent of pollutants, that is, localized or global, etc.

Pollutant removal ranges from removing toxic and hazardous materials, for example, oils and grease. The designer evaluates the need to remove pollutants from technical and cost-effectiveness point of view. This is informed through assessing the level of stormwater pollutant load and then decide on the choice of technology to remove the pollutant load. In many cases, treating stormwater is not economical due to the large volumes. Will you remove pollutants, is it economical to treat? Or will you use a different method? These are typical questions to answer.

Stormwater quality indicators that are commonly used to predetermine the effectiveness and efficiency of a particular SuDS are total suspended solids (TSS), total phosphorus (TP), and total nitrogen (TN) content. The indicators are evaluated for a given location by taking samples during or immediately after it rained. Based on the results, the suitability of a particular SuDS to perform effectively is evaluated and the SuDS planning and design are done accordingly.

Pollutant removal and stormwater runoff quantity control is decided based on site-specific conditions. Section 3.6 provides site-specific conditions that influence the suitability of particular SuDS. For example, infiltration systems are good at scaling down the stormwater flow volumes thus enabling interconnected storm sewers not to be overworked. However, the topography, water table, and native soils are primary factors that affect the suitability of infiltration systems. Also, sediment and grit control may be a primary goal. In that case, sediment basins or sediment traps may be designed in the drainage network. In many cases, sediment traps are specified where sediment basins are not feasible.

2.5.1.3 Peak flow reduction

This can be accomplished integrating ponds, such as dry ponds and wet ponds in the drainage networks. In many instances, peak flow reduction and velocity reduction work hand in hand especially on steeply sloping sewers or drains. Integrated drainage planning benchmarks on incorporating SuDS such as bio-swales with check dams, ponds, backdrop manholes to serve multiple purposes.

Also, steep areas warranty velocity reduction techniques employed such as backdrop manholes for pipes and check dams for drains. As an example, in the interest of planning for sustainability, vegetated swales may be incorporated in the drainage network and check dams made of wood, stone, or concrete are often employed in vegetated swales to enhance infiltration capacity, decrease runoff volume, rate, and velocity. Bio-swales specifically designed to attenuate and treat stormwater runoff for a defined stormwater volume.

These techniques are developed negotiating slopes in a stepwise manner, presented as flood wave movement control systems. In that case, the energy is significantly reduced as the flood wave propagates through the channel. See Figure 2.2 for a drain with controls and Figure 2.3 for backdrop manholes used to avoid steeply sloping storm sewers. The check dams in Figure 2.2 may be designed to help detain the runoff hence serving as a minor detention system. The check dams can be designed with an outlet at the bottom so that after the storm stops, the collected runoff is released at the appropriate time.

High velocities cause erosion and corrosion of pipes. In the recent past, erosion-resistant materials such as cast iron were used in areas susceptible to erosion and corrosion of pipes. Pipes are eroded where high velocities are encountered, especially velocities above 10 m/s. Corrosion is the primary factor affecting the longevity and reliability of buried pipelines throughout the world. Therefore, epoxy coatings are extensively used offering excellent resistance to high temperatures, chemicals, and corrosion.

High velocities are usually negotiated stepwise as shown in Figure 2.4. As an example, where a reduction in velocity, removal of pollutants, and peak flow reduction coincide as primary goals, a grass swale may prove a competitive option replacing standard gray kerb and

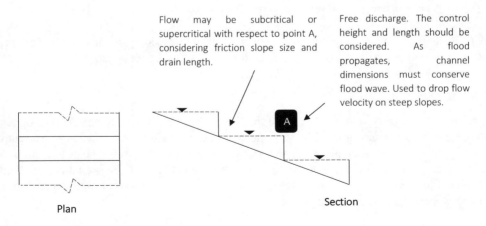

Figure 2.2 Drain with controls.

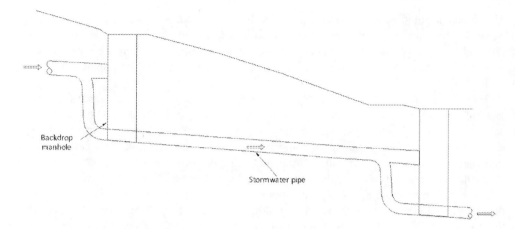

Figure 2.3 Backdrop manholes on a steep terrain.

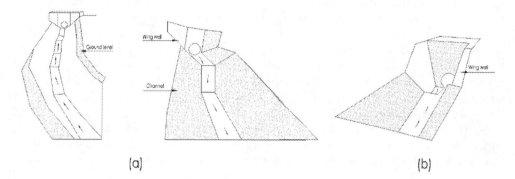

Figure 2.4 Steep slope negotiated stepwise to reduce flow velocity.

gutter system as shown in Figure 2.5. However, this is influenced by the quantity of flow, the site-specific conditions such as native plants, topography, and space constraints.

2.5.1.4 Reduced costs

A reliable drainage system would reduce the lifecycle costs for the project. As a designer, this might be a primary goal to ensure the planning fits within the financial constraints of the entity. A combination of capital and operational expenditures would be generated from different appropriate drainage options to weigh the associated costs before taking the final decision. Financial constraints influence choices. Lifecycle costs also predetermine the SuDS and corresponding gray infrastructure.

2.5.1.5 Low-carbon design

Following the need for each project to contribute to net-zero targets, the planner is tasked to provide a carbon offset sheet for each project. The carbon footprint calculation is a must

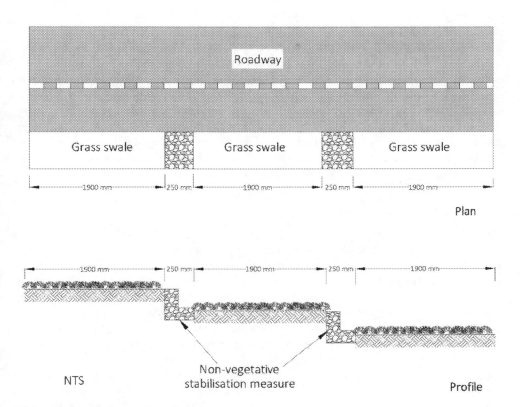

Figure 2.5 Grass swale along a highway.

have to approve a project as sustainable. This means optimal designs are necessary following a back-and-forward approach to minimize concrete use and other materials in building gray infrastructure (sewer pipework standard solutions). Again, this means design iterations would be conducted scaling down the size of concrete sewers and drains as much as possible.

In Section 2.7, playing with the Froude numbers for approach and exit flows at a given point together with mild/steep construction slopes always evaluates the available energy and localizes constraints leading to optimal sewer sizes and thereby reducing concrete use as much as possible. The goal of negotiating slopes is to ensure that we nearly get close to critical slopes. The reduction in sewer sizes is quickly made possible when supplemented with SuDS that absorb most of the runoff especially integrated infiltrations systems, bioretention systems, and detention/retention systems as appropriate.

Low-carbon design means reduced cost, as the site development tries to minimize impact on area hydrology. There could be a need to implement pollution prevention, proper maintenance, and public education programs in order to conserve natural areas wherever possible. The implication is that pave only areas you need to and leave other areas green. Achieving low-carbon design necessitates to scatter integrated SuDS across the site. Integrated SuDS are decentralized, microscale controls that infiltrate, store, evaporate, and/or detain runoff close to the source. For example, a site may be designed to have bioretention gardens, bio-swales, infiltration systems, etc., significantly reducing peak flows, quantity of flows, and effectively treat the stormwater runoff. Consequently, the stormwater sewers and drains receive the minimum flows, and this reduces capital carbon in concrete and associated materials used in gray infrastructure networks.

Among primary goals while planning for a low-carbon design, protecting the ecosystem is included. This includes protecting sacred sites and sanctified places that require careful planning. River restoration programs would be a primary goal where high volumes of stormwater runoff are anticipated thereby necessitating incorporation of SuDS following national statutes and regulations aimed at producing resilient communities. For example, creating stream buffers along the receiving stream or river would be essential so that they protect the stream or river to function properly. This greatly reduces the impact of human activity hence promoting the ecosystem while contributing to sustainability going forward. Buffers can differ greatly, ranging from flat floodplains to steep gorges.

2.5.1.6 Construction materials

The economics, durability, reliability or strength, and sustainability influence the choice of construction materials. The choice of materials greatly influences sustainability going forward. For example, preference for reinforced concrete pipe (RCP) or corrugated metal pipe (CPM) weighs between the economics and environmental/site conditions, especially waterlogged areas, may call for CMP in order to resist differential settlement. In waterlogged areas, concrete pipes usually experience premature structural failures. In some areas, subsurface detention facilities normally use CMP over high-density-polyethylene (HDPE) because CMPs resist differential settlement and have a higher flexural strength. Section 8.9 provides different types of materials for storm sewers. It also outlines the structural aspects of sewers.

Construction materials for use building storm sewers must be evaluated at the start to ascertain the availability, reliability, durability, and sustainability and to plan for the challenges that might be faced in construction. The available construction material in the locale inform the hydraulic efficiency of the drainage networks. For example, HDPE pipes are used because it is more expensive, and time-consuming, to rehabilitate RCPs than it is to repair HDPE pipes. HDPE joints are known for being watertight as the pipe. Their joints are fusion welded together, needing no extra couplers, grout, or other sealants to install.

2.5.1.7 Land stabilization and embankments

Sometimes it is essential to evaluate the use of geotextiles during the evaluation processes. Geotextiles are used for soil separation, soil stabilization, and for permeability purposes. Embankments may be used on sewers to be laid on the original ground.

2.5.1.8 Hydrologic and hydraulics methods

The hydrology and hydraulics techniques, standards and codes of practice, statutes, regulations and policies, and design software for use should be evaluated and goals set. The planning must comply with and conform to the statutory laws and regulations. For example, in the UK standards and guidance manuals such as Sewers for Adoption, design manual for roads and bridges (DMRB), and CIRIA C753 must be consulted and used in the integrated drainage solutions.

It is vital to note that flow routing techniques influence costs; depending on which method is used it may result in a large hydraulic radius, which is costly implying much carbon too!

What technique will you use to design pipework solutions (gray infrastructure) is a question that must be answered while setting goals. At planning stage, the designer decides which routing technique will be used. There are two commonest methods: the dynamic wave and kinematic wave. The dynamic wave method solves the complete St. Venant equations for shallow water.

The dynamic wave routing is used in complicated networks such as looped networks, pressurized networks, or tailwater conditions. Kinematic wave routing is a simplified form of the momentum equation found in each channel or pipe. If you have pressure or tailwater conditions in the network, it is not advised to use this routing method[2] (AutoDesk Civil 3D).

SuDS, such as detention ponds, are sometimes interconnected with standard pipework storm sewers and drains. Therefore, modeling the drainage system using different methods produces different results. In setting primary goals aimed at sustainability, SuDS would be integrated in the stormwater sewer networks, for example, sewers and drains interconnecting with detention ponds. Sometimes, stormwater sewer networks only would be ideal. In other instances, the subdivision of drainage systems would be necessary. Other considerations in goal setting would include sizing and designing of detention/retention ponds and outlet structures, highway drainage systems (including kerb and gutter inlets), bridge and culverts, including roadway overtopping and water quality analysis.

2.5.1.9 Other constraints

Several other constraints that would influence the drainage scheme routes, such as presence of rocks, and alignment would be evaluated to set appropriate goals.

2.5.2 Defining drainage networks

Defining drainage networks considers several factors that include the contributory basins/catchments. The designer identifies where junctions and manholes will be located as a preliminary step. Manholes can be made of masonry or reinforced cement concrete (RCC) chambers constructed at suitable intervals along the alignment of sewers.

The functions of manholes and junctions include the following:

a. They reduce the energy of flow on a steep terrain.
b. They help in cleaning up the sewerage system—debris can also be removed at a manhole by inserting an ell section in the outlet with the open end turned upward.
c. They provide access to the sewers for the purpose of inspection, testing, and removal of obstructions from the sewer lines.
d. They can boost the available energy for the next link (pipe or drain).
e. The more the number of manholes, mini-manholes, and junction boxes in a sewer network, the better especially in storm sewers of relatively small sizes, for example, 600-mm diameter and below. On straight reaches of sewers, manholes are provided at regular spacing that depends on the size of the sewers. The larger the diameter of the sewer, the greater may be the spacing between two manholes. The spacing between manholes also depends on the type of equipment to be used for cleaning sewers.

f. Whenever there is a change in direction of flow, the designer may propose a manhole, junction box, or a mini-manhole. As a rule of thumb, manholes may be provided at every change in alignment of sewers, at every change in gradient of sewers, at every junction of two or more sewers, at head of all sewers or branches, and wherever there is a change in size of sewer.

In some instances, urban sewers rarely span more than 100 m before connecting on a junction or manhole or meeting some other kind of controls. The decision to plan for controls may be decided during the evaluation stage, for example, weirs or flumes control the flow in a channel. Also, long sewers are control systems because flow tends toward normal flow. The first-line technique in urban drainage is to identify drain or sewer network junctions first. Sketch the way they should look like. Imagine the type of flows that would happen at the type of junction envisioned: normal flow, critical flow, subcritical flow, and supercritical flow—of course together with the bed slopes of the adjoining culverts/drains. For example, you can locate the outfalls first. Then locate the junctions (manholes, mini-manholes, and drop inlets). You can envision the flow profiles at specific points in the network as you move upstream. Integrated drainage systems planning, design, and development incorporate SuDS as nodes or link. Node SuDS may include detention and retention ponds, bioretention gardens, tanks, and infiltrations ponds/basins. Link SuDS may include swales, vegetated open-channel streams, and downspout disconnection.

In defining drainage networks, the primary consideration in coming up with an optimal hydraulic design is checking for surcharging and flooding of sewers. Therefore, an optimal design considers sewers not to flow pressurized and never to flood, considering the average recurrence interval (return period), which is normally considered based on the predicted level of risk.

Slopes are a primary consideration when evaluating the process to design effective stormwater runoff conveyance systems. The slopes significantly influence flow velocity, available energy, and flood wave routing time. Areas with varying slopes, i.e., steep, mild, flat, would influence the pipe sizes, pipe material, bending of routes, and position of junctions and manholes.

Preliminary stages would indicate steep slopes which risk high velocities and where supercritical flow would dominate. Similarly, subcritical flow dominates where mild slopes are predominant. The primary goal of forecasting the likely flow patterns is to ensure that flood waves are perfectly conserved in the drainage system at an optimal design that is reliable, cost-effective, and sustainable. This is because overdesigning leads to increased costs, which leads to more embedded carbon. Therefore, an optimal hydraulic radius is always arrived at through preliminary data gathering which include evaluating the available easements to ascertain the headroom that can support various or specific styles of hydraulic radius (geometric shapes), and through preliminary hydraulic radius forecasts, a back-and-forth analysis of the carbon costs and lifecycle costs for competing materials can be done together with extensive hydraulic modeling.

Therefore, defining drainage networks would juggle constraints to make normal flow equivalent to critical flow leveraging minimum energy as much as possible in order to minimize the size of hydraulic structure. In that case, the entire stormwater drainage sewer system would reduce capital carbon significantly. Figure 2.6 provides a conceptual diagram showing how links, that is, drain or conduit could connect with nodes, that is, junctions, inlets, or manholes while defining drainage networks.

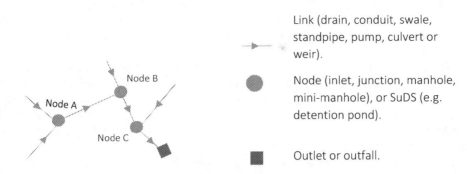

Figure 2.6 Conceptual diagram for a drainage network.

In defining drainage networks, it should be noted that constraints influence decisions on routes and structure types. For example, buildings, structures, utility lines, soils and rocks, socioeconomic factors, and safety factors affect scheme routes and type of drainage infrastructure asset to develop in the area. Haphazard housing units greatly affect drainage routes. Where underground utilities are so great, it would need relocation or avoiding by finding an alternative drainage route. The financial resources to carry project through implementation must be evaluated while defining drainage networks.

Conduit or drain? The available headroom influences the shape of the hydraulic structure, box or circular, U or arch, among others. However, a circular culvert is the best geometric section among all in terms of hydraulic efficiency. Adopting U-channels answers many questions surrounding several constraints. This is because U-channels' geometry provides a versatile situation in easing calculations (hydraulic radius) and also fitting within local site constraints, for example, available easements for stormwater sewer lines. This answers many sustainability goals going forward. For hydraulic performance, a circular pipe is best geometric section among all.

Central business districts (CBD) and ultra-urban are usually congested and generate increased quantity of flow. Underground stormwater pipelines are warranted when the scheme seems more feasible when more direct. Tunneling under buildings and roads is also proposed where space is a constraint, and the scheme is good when left in a straight line. Sometimes large detention ponds are built underground, in ultra-urban environments. From Figure 2.6, constraints influence choices on scheme layouts and drainage alignment. Above ground, micro-tunneling, and/or underground are all warranted juggling constraints posed by the environment, space, cost, safety, complexity of alternative routes, avoiding interruption to service delivery, avoiding traffic jam, among others. Considering safety and cost, the choice between underground and open-cut methods is done. Above the ground pipelines are dictated by the environmental conditions, such as crossing wetlands, and congested underground utilities may warranty anchoring drainage pipelines, to move above the ground especially combined systems.

Above the ground storm sewers are not very common in stormwater conveyance but are warranted when the stormwater runoff requires treatment and has to cross rivers, streams, railways, and canals before being discharged. In some commercial and industrial areas, stormwater may mix with spillage from tank vehicles and processing plants. This may necessitate designing a partially separate system which calls for draining stormwater to a treatment plant. In that case crossing rivers or streams might need raising pipes above the ground by resting them on piers, bridge works, and embankments. Lower pipes may be constructed from mass concrete or brickwork, while reinforced concrete piers are used for higher piers.

In some areas, a partially combined system may be warranted, for example, where rainwater mixes with a working area contaminated with grease and oils. These include garages, warehouses, and fuel stations. This requires the stormwater runoff to pass through grease traps, oil separators, and grit separation devices before discharging to public sewers. Also, wastes from kitchen sinks, hotels, and restaurant floor areas may mix with stormwater and would need a partially combined system.

Cast iron and steel pipes are the commonest pipelines used. However, cast iron presents an advantage over steel due to its strength in resisting corrosion. However, both pipes can span longer length. Manufacturers usually specify the length the pipes can span. Cast iron is usually manufactured with spigot-and-socket end.

In defining drainage networks, manholes are provided where there is change of direction, slope, and pipe diameter. In many cases, storm sewers are laid in straight lines, at least 110 m apart on long straight length. Except sewers large enough where humans can work in, the length of 100 m can be increased. Accessibility in manholes is key for trucks and humans working to easily access the locations. Inspection of manholes need to be made as easy as possible. Other vital considerations include the following:

- Shallow foundations might need concrete surroundings on sewers especially under roads.
- Inspection chambers are vital for cleaning and sewer maintenance activities.
- Manhole spacing should be considered for easy inspection.
- Considering curved lines over straight lines has gained popularity in the recent past.
- Sometimes engineers may space manholes as close as possible where slopes greatly vary to about 80 m.

Will channels receive flow along their length? Is there a need to plan for controls? What could be the reasons to warrant receiving flow and discharging flow along the channel lengths? In simplified cases, lateral inflows are usually neglected. However, where lateral inflows are predicted, hydraulic methods, for example, momentum principle and flow routing techniques, must be properly selected and applied. This means lateral inflows warrant the level of simplifications of these techniques as a key consideration.

2.5.3 Cost of SuDS and gray infrastructure

The cost of SuDS varies widely considering site-specific constraints that necessitate different approaches for design and implementation. In many cases, integrated management practices such as the bioretention gardens, which sometimes serve as infiltration systems, may call for high initial cost investments. However, to reduce cost, the designer targets applicable SuDS that are site-specific. This is derived from the local and global constrains such as available easement/land, topography, climate, native vegetation, soils, and the geological conditions of the place.

2.6 SOFTWARE APPLICATIONS

Software applications are increasingly being used for design purposes for several reasons, which include 3-D modeling simplifications while interpreting information to management, public authorities, and relevant stakeholders. Smart stormwater management and drainage begins with smart planning and design. Several software can ably optimize the performance and effectiveness of drainage systems, including SuDS—both Greenfield and retrofit. Software applications for a drainage engineer includes the following:

- AutoCAD CIVIL 3D, AutoCAD Storm, and sanitary analysis produced by AutoDesk
- GIS
- EPA SWWM
- HEC
- MicroDrainage from Innovyze®
- InfoDraiange from Innovyze®
- Innovyze®
- Bentley StormCAD
- MATLAB
- MS Excel®

The drainage engineer plans which software to use, imports survey data, import images and geo-references as appropriate. Modeling drainage systems in software saves time and cost. It improves the design precision—errors are almost inevitable where software is not used. In many cases, design oversights are inevitable where hand calculations are performed. After careful evaluation exercise, the drainage engineer can delegate less skillful engineers to perform some parts of the design with ease, using the software. This speeds up the exercise.

However, the most important issue is that the modeler must customize the software application to the regional policies, design standards, and codes of practice just in case it is of foreign jurisdiction. Knowledge and understanding of the software are essential for correct use. There is a big challenge in emerging economies where engineers greatly rely on software applications from North America and Europe. This impacts on the accuracy of results especially where the software users are not well-conversant with the latest technologies, lacking knowledge of customization procedures, applicable international and regional standards, and latest developments in the field.

2.7 THE DECISION-MAKING PROCESS

A multidisciplinary approach to drainage planning would generate holistic interventions that would be sustainable in the long run. Following economic comparisons conducted in line with environmental and social impact assessments, a decision is made on which drainage scheme or route is more feasible. Weighing the benefits and costs of different alternatives within the region's institutional framework, legal/policy framework is essential while taking decision on drainage and sustainable stormwater management solutions.

In addition, the decision on construction of drainage systems with available materials, funding, and equipment/machinery considers the whole lifecycle phases of the project. This means building reliability in the project from inception, planning, design, construction, operation and maintenance, and decommission/repairing/expansion. After goals are set as shown in Section 2.5.1, it is essential to consider operability and maintainability as independent variables that inform reliability. However, optimization is the backbone of sustainability. This is paramount in taking the final decision.

The design of hydraulic structures requires consideration of different but interrelated fields of planning, specifications analysis, hydrology, hydraulics, structures, installation, durability, maintenance, and economics. Usually most of the aforementioned fields are considered except two items, that is, the durability and economic aspects, which are generally not given proper consideration in drainage projects. Several drainage schemes select pipe materials or systems based on an initial (or capital) cost basis only. However, lifecycle (least cost) analysis is not given the attention it deserves which puts the reliability and sustainability of the project at a somewhat risk. The application of least (lifecycle) cost analysis to drainage projects has increased dramatically in recent years. It is now essential to conduct lifecycle (least cost) analysis before deciding on materials to use in drainage schemes for both gray infrastructure and SuDS. Local and state governments have increasingly included economic analysis in their material selection process.

In a reliable integrated drainage system, materials and systems that are more cost-effective going by the lifecycle (least cost) analysis are selected and parameters used in the hydraulics and structural design of the storm sewers. *Sustainable drainage projects' feasibility, economy, efficiency, effectiveness, and reliability are approached from a cost-effectiveness approach unlike other infrastructure projects.* Which combination of materials and technology provides the most cost-effective lifecycle cost analysis (LCCA)? That is the question to answer. The parameters to consider include Manning's roughness factors, strength to carry dynamic and static loads, resistance to thermal expansion, resistance to chemical attacks, and sustainability issues, that is, how much carbon is associated in the design and development of the material and the associated construction process. The choice to specify HDPE, CMPs, or concrete pipes should be based on LCCA and durability first before going into the hydraulics and structural design. A back-and-forth criterion may be necessary to arrive at the optimal solution that is applicable to solve a drainage problem following sets of circumstances, conditions, or factors. For more information about lifecycle (least cost) analysis, see Section 9.3.4.

During the evaluation process, the loading cases on storm sewer pipes need to be envisioned and assessed. Whether the pipe will be laid in normal trench conditions or wide-trench embankment, the designer needs to evaluate options following standards of practice such as BS 9295:2020 Guide to the structural design of buried pipes[3] (BS 9295:2020 Guide). This is because loading cases influence the reliability of the storm sewers in terms of bedding classes and strength of the pipe. This consequently influences the LCCA and carbon estimation. Chapter 8 presents detailed account of the loading cases for further reference.

A back-and-forth design goal to achieve sustainability in integrated drainage systems is achieved when the project is considered through all the lifecycle phases, that is, from planning, design, construction, operation and maintenance, and decommissioning. Several gray infrastructure may require repairing and replacing before the design life ends. Sometimes the service life may be longer than the design life. In case a material, system, or structure has a service life greater than the project design life, then it would have a residual value. Therefore, comparing lifecycle costs for alternative materials or drainage systems with their hydraulic efficiency, effectiveness, and sustainability metrics, the optimal solution is obtained. This is done by creating flowcharts where the designer iteratively drops integrated solutions that do not address the drainage problem optimally, reliably, and sustainably.

The back-and-forth criterion is achieved by juggling different types of SuDS suitable for an area. This is done considering a number of factors, that is, the lifecycle costs, amount of carbon emitted throughout the life span and that absorbed, site-specific factors, and availability of maintenance teams and maintenance funds. The goals set in Section 2.5.1 might require runoff velocity reduction, quantity of flow reduction, or peak flow reduction; this influences the nature of SuDS and the final decision.

Drainage prioritization is the work of planner or engineer. Drainage networks constantly have primary and secondary drainage networks. In some cases, they are called major and minor drainage networks. That means specifying what you call minor and major drainage systems is solely a responsibility for a drainage planner or engineer to advise the public authorities. The minor system may be designed to accommodate storms of short average recurrence intervals typically between two and five years which primarily depends on the nature of the area. The major system should be adequate to carry flows resulting from storms of long average recurrence interval of at least 25 years.

The planner delineates the types, that is, minor or major well for proper planning, operation, and maintenance. Catchment delineation and annual exceedance probability measures must conform to what is required, either primary sewer or secondary sewer. In the prioritization, the planner assigns a risk per scheme that is complementary with annual exceedance probability. In this case, the area served by the drainage scheme is vital; the rainfall received and activities conducted in the area are influential factors in drainage planning.

The existing condition of the drainage system, if any, is evaluated both qualitatively and quantitatively. The results would include the nature and size of hydraulic structures, the materials used to build the existing structures, the maintenance practices carried out on the hydraulic structures, the institutional and policy frameworks that guide maintenance, and new developments.

The decision to remodel, rehabilitate, or replace the existing structure with a new hydraulic structure would be reached at after setting goals as provide in Section 2.5.1. These goals result from the analysis of the extent of the drainage problem and of course fitting within the public authority's vision. Today's conventional drainage planning approaches prioritize sustainability over and above any other vision, globally.

The extent of the drainage problem would indicate whether the existing technology used to collect, convey, store, or treat stormwater and stormwater runoff has become obsolete, worn-out, silted, of inadequate capacity, or outdated. The available resources, therefore, would influence remolding the existing drainage infrastructure or developing new drainage systems.

The decision is highly based on whether the drainage system will be safe, economic, and sustainable. In that regard, public participation/engaging community is essential through stakeholder consultations. Taking a decision from a professional judgment point of view weighs the cost, time, safety, and social and environmental aspects clearly defining sustainability—embedded and operational carbon.

As a means of financing and funding, traditional public sector financing (TPSF) is commonly used, which includes loans and grants, revenue and taxes collected, which could include own source revenue. Private financing initiatives, that is, public–private partnerships (PPP) include NGOs and individuals partnering with public authorities. The decision to take which financing alternative is guided by sustainability goals and availability of the funds.

Traffic management during construction, closure of lanes, open-cut method versus micro-tunneling, time, and budget are primary factors when evaluating the decisions on drainage and stormwater management projects. The available machinery and equipment, manpower, and timelines are decisive factors that highly influence drainage projects.

Construction time is an essential component of drainage planning. In some cases, where the water table is high, micro-tunneling may not be feasible except when it is completely unavoidable. In that case, techniques such as ground freezing, pumping, and well-point dewatering and chemical injection may be applied. This may require hiring experienced specialist contractors. For instance, where soil needs to be stabilized to prevent collapsing next to excavations, or to prevent contaminants spilled into soil from being leached away, ground freezing is applied. Ground freezing is a construction technique that has been used worldwide for at least 100 years.

In terms of sustainability, a scheme route is feasible only when it strikes a balance between social, economic, and environmental factors. A scheme route would be feasible following sustainability guidelines only when it strikes a balance between social, economic, and environmental factors. That means the decision to adopt a scheme route as the best choice would consider the total lifecycle costs. The costs would include the carbon costs resulting from the amount of energy used in construction, operation, and maintenance phases. In the bid to reduce carbon emissions, SuDS would be adopted to supplement the standard pipework solutions in a back-and-forth approach to find the optimal point.

Figure 2.7 is a typical site layout example of how SuDS can be incorporated on the site to minimize the impact on the environment by mimicking the natural environment. Before deciding on the type and capacity of sewers and drains to remove excess stormwater runoff, several SuDS are integrated on the site. The figure shows a network of SuDS, that is, bioretention gardens, swale, permeable pavers, tree trench filters, rainwater tank, and tree planting (reforestation). Depending on the area hydrology, the amount of stormwater to be captured by a particular SuDS is evaluated following local regulations and standards of practice. The local manuals can guide the planner or designer on the quantity of stormwater runoff controlled and managed by a SuDS per unit area of the catchment.

Section 3.8 provides an account of how SuDS can be planned and their design criteria. It provides an account of key design factors. Following an estate planned for stormwater control and management, sustainability can be achieved when juggling constraints, back and forth, thereby minimizing runoff ending in storm sewers, reducing heat in the estate, capturing heavy pollutants, and finally reduce the risk of flooding.

As you conduct a back-and-forth analysis, you find that the proposal may not work because of site constraints, for example, slope, nature of soils, cost or technology constraints, etc. This may necessitate revision to adopt the next most appropriate type of SuDS. Sometimes, an integrated stormwater management strategy is preferred—one that reduces the quantity of flow, reduces velocity, captures pollutants, and significantly drops the peak flow discharge. That is the reason bioretention gardens are proving essential SuDS than any other type.

Figure 2.7 shows a site planned to significantly reduce the impact of the built environment to the predevelopment site hydrology in a suburban environment. It shows bioretention gardens, vegetated swales, cistern, green roof, reforestation, and tree filters incorporated in the project development. The main storm sewer receives the excess follows from the project site. The underdrains from the bioretention gardens, swale, and permeable pavers transport the excess flow to the main storm sewer. Understanding the area hydrology informs the

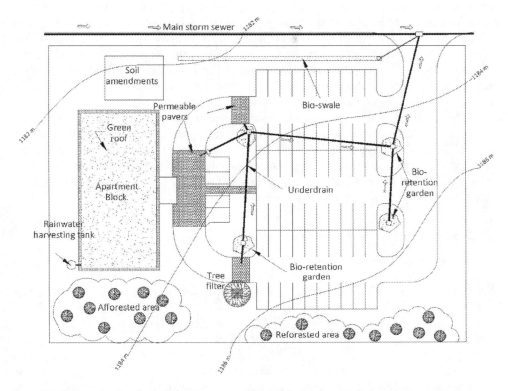

Figure 2.7 Site planned to have little impact on site hydrology in a suburban environment.

appropriate design of the SuDS and capacity of the main storm sewer is reduced significantly. See Section 3.8 for specialized key design factors and approaches.

Decision to extensively plan and design for SuDS may be influenced by topography of the area. Highly steep areas warranty runoff velocity reduction strategies, lowering time of travels and influencing the time of concentration downstream to reduce the risk of flooding. It is prudent while planning to imagine variances in catchments with, for example, sacred sites, hospitals, schools, golf courses, recreation centers, commercial and residential estates, roads, and other infrastructure of any kind. This variation in land use will necessitate the planner to holistically plan while considering sustainability goals, lifecycle costs, reliability, safety, and health. The choice to specify SuDS such as bioretention gardens, infiltration systems, swales, and others should be based on LCCA and durability first before going into the hydraulics hydrology. This is an integrated drainage design which includes both gray and green infrastructure to contribute to sustainable development. For more information about lifecycle (least cost) analysis, see Section 9.3.4.

2.7.1 Reliability and sustainability

Reliability is a function of operability and maintainability. Sustainability is about economizing resources while contributing to a carbon-neutral economy. This means planning must emphasize low-carbon design as much as possible while upholding the desired operability, maintainability, and functionality of drainage scheme. Chapter 9 presents a procedure to conduct environmental lifecycle (least cost) analysis (LCCA). Reliability of the drainage scheme is evaluated from hydrologic and hydraulics models first, and then through the LCCA. The LCCA considers the cost of the desired structural strength especially the bedding classes and storm sewer concrete crushing strength. Chapter 8 presents the structural aspects of sewers, loading cases, bedding classes, and crushing strength of concrete pipes. Therefore, building reliability in the drainage system looks at the storm sewer structural strength, hydrology, and hydraulics. In addition, the topographical survey must be properly done because it influences cuts and fills, invert levels and pipe covers, and the overall earthworks. Not forgetting deploying competent personnel through the design, implementation and maintenance phases are also important. By minimizing the materials and resources in production and operation and maintenance phases, sustainability is achieved.

2.7.2 Hydraulics and flow routing

The Froude number choice depends on evaluation process, that is, channel/drain or storm sewer length, site steepness, control sizes, flow quantity, and other localized constraints. The design process for drainage projects, unlike auditing drainage schemes, will analyze the Froude number based on site constraints.

The Froude number choice forecasts the flow profile during design. It predicts how a flood wave would propagate through the sewer or drain. The Froude number ties the energy and momentum equation together obeying program conditions during design to properly conserve the flood waves. It is essential to have an appreciation of how the flow will propagate through the sewer depending on the goals set in Section 2.5.1. Selection of a Froude number would consequently influence the flow routing technique. A Froude number equal to 1 predicts that flow would be normal and steady. Therefore, in case the designer opts for a dynamic wave flow routing technique, setting Froude number equal to 1 would reduce the dynamic wave equation to steady flow routing technique by reducing the temporal and spatial acceleration.

Important: The Froude number is the measure of the ratio of inertial to gravitational forces. In simplicity, Froude number is the ratio of flow velocity to the celerity of the surface waves. For rectangular channels which are the perfect candidates for sustainability goals going forward, the wave celerity is given as shown in equation (2.1) (see Section 6.7.1):

$$c = \sqrt{gy} \tag{2.1}$$

Where:

c : Wave celerity
g : Acceleration due to gravity
y : Flow depth

Note that wave celerity is not constant, as depth varies so does the wave celerity. The flow velocity can increase or decrease beyond the wave celerity. Thus, where the flow velocity is larger than the celerity, supercritical flow occurs. Where the flow velocity is lower than the celerity, subcritical flow occurs. The Froude number is given by:

$$F_r = \frac{V}{c} \quad (2.2)$$

Where:

F_r : Froude number
V : Flow velocity
c : Wave celerity

Note:
$F_r < 1$ Subcritical flow
$F_r > 1$ Supercritical flow
$F_r \approx 1$ Critical flow

To minimize earthworks/construction cost and easement cost, enough care must be exercised about the optimal hydraulic radius, scaling b as shown in Figure 2.8. Reducing earthworks, embodied carbon from construction plants and trucks is reduced significantly. Also, where possible, narrowing structures for the flow to flow at critical flow significantly reduces capital carbon especially in concrete structures. Therefore, hydraulic radius is a key component in reducing carbon. For a channel of a given cross-sectional area and slope, the discharge is maximum for maximum hydraulic radius.

Hydraulic radius, $R = \frac{A}{P}$, where A is the cross-sectional area and P is the wetted perimeter.

$$\Rightarrow R_{max} = \frac{A}{P_{min}} \quad (2.3)$$

Therefore finding R_{max} maximizes discharge and minimizes the cost of construction. Reducing earthworks means reducing capital carbon during construction from vehicles such as backhoes, front-end loaders, wheel loaders, and trucks hauling the excavated materials. Furthermore, finding the maximum hydraulic radius will significantly reduce the capital carbon embedded in concrete and other materials used to build gray structures because of reduced sizes.

Note: a critical flow, the flow has maximum discharge minimum energy.

A rectangular section is a special case which produces the best hydraulic section. During design, a rectangular section should be proportioned as much as possible so that the depth of flow is one-half the width. This is because,

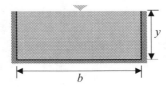

Figure 2.8 U-drain.

$$R_{max} = y/2$$
$$\Rightarrow b = 2y \qquad (2.4)$$

It is essential to note that RCC pipes and U-concrete channels are usually the best choices. This is because U-channels' geometry provides a versatile situation in easing calculations (hydraulic radius) and also fitting within local site constraints, for example, available easements for stormwater sewer lines. For hydraulic performance, a circular pipe is best geometric section among all.

Deciding on materials to use depends on cost, availability, and hydraulic efficiency desired. However, also sustainability goals influence the final decisions going forward. The roughness factors influence the flow velocity significantly hence determining the hydraulic efficiency. For example, the friction resistance is lower in concrete channels than in stone-pitched channels. This influences the hydraulic efficiency. However, using cobblestones and leaving concrete may be warranted where hydraulic structures lie in the municipal environs and not in the city center, or peri-urban environments, and when it is readily available. This would reduce the construction cost on concrete works thereby reducing carbon too.

Due to land constraints, existing roads and buildings, it is always necessary to check for available headroom because they influence the type and geometry of the structure. Also, safety issues are an essential aspect of sewers. **Slopes** influence the velocity of runoff and consequently influence the available specific energy as given by equation (2.5):

$$E = y + \left(\frac{\alpha V^2}{2g}\right) \qquad (2.5)$$

Where:

E : Specific energy
y : Depth of flow
V : Velocity
g : Acceleration due to gravity
α : Kinetic energy velocity distribution coefficient

In the Figure 2.9, links 1 and 2 feed link 3 through a node A, which can be a manhole or junction box. The slopes at which links 1, 2, and 3 are laid influence the flow regime. Whether the slopes are mild or steep, it influences the flow profile in all links. In the case of subcritical flow in links 1 or 2, backwaters will occur at the end of links 1 and 2 or at the start of link 3

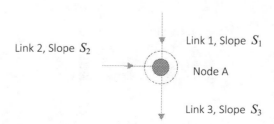

Figure 2.9 Two links feed into a node and one link routes the flow away from the node.

due to the creation of sort of a hydraulic jump caused by an abrupt change from supercritical flow to subcritical flow.

During design, the Froude number for the approach flows is usually fixed a constant either 1, < 1, or > 1 to produce a uniform structure link (pipe or drain) running from one point to another without unnecessary controls unless the structure is required to have controls amid sections of the link. This is justified especially when lateral inflows are neglected. The goal is to ensure that the flow waves are conserved in the hydraulic structure at the maximum hydraulic radius.

The fixing of the Froude number influences the dimensions output from the Manning's equation governed by the overreaching goal of reducing materials to meet sustainability goals through maximizing the hydraulic radius. However, in reality, as flows propagate through the structures, flow profiles keep changing depending on the slopes and length of the hydraulic structure (pipe or drain). Therefore, catering for nonuniform flow, the approach velocity for each hydraulic structure (drain or conduit) in the drainage network is assigned an independent Froude number that conforms to the routing technique used. Note that the kinematic wave routing method equates the friction slope to the bed slope. This risks obtaining wider drains and conduits, which does not promote sustainability goals going forward. That is the primary reason the kinematic wave routing method is not recommended for looped drainage networks.

An iteration of Froude numbers for approach flows is considered until when an optimal design that is environmentally sustainable and cost-effective is obtained. In subcritical flows, which tend to cause backwaters, the power dissipated in the hydraulic jump so caused is related to the Froude number.

In many cases, for overland flow drainage, $S_3 < S_1$ and $S_3 < S_2$. However, it is not the usual norm—sometimes mild slopes follow steep slopes. This might prompt the designer to reduce the structure size preferably a pipe in link 3 due to the steeper slope to meet sustainability goals going forward. However, a reduction in pipe size in link 3 creates pressurized flow in either link 1 or 2 or both leading the two pipes in links 1 and 2 to surcharge. This is because stormwater runoff flow tends to flow backward at the node especially where the invert level at the node manhole for links 1 and 2 are slightly above the invert of the manhole creating critical flow at the exit of pipe links 1 and 2.

The Figure 2.10 illustrates the backwater effects and pressurized flow. Energy losses at manhole are commonly neglected. The pipes are of the same diameter from point A to point D. Consequently, the critical depth is uniform throughout. The pipes in link CD are laid at a steep slope while the pipes in links AB and BC are laid at mild slopes. The flow at C is critical flow and the flow upstream of C is subcritical and the steep reach above A is delivering supercritical flow into link BC. An abrupt change from supercritical flow to subcritical flow exists in the link BC creating a hydraulic jump, leading to backwater effects in link BC.

Reducing the pipe size in link CD in a bid to take advantage of a steep slope results in water flowing backward in link BC and some energy is lost in the hydraulic jump as shown in Figure 2.11. This is negotiated by limiting the approach flow Froude number because it is related to the power dissipated in the hydraulic jump at the foot of the pipe. This forces the pipe in link BC to surcharge in order to provide the necessary specific energy to compensate for the lost energy in the hydraulic jump so that the flow gains enough momentum to flow through the pipe (link CD). The problem with this is that it causes pounding of the pipes and erosion problems. Also, it might call for more robust haunching and heavy concrete

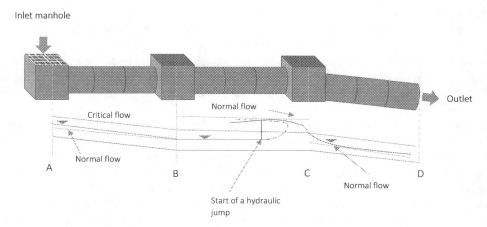

Figure 2.10 Backwater effects and pressurized flow.

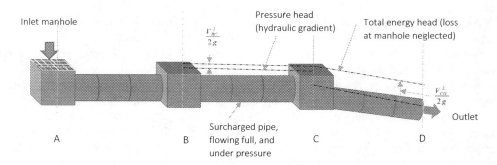

Figure 2.11 Pipe is reduced in link CD.

surroundings to firmly install the pipes to avoid movements resulting from pressurized flow. It can also influence the thickness of pipes. The internal water load in the pipes flowing under pressure increases because stresses additional to those due to fill-loading are created.

To avoid extra unnecessary costs, pipes of adequate sizes are designed by locating an optimal solution weighing embedded and operational carbon through the project lifecycle phases. Keeping pipe sizes uniform through links AB, BC, and CD and to never reduce the size of pipes in the flow direction have always been the rule of thumb. Trying to deviate from the rule of thumb to answer sustainability goals, the designer has to weigh the cost of hydraulic efficiency with carbon costs!

In the Figure 2.9, the angles at which the link 2 meets links 1 and 3 may necessarily not be at a right angle due to site constraints and the orientation of the drainage networks. It could be a bend with a varying radius of curvature r. In case flows are supercritical, approaching the bend, the radius of curvature of the bend presents an interesting scenario. In some cases, the flow may cause rotation of the pipe cross-section tightfitting on the pipe walls and leaving a core in the middle, and in other cases the flow may cause superelevation creating a train of standing waves. However, this has no effects such as backwaters, and the flow remains supercritical in both cases, which does not warranty the use of a larger-diameter pipe.

The governing formula to conduct the checks is given by equation (2.4):

$$\frac{V^2}{gr} \qquad (2.4)$$

Where V is the velocity, g is the acceleration due to gravity, and r is the radius of curvature. Rotation of the bend would occur when this formula provides a value greater than 1 and a train of standing waves occurs when the value is less than 1. The goal is to avoid sharp corners because they are more likely to cause erosion. Attempt to localize the hydraulic jumps at node manholes and junctions is well catered for through the hydrodynamic method of routing flows through the drainage network. This solves the backwater effects that cause pressurized flow or the surcharge of pipes. This automatically influences the pipe sizing. The dynamic wave method solves the full St. Venant equation catering for backwater effects and pressurized flow through the convective terms and accelerations terms. The dynamic wave routing method clearly accounts for subcritical flow.

The primary purpose of routing flows in a drainage network is to estimate the duration and discharge at nodes to adequately connect adjoining structures (pipes and drains) of adequate size without surcharging and flooding. This influences the sizing of the node manhole or junction box because the rim elevation of the node manhole influences the available headwater at a node that determines the available energy for the flow in the exit link. Invert levels take into consideration the pipe cover. For example, the interconnectivity at node A necessitates the routing of flows through links 1 and 2 to determine the time to peak at node A. The peak flow at node A influences the sizing of the exit link 3. When the control volume is considered as small as possible, specific force and specific energy techniques can be used to model node manholes and junction boxes especially where boundary and gravity forces are neglected. See detailed information from specialized hydraulics handbooks.

2.8 CASE STUDIES

2.8.1 Case study 2.1: Kabale municipality drainage master planning, Uganda, 2016

Kabale municipality is located at coordinates of 01 15 00S and 29 59 24E in the western region of Uganda. The region experiences bimodal rainfall and a tropical climate. The size of Kabale municipality is 44 km^2. This necessitated delineating major basins and subbasins when conducting the drainage master planning for the municipality. The municipality was divided into eight major basins following contours on the topographical survey produced. The drainage master planning for Kabale municipality in western Uganda necessitated incorporation of SuDS to reduce huge runoff volumes on a very steep terrain. In developing countries where financial constraints are among the top factors influencing choice of SuDS, rainwater cisterns were found more suitable to reduce stormwater runoff to public sewers. Rethinking sustainability and resilient communities necessitated an incorporation of stormwater best management practices.

The steep terrain warranted great reduction in stormwater runoff quantities. Kabale CBD is in a valley surrounded by hills as shown in Figure 2.12. It is the most prone area to floods

46 Integrated Drainage Systems Planning and Design for Municipal Engineers

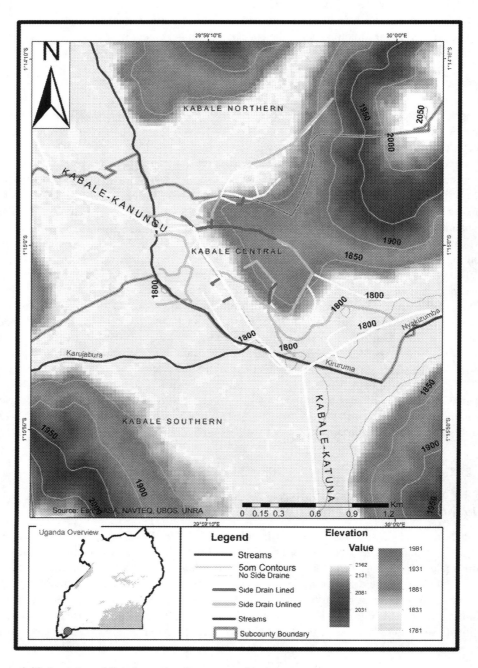

Figure 2.12 A section of Kabale municpality topographical map.

within the municipality. The implications of the steep terrain include possibility landslides, excessively high velocity of runoffs for which flash floods can become disastrous, and sediment and erosion challenges, which ought to be dealt with. This necessitated backdrop manholes and check dams installed at many sections in the drainage routes.

The highest point in the municipality is at 2050 m ASL, and the lowest point in the municipal is at 1800 m ASL, an average change in elevation of about 250 m. The current level of urbanization is still low but risked growing exponentially in the next 10 to 30 years depending on the region's economic indicators. In that case, physical plans communicated essential information used to forecast imperviousness.

To reduce the risk of flooding in the CBD, rainwater cisterns were proposed as SuDS to supplement the gray infrastructure. Other SuDS proposed included infiltration trenches for the golf course, tree planting, retrofitting many existing areas, and creating more open green spaces. These sustainable drainage practices estimated to save the municipal council about 25 tons of CO_2 annually.

In terms of drainage prioritization, the CBD was the first priority, because it is the most prone to severe floods and it is at the heart of central business activities. It was mapped as the flood-prone zone. The flood-prone CBD DEM is shown in Figure 2.13. Through engaging several stakeholders, the drainage master plan was developed as a live document updated in every five years with extra emphasis on enforcing SuDS.

During the planning, evaluation, and process selection, enormous data were collected in order to determine hydraulic structures that are most suitable for council and could fit within council's financial constraints. The data collected included, but not limited to, meteorological data, R. Kiruruma flow data (50 years), national development plan, underground utility networks, aerial photographs, geological maps, topographical maps, satellite imagery, GIS

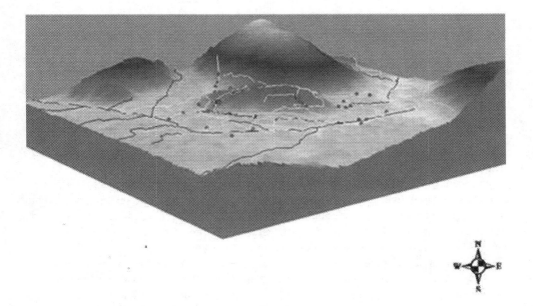

Figure 2.13 Digital elevation model for Kabale central business district.

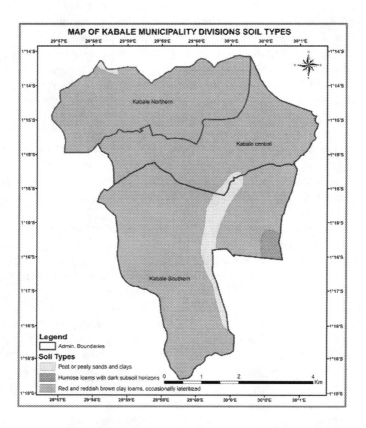

Figure 2.14 Kabale municipality soil types.

database, council physical plan, past study reports, as-built drawings, and demographics. River Kiruruma is the receiving stream for most of the stormwater runoff in the municipality. The map in Figure 2.14 shows the soil types within Kabale municipality.

From data collected, constraints were juggled to select drainage routes, materials, and type of structures suitable for a location. Such localized constraints included drainage easements, built environment, roads, buried utilities, landfills, wetlands, and topography (steep/mild/flat).

The extent of constraints would influence drainage route direction, underground or open drainage structures, hydraulic radius, and structure type. For example, several underground conduits were proposed in CBD, tunneling under roads/buildings. Weighing between safety, environmental impact, easement/land cost, earthworks, proposed structure hydraulic efficiency, and overall cost, preliminary designs were produced.

By judgment, RCC pipes of 600-, 900-, 1200-, and 1500 mm were readily available in municipal and neighborhoods. Therefore, considering time and cost, these were opted for their sizes/material except for waterlogged areas for which CMPs were the best alternative to resist differential settlement. From a global perspective, the minimization of carbon emissions from vehicles ferrying materials during construction by specifying local products was essential. Stone-pitched drains could as well reduce on cement and steel needed where possible. The World Bank (Funder), for example, currently emphasizes economical use of cement and steel because their production comes with enormous amounts of carbon emissions. Therefore, designers owed a duty to reduce global warming in that sense. Designers categorized and prioritized structures as primary and secondary. In that case, the residential and council structures took on between 5- to 25-year ARI, respectively—residential structures such as ditches and small sewers considered as the *minor system* and the council structures considered as the *major system*. River Kiruruma which receives most of the municipal stormwater runoff was considered for 100-year ARI. See Section 4.4.1 for detailed information on average recurrence intervals.

In a holistic view, sewers, drains, and SuDS were integrated in the drainage networks. It was necessary to visualize modern designs, but constraints always pulled for a cost-effective, reliable, safe, and sustainable design. The fact that sustainability goals guided the designers, proposals had to be evaluated back and forth to fit within the financial envelope. Therefore, designer's concepts would not overstretch team and available budgets without proper justification. Otherwise, fitting within municipal limited budgets, thoughtfulness on the feasibility of stormwater management proposals was essential. In today's flood risk management approaches, and need to make decisions that consider sustainability as overarching goal, recommendations were made for leaders to scrutinize. More than ever, the political actors are now required to be vigilant on green policies.

The detailed drainage master plan was produced adopting the dynamic wave routing method that provided the optimal hydraulic radiuses for different hydraulic structures (conduits and channels). Micro-tunneling under buildings and roads was proposed where the scheme route would be more feasible when left in a straight line and where space was limited.

2.9 CONCLUSION

The evaluation process for integrated drainage systems planning is holistic and considers back-and-forth approaches aimed at meeting sustainability goals. Deciding on the most appropriate SuDS to supplement gray infrastructure is based on site-specific constraints. When site-specific constraints and other constraints are juggled, the designer arrives at the optimal solution to solve a drainage problem. Integrated drainage systems planning considers how to minimize surface runoff, reduce risk of flooding, building resilient communities, and fostering a change in how the community views stormwater management programs.

References

1. Zhao, J. Z., Fonseca, C., & Zeerak, R. (2019). *Stormwater utility fees and credits: A funding strategy for sustainability*. Humphrey School of Public Affairs- University of Minnesota, 301 19th Ave S, Minneapolis, MN 55455, USA. *Sustainability* **2019**, *11*(7), 1913; https://doi.org/10.3390/su11071913
2. AutoDesk Civil 3D Support and learning. The difference between Kinematic wave or Hydrodynamic link routing in Storm Sanitary Analysis in Civil 3D. https://knowledge.autodesk.com/support/civil-3d/troubleshooting/caas/sfdcarticles/sfdcarticles/SSA-difference-between-Kinematic-wave-and-Hydrodynamic-link-routing.html
3. BS 9295:2020 Guide to the structural design of buried pipes.

Further reading

The Construction (Design and Management) Regulations 2015. UK Statutory Instruments, 2015 No. 51. www.legislation.gov.uk/uksi/2015/51/contents/made

Health and Safety at work Act (1974), UK.

Chapter 3

Planning for sustainability

3.1 INTRODUCTION

Sustainable development is development that meets the needs of the present without compromising the ability of future generations to meet their own needs[1] (UN Brundtland report, 1987). The 2015 Paris Summit defined three pillars of sustainable development, that is, social, economic, and environment, as shown in the Venn diagram (Figure 3.1). Since then, every construction project is scrutinized to ensure that it is sustainable and saves the planet. Greenhouse gas (GHG) emissions must be minimized from all construction activities, throughout the lifecycle phases of the projects. Global warming is caused by abruptly increased GHGs' emission by human activities. In the construction industry, carbon dioxide (CO_2) emission majorly comes from cement production. All project stakeholders have a role to play ensuring that the project's carbon footprint is reduced as much as possible contributing to net-zero targets. Combating climate change has become the overarching goal of project delivery mechanisms.

From a global perspective, Earth lost 28 trillion tonnes of ice between 1994 and 2017. Arctic sea ice (7.6 trillion tonnes), Antarctic ice shelves (6.5 trillion tonnes), mountain glaciers (6.1 trillion tonnes), the Greenland ice sheet (3.8 trillion tonnes), the Antarctic ice sheet (2.5 trillion tonnes), and Southern Ocean sea ice (0.9 trillion tonnes) have all decreased in mass. Just over half (58%) of the ice loss was from the Northern Hemisphere, and the remainder (42%) was from the Southern Hemisphere. The rate of ice loss has risen by 57% since the 1990s—from 0.8 to 1.2 trillion tonnes per year—owing to increased losses from mountain glaciers, Antarctica, Greenland, and from Antarctic ice shelves.[2]

Fluctuations in Earth's ice cover have been driven by changes in the planetary radiative forcing,[3] affecting global sea level.[4] Increased ice loss is due to global warming. If emissions continue to rise unchecked, the Arctic could be ice-free in the summer by 2040.

Radiative forcing is what happens when the amount of energy that enters the Earth's atmosphere is different from the amount of energy that leaves it. Energy travels in the form of radiation: solar radiation entering the atmosphere from the sun, and infrared radiation existing as heat[5] (worldwildlife.org, 2022). The melting sea ice affects the planet in several ways outlined below:

a. **Temperatures.** The Arctic and Antarctic are the world's refrigerator because they are covered in white snow and ice that reflect heat back into space. They help to balance out other parts of the world that absorb heat. When ice melts, less ice means less reflected heat, meaning more intense heatwaves worldwide. It also means more

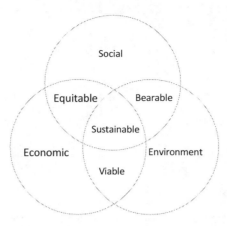

Figure 3.1 Sustainable development pillars.

    extreme winters: as the polar jet stream—a high-pressure wind that circles the Arctic region—is destabilized by warmer air, it can dip south, bringing bitter cold with it.
  b. **Permafrost.** Arctic ice and permafrost (ground that is permanently frozen throughout the year occurring mainly in the polar regions) store large amounts of methane, which is a greenhouse gas that contributes to climate change. When it thaws or melts, that methane is released, increasing the rate of global warming. This, in turn, causes more ice and permafrost to thaw or melt, releasing more methane, causing more melting.
  c. **Wildlife.** When there's less sea ice, animals that depend on it for survival must adapt or perish. Loss of ice and melting permafrost spells trouble for polar bears, walruses, arctic foxes, snowy owls, reindeer, and many other species. As they are affected, so too are the other species that depend on them, in addition to people. Wildlife and people are coming into more frequent contact—and often conflict—as wildlife encroach on Arctic communities, looking for refuge as their sea ice habitat disappears.
  d. **Coastal communities.** Global average sea level has risen by about 177.8 to 203.2 mm since 1900 and it's becoming worse. Rising seas levels endanger coastal cities and small island nations by exacerbating coastal flooding and storm surge, making dangerous weather events even more so. **Note:** Glacial melt of the Greenland ice sheet is a major predictor of future sea level rise; if it melts entirely, global sea levels could rise 6.1 m.
  e. **Food.** Polar vortexes, increased heat waves, and unpredictability of weather caused by ice loss are already causing significant damage to crops on which global food systems depend. This instability will continue to mean higher prices and growing crises for the world's most vulnerable.
  f. **Shipping.** As ice melts, new shipping routes open up in the Arctic. These routes can be tempting time-savers, but extremely dangerous. Imagine more shipwrecks or oil spills like the Exxon-Valdez in areas that are inaccessible to rescue or clean up crews.

With the above insight, planning for infrastructure development has now shifted goals to prioritize development that has a low impact on the environment. That means planners and engineers are now tasked to come up with techniques that would help to reduce emissions as much as practicable. In designing and developing drainage systems, the need to save our planet Earth outweighs any other option. Therefore, sustainable drainage systems (SuDS) are found competitive because they mimic the natural environment and act as a carbon sink which is the primary goal while fostering sustainable communities worldwide. Stormwater can safely be controlled at the source. In doing so, the size of conveyance systems such as storm sewers and drains reduces drastically which helps in reducing embedded carbon in building these gray drainage systems.

In many cases, SuDS outperform standard storm sewer systems in environmental and social aspects. However, their high maintenance costs and shorter life expectancy hinder SuDS' economic feasibility in some cases. Therefore, drainage schemes for urban areas must focus on integrated designs that influence each other in a complementary way whereby the strengths of both drainage alternatives, that is, SuDS and gray infrastructure assets are combined in order to achieve sustainable development.

The UN set 17 SDGs to achieve by 2030 and combating climate change is Number 13 (Climate action). Article 1(a) of the Paris Agreement states,

> Holding the increase in the global average temperature to well below 2C above pre-industrial levels and pursuing efforts to limit the temperature increase to 1.5°C above preindustrial levels, recognizing that this would significantly reduce the risks and impacts of climate change.

Global activities that contribute to the increased amounts of GHGs in the atmosphere perpetuate the global temperature rise. As a result of the temperature rise, extreme heat, and huge storms constantly hit different parts of the world. In that regard, planners and engineers are at the forefront of mitigating the worst case. Sometimes, flooding exacerbated by climate change puts critical infrastructure at risk; flash floods, rivers topping their banks, and coastal flooding, and storm surge present drainage engineers with more work to do than ever. Every aspect of the construction industry must reduce carbon from the conceptual design stage through delivery. Drainage engineers are tasked to produce low-carbon designs to address climate change.

Due to climate change, floods have hit several communities globally interrupting service delivery, destroying farmlands and roads, displacing people and destroying homes, etc. City and municipal planners have since changed the conceptual frameworks in compliance with UN's call for designing and implementing sustainable cities globally. In that regard, all projects require a carbon footprint and it is the role of planners to create a roadmap for the cities. The construction industry and associated agencies worldwide promote initiatives to support drainage engineers and planners to achieve net-zero emissions in their practices.

These initiatives are governed by one special interest of sustainable growth and development by fostering low-impact development (LID) approaches. The LID approaches use nature as a model and manage stormwater at the source. Therefore, by mimicking the natural environment, runoff is minimized as much as possible. These methods are sequenced in a way that runoff prevention/mitigation strategies are the first priority; and lastly the rainfall that converts to runoff is treated to remove pollutants.

However, mitigation measures through effective drainage systems' design and implementation programs do not guarantee that heavy storms cannot happen. Several regions worldwide have experienced severe floods regardless of sound planning. The primary strategy has since shifted to climate adaptation strategies and these include best management practices (BMPs) that make societies more resilient and adaptable to climate change.

The primary goal of LID approaches is to reduce as much as possible impervious areas and to only develop and pave areas that are a must optimally meeting the demand. This means paving only areas that are a must to be paved and conserve natural areas to the extent possible, or developing them in a way that mimics nature. LID approaches also aim to reduce impact on hydrology and to implement pollution prevention strategies. Through proper maintenance and public education programs, LIDs foster growth of sustainable communities.

Although the economics of a drainage scheme greatly influences the decision to undertake the project, the position is slowly changing. Environmental and social concerns are now equally considered in order to have an inclusive project outcome. Environmental and social impact assessments have increasingly become essential elements in determining SuDS globally.

SuDS would significantly reduce the size of channels and sewers because the runoffs would be reduced due to reduction in impervious areas. The estimation of capital carbon reduction is conducted by comparing fully impervious conditions to fractional imperviousness. Strom sewers are predominantly constructed with concrete.

3.2 GREENFIELD AND RETROFITTING

Planning for SuDS development will always be done on either new areas or old landscapes. With the world increasingly anxious about drainage problems, stormwater, and flood management and the sustainability of life as we know it on earth, it's a great time to consider how to develop potential solutions for the future of drainage, stormwater, and flood management. Smart drainage starts with smart design considering optimizing the performance and effectiveness of drainage systems, including SuDS—both the Greenfield and retrofit.

The previously undeveloped sites for residential, industrial, and commercial development would always call for fresh SuDS planning. In that case, the planning for LIDs is always flexible considering several alternatives going by site-specific conditions and constraints. However, retrofitting always taps into available opportunities. Such retrofit opportunities might include the following:

a. Retrofitting parking facilities to be greener.
b. Design traffic safety features to manage stormwater and improve aesthetics.
c. Build in green features during routine right-of-way maintenance and operations.
d. Building rain gardens in public places.
   1. Creating stormwater micro parks.

3.3 WHAT ARE SUSTAINABLE DRAINAGE SYSTEMS AND HOW DO THEY WORK?

SuDS refer to a decentralized network of site-specific stormwater management practices. SuDS are implemented to reduce the volume of stormwater runoff entering the storm sewer system while also restoring the natural hydrologic cycle.

SuDS aim to serve three primary goals: stormwater runoff quantity reduction, runoff quality control, and sinking carbon. This is achieved through physical, chemical, and biological processes that SuDS employ. Stormwater is collected, stored, conveyed, and treated in a way that mimics the natural environment. Mimicking the natural environment is a good opportunity to promote sustainability.

SuDS employ the following processes that mimic predevelopment conditions: evaporation/transpiration using native vegetation, infiltration (allowing water to slowly sink into the soil), and rainwater capture and reuse (storing runoff to water plants, flush toilets, feeding animals, etc.).

SuDS are essential in urban areas with combined sewers or partially combined sewers, where during wet weather events, combined sewer overflows (CSOs) result in untreated combined sewage being discharged directly into water bodies. These CSO events can significantly impact downstream water quality. As opposed to the costly traditional standard pipework drainage solutions, SuDS manage stormwater through a variety of small, cost-effective landscape features located on-site.

3.4 APPROACHES TO DEVELOP SUDS

There are several steps to consider before undertaking site-specific SuDS projects for both the Greenfield and retrofit. Figure 3.2 summarizes the steps to consider when planning SuDS and more details on specific actions that are needed to begin planning, installing, and maintaining SuDS are provided henceforth.

3.4.1 Review standards and codes of practice

This involves desk reviews for applicable statutes, development codes, rules and regulations, codes of practice, and standards to start planning developing Greenfield SuDS or retrofits. An extensive review of planning codes and documents would identify potential barriers to implementing SuDS projects. The review would help determine constraints and barriers and how to overcome them. Several regions develop their standards and codes of practice, technical specifications, and design templates for developing SuDS on public and private properties.

A number of planning and development codes need to be visited to conform to regional applicable laws. In the interest of planning for sustainability, the global frameworks should be complied with from which many national statutes, sustainable development codes, specifications, and rules and regulations are derived.

After a municipal authority or any other public authority establishes that SuDS are applicable in the projects' development and redevelopment, then it is essential to also establish whether SuDS can be incorporated into already planned maintenance projects as a way to reduce costs or retrofit an area. It is also good to include small-scale SuDS projects in individual/private municipal

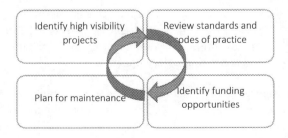

Figure 3.2 A summary of steps to consider while planning for SuDS.

projects that are currently in the planning stages. Where possible, integrate SuDS requirements into competitive bid packages for projects' portions that hired contractors will complete.

It is now common practice that almost all projects developed have a green component working toward a sustainable built environment. Projects such as roads and highways, pipelines, and buildings all have a green infrastructure component working toward sustainability. In that case, spotting opportunities planning for the council, all municipal projects would allocate a minimum amount of the total project cost to SuDS components with a minimum of such projects as planting trees, vegetated curbs, vegetated swales, etc. The municipal can think about incorporating SuDS into projects that have already been approved or are in the planning stages.

Retrofitting SuDS while repairing or replacing curbs presents a great opportunity. A grass swale in the median can potentially become a bioretention area. While repairing sidewalks from tree root damage, or a failing street tree is slated for replacement, the municipal can plan minor modifications to include a tree trench box that filters stormwater as appropriate. Greenfields largely call for planning site-specific SuDS. Section 3.6 provides site-specific constraints and considerations to develop the most appropriate SuDS.

3.4.2 Identify funding opportunities to install and maintain SuDS

Funding opportunities include conditional and unconditional grants, local revenues, loans, and private finance initiatives. The cost of SuDS varies widely based on the setting from a few thousands of dollars to millions of dollars. Greenfield projects can be less expensive than those in redevelopment or retrofit situations. Usually, SuDS redevelopment or retrofitting encounters utility conflicts, space constraints, compacted soils, and other site-specific factors that may require specialized technology to design and develop.

Large-scale SuDS projects with large catchment areas typically call for huge capital investments compared with small-scale projects such as downspout disconnection and curbs to direct flows to an open green space.

Public authorities such as municipalities have several ways of sourcing SuDS financing ranging from local revenue streams, state and NGO grants, loans, and private sector financing. It is an excellent practice for public authorities to work with partners such as nongovernmental organizations (NGOs) and volunteer organizations to effectively design, construct, and maintain SuDS. This promotes the sustainability of SuDS projects blended to reflect a mix of local, state, and NGO funding mechanisms.

- **Green Bonds**
  These bonds are now common in the USA and UK and are a new source of funding for climate and environmental projects. They are typically asset-linked and backed by the issuing entity's balance sheet. They carry the same credit rating as their issuers' other debt obligations. Since environmentally friendly projects are fast becoming a model for delivering sustainable projects development globally, green bonds are a means of borrowing money to finance SuDS development.
- **Loans**
  Through well-engineered sustainable drainage projects, public authorities are usually eligible to source for a funder or a soft loan from development banks. Concessional loans and low interest loans may be secured but are generally used for planning and capital-intensive projects.

- **Conditional and unconditional grants**
  Grants and other monetary transfers to local governments from the state or NGOs are either conditional or unconditional. These are state and federal state and NGO grants that provide additional funding for water quality improvements in specific local governments, cities, municipalities, town councils, and other zones. Conditional grants are designated for a specific purpose and may not be used for another project, while unconditional grants may be used for any purpose the recipient local government finds appropriate. SuDS planning and development may receive conditional or unconditional grant financing depending on the public authority's voting decisions.
- **Private sector financing initiatives**
  These are public–private partnerships (PPP) that allow private sector participation in SuDS financing, planning, design, construction, and maintenance through signing contract agreements with the public agencies.
- **Locally generated revenue (own source revenue)**
  These include fees, taxes, and stormwater utility revenues. Funds may be raised through collecting developer impact fees which are one-time charges linked with new development, for example, legal point of stormwater discharge (LPSD) fees, inspections fees, and issuing permits. Stormwater utility generates revenue through fees going into a designated fund specifically for stormwater services. Taxes can be raised from property sales and property incomes. Special fees may also be collected for permit and plan reviews and inspections and should be directly linked to stormwater management programs. LPSD and other special developer impact fees are usually a one-time fixed fee charged for new development. Pollution prevention would be based on appropriate policies and laws on pollution and the polluter would pay damages following the 'polluter must pay sustainable development principle'.

  In some areas, a fee-based stormwater utility may be developed imposing fees from organizations, schools, commercial buildings, industrial parks, etc. prorated by land area, imperviousness, roof area, or other similar parameters. This stormwater fee would be charged annually or monthly or periodically as appropriate contributing to the stormwater management fund for capital cost funding, operating, and maintaining SuDS and associated gray infrastructure.

3.4.3 Plan for maintenance

Planning for maintenance considers the equipment required, the lifecycle costs built based on the capital and operational expenditure. Some SuDS require constant watering through dry spells and during establishment, weeding, and pruning. Other SuDS require routine removal of leaf litter, trash, and debris from time to time. Integrated SuDS and pipework drainage solutions might need heavy equipment such as backhoes and front-end loaders to remove large volumes of sediment load. The required personnel to maintain the SuDS should be identified and trained. It is essential to establish whether the SuDS project would require a specialist contractor to maintain it.

3.4.4 Identify high-visibility projects

This is a critical step in planning SuDS development and retrofitting. There is a need for identification of high-visibility projects to garner support for developing SuDS across the region. The public authority should identify places that are clearly visible in the area to locate SuDS. The purpose of this is to promote SuDS by allowing the residents to enjoy the SuDS benefits and educate the community about specific SuDS hence getting more support for more SuDS development in the area. Areas such as highly trafficked places, medians and islands, roundabouts, cafeteria and restaurants, and recreation centers are good spots for SuDS development.

3.5 THE ROLE OF SUSTAINABLE DRAINAGE SYSTEMS

Planning for drainage infrastructure's sustainability is based on three main factors:

- Reducing the **quantity** of flow (discharge rates or volume).
- Improving the **quality** of flow (absorbing dissolved particles, turbidity control, etc.).
- Control and management of **GHGs**.
- **Biodiversity and ecological** systems preservation.

Understanding SuDS is to look at it this way, the impervious surfaces greatly produce runoff because little-to-no stormwater infiltrates into the ground. Where urbanization grows rapidly, infrastructure such as roads and building greatly produce runoffs. The role of SuDS is to capture the runoffs and stormwater and deal with it in a way that mimics the natural water cycle, which is usually to infiltrate the water into the soil, store it, purify it, and evaporate back into the atmosphere. For example, combined kerbs, drainage can be designed with outlets to a vegetated swale, which is a very good example of an achievable SuDS.

In designing and developing infrastructure, the SuDS techniques would be integrated as appropriate, so that as water collects on the impervious surfaces, it runs to the immediate SuDS facilities. In this way, it is dealt with in a way that mimics the natural environment.

Sometimes, green roofs are developed to make the rooftops aesthetically appealing, but it serves multiple benefits, including infiltrating, detaining stormwater for some time and later release it slowly through underdrain pipes. These green roofs are characterized by flowerpots, trees, and ancillaries that make them appear more aesthetically appealing.

SuDS can always be integrated differently depending on the purpose they aim to serve. If the designer thinks stormwater is required to be detained for some time, and later release it, detention facilities in form of underdrain pipes can be designed on top of an infiltration basin or trench. Bioretention gardens can also serve as infiltration systems and as partially retention systems. The choice is for the designer. The primary goal is to design systems that mimic the concepts of water infiltration, percolation, groundwater recharge, evaporation, storage, treatment, conveyance, and collection.

3.6 ACHIEVABLE SUDS (SITE-SPECIFIC SUDS)

A site-specific work or activity such as SuDS is **designed for a specific location**, if removed from that location, it loses all or a substantial part of its meaning. The term site-specific is often used in relation to installation of art, as in site-specific installation; and land art is site-specific almost by definition. Therefore, site-specific SuDS are synonymous with land art, because they are developed largely mimicking the natural environment that involves art and engineering skills. With the variability of **soils, topography**, and **other design constraints** that occur between sites, it is often difficult to determine which SuDS will be the most appropriate and cost-effective solution. Several site-specific factors predetermine the type of SuDS to develop for a particular site. In some instances, SuDS are found inapplicable, depending on the site constraints. Also, site-specific flood investigations are a vital step in selecting the most suitable SuDS. Site-specific factors considered generally include the following:

3.6.1 Topography/slope

This influences many types of SuDS, especially infiltration systems. Infiltration systems do not perform effectively on steep slopes because water fails to infiltrate into the subsoils but instead converts into the runoff due to high velocities that steep slopes cause. Therefore, steep terrains warranty velocity reduction techniques. Sloping grounds might need embankments. Again, vegetated swales and vegetated kerb extensions work effectively on roads, and roads need trees, and bioretentions are largely developed in depressions. Parking lots need infiltration systems. Evaluating the area slope as mild, flat, or steep would enable the designer to know the type of SuDS most appropriate to develop.

3.6.2 Depth to water table and to bedrock

The water table is a key consideration. Some SuDS are designed to infiltrate water into the subsoil, and where the water table is high, they cannot work because the subsoil is already saturated. For example, infiltration systems require areas whose water table is low enough so that their primary function is achieved. The water table is the upper surface of the zone of saturation. The zone of saturation is where the pores and fractures of the ground are saturated with water. It can also be simply explained as the depth below which the ground is saturated. A water table is surface of the aquifer which is the surface where the water pressure head is equal to the atmospheric pressure. Below the water table, the aquifer is completely saturated with water. The distance down to the groundwater table may vary from zero meter to several tens of meters, depending on whether it is being drilled in the valley or on higher hill. The height of water table also varies with climate and season.

On the other hand, sites with shallow bedrock are of great concern. Sites with shallow bedrock are defined as having bedrock within 1.80 m or less of the ground surface[6] (Minnesota Stormwater Manual, 2022). In some manuals such as the Minnesota Construction General Permit (CGP), when installing an infiltration SuDS, there must be at least 0.90 m of separation between the base of the SuDS and the bedrock. Bedrock at the 1.80-m depth is a trigger to perform a geotechnical investigation to determine the location of the bedrock in the area in and around the proposed SuDS to ensure that the 0.9144-m separation can be achieved.

The shallow depth to bedrock is a great concern because it limits the depth of SuDS, reduces the potential for subsurface infiltration, and reduces the depth over which treatment can occur. Sites with shallow depth to bedrock present challenges to stormwater management. However, these challenges can be managed through adequate designs. Several manuals present general guidelines for investigation and management. Special caution for steep slopes and fractured bedrock is urged.

It is essential to understand the general depth to bedrock over the entire project site, but more specifically, it is important to know the depth to bedrock in and around the area of the proposed SuDS. Geotechnical investigations are recommended for all proposed stormwater facilities located in regions with shallow bedrock. The purpose of the investigation is to identify subsurface conditions which can pose an environmental concern or a construction hazard to a proposed SuDS. Several guidelines for investigating all potential physical constraints to infiltration on a site can be obtained from regional manuals for SuDS, which sometimes vary, but the overarching goals remain almost constant to achieve efficiency and effectiveness of the SuDS. The size and complexity of the SuDS project greatly influence the extent of any subsurface investigation.

Subsurface material investigation is designed to determine the nature and thickness of subsurface materials, including depth to bedrock and to the water table. Subsurface data for depth to groundwater may be acquired by soil boring or backhoe investigation. These field data should be supplemented by geophysical investigation techniques deemed appropriate by a qualified professional, which will show the location of the geologic and groundwater formations under the surface. The data listed below should be acquired under the direct supervision of a qualified geologist, geotechnical engineer, or soil scientist who is experienced in conducting such studies. Pertinent site information should include the following:

a. Known groundwater depth or bedrock characteristics (type, geologic contacts, faults, geologic structure, and rock surface configuration).
b. Soil characteristics (type, thickness, and mapped unit).
c. Bedrock outcrop areas.

3.6.3 Soils

Areas with rocks might not work well with some SuDS that require water to infiltrate into the subsoil. Permeable native soils can support infiltration systems. Types of soils influence the suitability of SuDS; infiltration would require native soils to have a good infiltration rate, bioretention systems require particular soils to support the growth of particular plants, and at the same time are designed with either filtration or infiltration medium. Waterlogged areas may not support infiltration systems unless otherwise designed uniquely. Such areas may call for retention and detention ponds.

The determination of site infiltration rates (for facilities with infiltration and/or recharge) is necessary. For design purposes, there are two ways of determining the soil infiltration rate. The first, and preferred method, is to field-test the soil infiltration rate using appropriate methods. The other method uses the typical infiltration rate of the most restrictive underlying soil (determined during soil borings).

If infiltration rate measurements are made, a minimum of one infiltration test in a soil pit must be completed at the elevation from which exfiltration would occur (i.e., interface of gravel drainage layer and in situ soil). When the SuDS surface area is between 95 and 475 m^2, two soil pit measurements are needed. Between 475 and 950 m^2 of surface area, a total of three soil pit infiltration measurements should be made. Each additional 475 m^2 of surface area triggers an additional soil pit. Soil characteristics such as type and thickness must be examined through geotechnical investigations. Specifically for bioretention gardens, the following performance specifications are applicable:

a. All bioretention growing media must have a field-tested infiltration rate between 25 and 205 mm/h. The growing media with slower infiltration rates can potentially clog over time and may not meet drawdown requirements. The target infiltration rates should be no more than 8 in/h to allow for adequate water retention for vegetation as well as adequate retention time for pollutant removal.

b. Temperature: Slower rates are preferable (< 50 mm/h).
c. Metals: Any rate is sufficient, 50 to 150 mm recommended.
d. Pathogens: Any rate is sufficient, 50 to 150 mm recommended.
e. Soluble salts (soil/water 1:2) should not exceed 500 parts per million.
f. The following infiltration rates should be achieved if specific pollutants are targeted in a watershed. Total suspended solids: Any rate is sufficient, 50 to 150 mm recommended.
g. The pH range (soil/water 1:1) is 6.0 to 8.5.
h. Total nitrogen (TN): 25 to 50 mm/h, with 1 in/h recommended.
i. Growing media must be suitable for supporting vigorous growth of selected plant species.
j. Total phosphorus (TP): 50 mm/h.

3.6.4 Potential hotspots

Areas with a high pollution levels call for SuDS with a high level of pollution removal potential without treatment. Therefore, in that case, weighing between several SuDS, the option to take on a particular SuDS without treatment comes first before that requiring treatment. For example, infiltration systems and bioretention systems when designed effectively can work properly in pollution hotspots with limited treatment.

3.6.5 Maximum drainage area

Some SuDS have a maximum drainage area that they can be designed to serve. For example, infiltration systems can serve an area of up to 5 acres. Also, the recommended maximum drainage area for bioretention gardens is typically 5 acres but can be greater if the discharge to the basin has received adequate pretreatment and the basin is properly designed, constructed, and maintained. For larger sites, multiple bioretention gardens and infiltration systems may be developed to treat and infiltrate site runoff provided appropriate grading is present to convey flows.

3.6.6 Climate of the area

Some SuDS that work well in the Mediterranean climate may not work the same way in the tropical climate. The six major climate regions are polar, temperate, arid, tropical, Mediterranean, and tundra. These call for variations in the application of sustainable drainage systems. The climate of an area is an important consideration in determining the feasibility of a particular SuDS; meteorological data are always collected. Climate change allowance is also applied to rainfall intensities within design calculations.

3.6.7 Native species/vegetation

Native vegetation influences the development of bioretention gardens, tree trenches, bio-sales, and other SuDS. Native species are grown in the bio-swales and bioretention gardens. The advantage they present is that they are already accustomed to the environmental conditions of the place and their sustainability and longevity is guaranteed.

3.6.8 Space

Some SuDS require large spaces which may not be available, for example, retention and detention ponds require space. Space constraints dictate the suitability of dry ponds and bioretention systems. They also influence the egress and entry points. Space constrains would influence an above-the-ground or underground SuDS, for example, rainwater harvesting jars (RWH jars). RWH cisterns can be of two types: underground and above the ground; space constraints sometimes dictate to have underground or on-the-ground system.

3.6.9 Sacred sites and culture

Sacred sites associated with cultural activities call for sustainability metrics incorporated in the design and development of SuDS. Some species may be debarred in some sanctified places, while some SuDS like bioretention may not be suitable. Open and green spaces associated with the work of art are largely the primary SuDS that would make the sacred sites aesthetically appealing.

3.6.10 Integrated factors

The physical plan communicates how the area will be developed. It links with the national or regional development plan because every aspect of development in the area follows what was proposed in the physical plan. In that case, SuDS integration in the city or municipal activities follows what the physical plan says together with what the national or regional development plan communicates. For example, for many years, bio-swales have been used along rural highways and residential streets to convey runoff. Like ditches, swales collect stormwater from roads, driveways, parking lots, and other hard surfaces. In this regard, bio-swales would be developed in those mentioned areas. Bioretention systems function effectively in depressions and in many cases developed on raised areas/uphill where the underdrain pipe would effectively drain the excess water. Vegetated swales may act as pretreatment method for stormwater that would later be infiltrated into the ground through the infiltration system. They have vegetation planted in the ditch with stones. Therefore, as an example, the physical plan and national or regional plan significantly influence the suitability of a particular SuDS—golf courses would call for extensive application of infiltration systems.

As a rule of thumb, when choosing which SuDS to include in a stormwater solution, consider the following questions[7]:

- What are the climate and weather conditions of the area?
- How much land is available, and is it suitable for construction? Look into factors like topography, soil type, and proximity to other bodies of water or drainage areas.
- How much maintenance will be required to keep the SuDS operating effectively?
- How will the SuDS impact the aesthetics of the property and/or community?
- What effect will the SuDS have on the surrounding environment (e.g., water quality, fish and wildlife, insect control, odor, etc.)?

3.7 SUDS APPLICATION AND FEASIBILITY POTENTIAL GOING BY AREA PHYSICAL PLAN

Section 3.6 provides the site factors considered to determine whether a particular SuDS will perform reliably and safely. Site factors would be evaluated in conjunction with the potentiality of a SuDS to fit in a location on a physical or a structural plan. Industrial, commercial, residential, streets, and roads and highways might not need the same SuDS. Table 3.1 provides areas that are most suitable for particular SuDS.

Table 3.1 Areas suitable for particular SuDS

Application area	Residential	Commercial	Ultra-urban	Industrial	Retrofit	Highway/road	Recreational	Public/private
Bioretention gardens	Yes	Yes	Limited	Yes	Yes	Yes	Yes	Yes
Infiltration systems	Yes	Yes	Limited	Limited	Yes	Limited	Yes	Yes
Tree trenches	Yes	Yes	Limited	Yes	Yes	Yes	Yes	Yes
Swales	Yes	Yes	Limited	Yes	Limited	Yes	Yes	Yes
Planter boxes	Yes	Yes	Yes	Limited	Yes	Limited	Limited	Yes
Green roofs	Limited	Yes	Yes	Yes	Yes	Yes	Yes	Yes
Rainwater harvesting systems	Yes	Yes	Yes	Yes	Yes	Yes	Yes	Yes
Porous/permeable pavement	Yes	Yes	Yes	Yes	Yes	Limited	Yes	Yes
Open/green spaces	Yes	Yes	Yes	Yes	Yes	Yes	Yes	Yes
Detention ponds	Limited	Yes	Yes	Yes	Limited	Yes	Yes	Yes
Retention ponds	Limited	Yes	Yes	Yes	Limited	Yes	Yes	Yes
Downspout disconnection	Yes	Yes	Limited	Yes	Limited	No	Yes	N/A

3.8 PLANNING, DESIGN, AND DEVELOPMENT OF SUDS

The planning of SuDS requires many considerations necessitating the designer to consider the SuDS lifecycle. From project initiation, planning, design, development, operation, and maintenance, the project must be evaluated as feasible. The design of SuDS is dictated by local stormwater quantity and quality control requirements. Local ordinances, standards, and guidelines typically dictate the approach to design and develop SuDS and require that postdevelopment peak flows be controlled to predevelopment levels. For example, in the UK, the Code of Adoption and Design and Construction Guide for SuDS, came into force on 1 April 2020, is accelerating change in the industry.

The cost of SuDS is an important aspect because it varies wildly from place to place, depending on site-specific conditions and the type of SuDS. The good news is that the cost of SuDS is usually competitive compared with standard pipework solutions. Similar to conventional methods of solving drainage problems, SuDS require the planner to evaluate the level of maintenance required to ensure optimal performance. For example, the lifecycle of the technology used in SuDS must be considered drawing up a maintenance plan. SuDS require structural and nonstructural maintenance. For example, structural practices may include repairing technologies and nonstructural practices include keeping the vegetation healthy, through pruning, weeding, slashing, and mulching.

The hydrologic characteristics of an area are always modified with new developments such as roads, buildings, and other infrastructure. The goal of the drainage planner is to try as much as possible to lower the impact of the built environment on the natural environment. The built environment greatly reduces the stormwater that infiltrates into soil and instead a considerable portion converts into runoff. The task of the drainage engineer is to design and develop a low-impact built environment, creating little effect on the hydrologic characteristics of the area before new developments.

In Figure 3.3, let z represent the total area for the predevelopment environment (Greenfield/unbuilt area) and x represent the built-up area, while y is the area that remains unbuilt at time t. Therefore, at any time t,

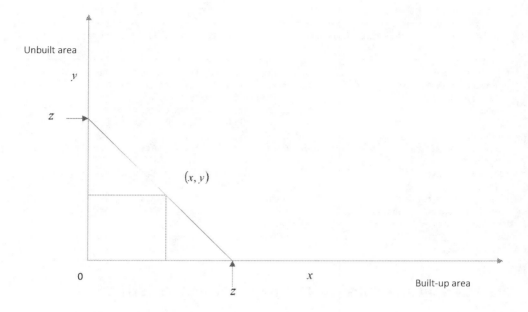

Figure 3.3 Simple site illustration to locate the optimal solution between built and unbuilt areas.

$$z = x + y \tag{3.1}$$

At any time t,

$$y = z - x.$$

Since z is a constant, the catchment can be modeled as appropriate incorporating SuDS. The goal is to minimize impervious area as much as possible. For example, consider a 4-acre entirely wooded Greenfield. In that case,

$$y = 4 - x.$$

The goal of the designer is to reduce the effect of x on the hydrologic condition of the place. That means predevelopment hydrology should not be significantly disturbed. This implies that the designer devises methods to develop x in a way that creates the least impact on the natural landscape is the best option. In that case, low-impact techniques are integrated mimicking the natural environment. LID would require the designer to evaluate options coming up with several assumptions.

Predetermining the amount of carbon offset is conducted by comparing the quantity of materials used to build stormwater runoff sewers when integrated with SuDS versus when no SuDS are incorporated. In that case, fully impervious areas generate more runoffs than partially impervious areas. This implies that the hydraulic capacity of sewers would vary depending on the peak flows generated. More details are presented in Chapter 9. The design of SuDS varies widely, and standard guidelines for designing SuDS are outlined. It largely

Planning for sustainability 65

depends on the type of stormwater management project, the overarching goals to be achieved, and the project constraints that dictate the type of SuDS required. However, the overarching universal goals are as follows:

- To reduce stormwater runoff minimizing flooding risk.
- To promote the ecosystem.
- To store stormwater for irrigation, toilet flushing, feeding animals, and washing.
- To infiltrate stormwater runoff into the ground facilitating groundwater recharge.
- To capture pollutants, for example, dust and CO.
- To control sediment load.
- To capture trash and debris, garbage.
- To absorb GHGs' emissions, for example, CO_2.
- To preserve base flow in streams.
- To reduce thermal impacts of runoff.

The above are the SuDS universal primary goals. The planning and design criteria presented in the sections below are simplified approaches followed to design and develop SuDS. A general overview of the planning and design criteria of SuDS is provided below.

3.8.1 Infiltration systems

3.8.1.1 Description and purpose

Infiltration systems are systems that capture and temporarily store the water quality volume before allowing it to infiltrate into the soil over approximately a two-day period. Infiltration systems can be in different categories such as bioretention gardens, infiltration trenches, infiltration basins, tree trenches, and permeable/porous pavers. Infiltration basins are well-known for clogging very easily, and therefore their success rely heavily on the site-specific conditions. They are used to infiltrate stormwater into the soil, thus minimizing surface runoff. The design of infiltration system varies widely, but the governing principle is determining the

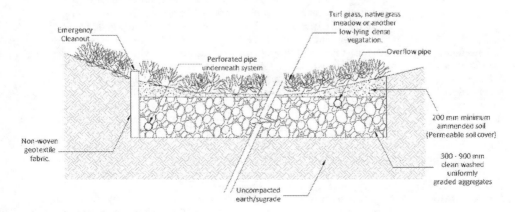

Figure 3.4 Typical infiltration system.

infiltration rates of the soils. It is highly recommended that all infiltration systems should be designed only after having an **infiltration test**. Infiltration systems are designed specifically to collect stormwater and infiltrate it into the groundwater table or surficial aquifer. Infiltration systems can be constructed under large open areas such as ball fields and parks. Properly designed infiltration systems can be effective and native (in situ) soils would support infiltration of a considerable amount of runoff into the soil layers. A typical infiltration system is provided in Figure 3.4. Infiltration systems' environmental benefits include water quality control, groundwater recharge, and improved aesthetics.

Infiltration systems can serve several benefits, which include reducing stormwater volumes, increasing groundwater recharge, improved aesthetics, reducing peak flow rates, thermal insulation, and multiple uses, among others. Table 3.2 provides a narrative of different types of infiltration systems.

Table 3.2 Different types of infiltration systems

Infiltration system	Narrative
Infiltration trenches	These are linear subsurface infiltration structures typically composed of a stone trench wrapped with nonwoven geotextile which is designed for both stormwater infiltration and conveyance in drainage areas less than 5 acres in size. They are rock-filled trenches with no outlet that receives stormwater runoff. The main pollutant removal mechanism of this type of infiltration system is filtering through the soil. The runoff is stored in the voids of the stones, and infiltrates through the bottom and into the soil matrix. Sometimes, the stormwater runoff passes through a combination of pretreatment measures, such as a swale and detention basin, and into the trench.
Infiltration berms	Berms may function independently in grassy areas or may be incorporated into the design of other SuDS facilities such as bioretention and constructed wetlands. These use a site's topography to manage stormwater and prevent erosion. Berms may also serve various stormwater drainage functions including creating a barrier to flow, retaining flow for volume control, and directing flows.
French drains	These are also referred to as seepage pits, dry wells, or Dutch drains. They are subsurface storage facilities that temporarily store and infiltrate stormwater runoff from rooftop structures. They are also used to remove excess water in waterlogged areas during road construction. They are structural chambers or excavated pits, backfilled with a coarse stone aggregate or alternative storage media. Because of their size, French drains are usually designed to handle stormwater runoff from smaller drainage areas, less than 1 acre in size.
Subsurface infiltration beds	These consist of a rock storage (or alternative) bed below surfaces such as parking lots, playfields, and lawns for temporary storage and infiltration of stormwater runoff with a maximum drainage area of 10 acres.
Infiltration basins	These are shallow surface impoundments that temporarily store, capture, and infiltrate runoff over a period of several days on a level and uncompacted surface. Infiltration basins are typically used for drainage areas of 5 to 50 acres with land slopes that are less than 20%.
Porous/ permeable/ pervious pavers	These infiltrate and store stormwater. The structural pavement consists of a permeable surface underlain by a storage/infiltration bed. Permeable pavement is well suited for parking lots, tennis courts, playgrounds, walking paths, sidewalks, plazas, and other similar uses.
Bioretention	A bioretention can function as an infiltration system. It employs biological, chemical, and physical processes to control the quality and volume of flow. See Section 3.8.2 for further discussion about the bioretention.

3.8.1.2 Potential limitations for infiltration systems

a. Infiltration systems do not function well on steep slopes. Therefore, areas with steep slopes are not recommended for developing infiltration systems.
b. Infiltration systems require pretreatment in order to prevent clogging.
c. The longevity of infiltration systems is less than five years without multiple pretreatment practices.
d. It may not be used if the contributing drainage area is hotspot.
e. Limited monitoring data are available and field longevity is not well documented.
f. Failure can occur due to improper siting, design, construction, and maintenance.
g. Systems are susceptible to clogging by sediment and organic debris.
h. There is a risk of groundwater contamination depending on subsurface conditions, land use, and aquifer susceptibility.
i. They are not ideal for stormwater runoff from land uses or activities with the potential for high sediment or pollutant loads.

3.8.1.3 How does an infiltration system work?

As the name suggests, an infiltration system works by collecting water and infiltrating it into the subsoil. It is typically constructed in permeable soils that capture, store, and infiltrate the volume of stormwater runoff through a stone-filled bed and then into surrounding soil as shown in Figure 3.4. It is designed with soil layers that support water movement down to lower soil layers. This is achieved through specifying soils with high infiltration rates. Although infiltration occurs more rapidly when the underlying soils are sandy, acceptable levels of infiltration can be achieved in clay soils with appropriate design modifications.

An infiltration system can be designed with vegetation cover on top or without vegetation cover. If the level of pollutant loads is considerably high, infiltration systems may be designed to have a well-specified vegetative cover. This helps to prefilter the water before it infiltrates into the subsoils. This is constantly kept in check to ensure it remains healthy by controlling weeds, insects, and to introduce nutrients such as nitrogen fertilization. The grass is kept at constant height by mowing.

Prefiltration allows debris and trash to remain on top and later removed during routine maintenance of the infiltration systems. The infiltration is designed with underdrain perforated pipes used as bypass so that when runoff exceeds the volume capacity, the excess can be passed over into drainage system. Similarly, infiltration systems are designed to accommodate a certain volume of water, and if this is exceeded, the overflow pipes would drain it off to the nearby sewers. Infiltration systems have wide applications in golf courses, urban islands, residential areas, and sports fields.

3.8.1.4 Infiltration systems' design criteria

Several key design features are considered for infiltration systems, that is, basins and trenches. These include pretreatment is recommended to avoid clogging, depth to water table or bedrock, soil types (infiltration rates, permeability, limiting layer, etc.), underdrain and overflow pipes, proximity to buildings, drinking water supplies, karst features, and other sensitive areas.

As mentioned in Section 3.6 (site-specific factors), site gradient highly affects the performance of an infiltration system. In some instances, maximum slope of 20% is considered safe for an infiltration system to perform effectively. A minimum depth to bedrock of 0.61 m and a minimum depth to seasonal high water table (SHWT) of 0.61 m are also necessary.

An infiltration designer or planner must identify sites that are well suited for the use of infiltration SuDS practice that particularly rely on appropriate in situ soils. In doing so, there is need to understand the primary design considerations especially the key design features and site factors. The nature of soils, water table, and other site constraints highly govern design.

The key design features of infiltration practices include the depth to water table or bedrock; pretreatment is often needed to prevent clogging, often required level infiltration surface, proximity to buildings, drinking water supplies, karst features, other sensitive areas, soil types (permeability, limiting layer, etc.), and provide positive overflow in most uses.

In addition, site factors to consider include the maximum site slope should not exceed 20%, the minimum depth to bedrock should be about 0.61 m, minimum depth to seasonally high water table should be about 0.61 m, potential hotspots: yes with pretreatment and/or impervious liner, hydrologic soil group (HSG) soil types A and B are preferred, C and D may require an underdrain, maximum drainage area not more than 5 acres.

Stages followed to design, evaluate, and analyze an infiltration system are provided below:

a. Stage 1. Determine the statutory and regulatory requirements

As a preliminary step, review all applicable statutes, statutory instruments, orders, regulations, and codes of practice (whether or not having the force of law) in force from time to time concerning the SuDS and particularly the infiltration systems. These requirements are usually nonnegotiable and must be complied with. Failure to comply with a legal requirement may result in a fine or penalty and possibly a custodial sentence for the person or persons responsible or organization for such failure. Both statutory requirements and regulatory requirements are those requirements that are required by law.

As a designer, you must learn how to find sources for design guidance and get acquainted with a better understanding of the use and limitations of infiltration practices for the management of stormwater quantity and quality from existing and proposed area to be developed.

b. Stage 2. Evaluation process

The evaluation process involves listing the advantages and disadvantages of the type of infiltration practice to undertake the project, applicability, suitability, and the feasibility of the infiltration practice. The feasibility criteria generally follow the following guidelines:

  i. The infiltration rate of the soil should be greater than 13 mm/h. In case there is no information about the infiltration rate of the soil, a geotechnical test must be conducted.
  ii. The silt/clay content of the soil should be less than 40%.
  iii. The clay content of the soil should be less than 20%.
  iv. The infiltration system should not be located on slopes greater than 20%.
  v. The infiltration system should be located at least 1.2 m above the high water table or bedrock.
  vi. All contributory drainage area must be stabilized before constructing the infiltration system.

vii. The maximum contributing area should not exceed 5 acres.
viii. The horizontal setback from a water supply well should not be less than 30 m.
ix. The horizontal setback down-gradient from structures should not be less than 7.5 m.

c. Stage 3. Soils and seasonal high water table evaluation
Soil investigation for infiltration purposes is examined for the composition. Soil texture, or the percentage of **sand, silt**, and **clay** in a soil, is the primary inherent factor affecting infiltration. As mentioned is Section 3.8.1.4(b), the amount of silt/clay content of the soil should be less than 40%. Specifically, the clay content of the soil should be less than 20%. Water moves more quickly through the large pores in sandy soil than it does through the small pores in clayey soil, especially if the clay is compacted and has little or no structure or aggregation.

The applicability of an infiltration practice heavily relies on the water table. A high water table results in low infiltration capacities of the soils and high surface runoff rates. The infiltration system should be located at least 1.2 m above the high water table or bedrock. SHWT is the highest zone of soil or rock that is seasonally or permanently saturated by a perched or shallow water table—a planar surface, below which all pores in rock or soil (whether primary or secondary) that is seasonally or permanently saturated. Usually, poorly drained soils have a blackish surface layer that is underlain by a gray subsoil that is mottled. These soils have a water table at or near the surface for a significant portion of the year.

SHWT may be determined in the field through identification of redoximorphic features in the soil profile, monitoring of the water table elevation, or modeling of predicted groundwater elevations. The SHWT is also detected by the mottling of the soil that results from mineral leaching. Soil mottling may indicate the presence of a seasonal high groundwater table. **Soil mottling** refers to the presence of irregular areas of different colors in the soil observed during a soil evaluation. Such mottling indicates poor aeration and impeded drainage characteristics, usually from seasonal saturation of the soil.

For proper functioning of an infiltration practice, soils must be examined for composition to ascertain the potential to successfully infiltrate stormwater by obtaining the infiltration rates and also evaluate the variability of the water table throughout the year.

d. Stage 4. Topography and other site constraints
Steep terrains typically with slopes greater than 20% are not feasible to locate an infiltration system. This is because steep terrains warranty high velocities of stormwater runoff and this leads to surface runoff to quickly route off the surface of the infiltration system. Other site constraints such as buried utilities, tree roots, available easements, and buildings may need to be evaluated.

e. Stage 5. Design criteria
Pretreatment and treatment practices are essential for the effective performance of the infiltration systems. This can either be done by the use of redundant pretreatment methods such as the washed bank run gravel as aggregate, grass channel, bottom sand layer, and grass filter strip.

Without using the redundant techniques for pretreatment of the runoff, other techniques include the sedimentation basin, stilling basin, and sump pit. As a rule of the thumb, 25% or more of the **water quality volume** (WQ_v) must be treated before it enters the infiltration system. And in case the infiltration rate of the soil is greater than 50 mm/h, at least 50% or more of (WQ_v) should be treated. In designing the pretreatment area, the emphasis is also put on designing to exfiltrate the WQ_v less pretreatment volume.

Water quality volume (WQ_v)

Hydrologic properties of watersheds usually change when natural land is cleared for development. When the land is cleared for development purposes, sediment loads carried by the runoff increases as compared to predevelopment conditions. Water quality volume is that stormwater runoff storage volume just enough to capture suspended sediment load before it is transported to the receiving waters. The technique requires a modest amount of information, including the watershed drainage area and impervious cover, stormwater runoff pollutant concentrations, and annual precipitation. The formula for calculating WQ_v is given in Equation (3.2).

$$WQ_V = 4.047 PAR_v \tag{3.2}$$

Where:

WQ_V : Water quality volume (m³)
P : Precipitation depth (mm)
A : Watershed area (acres)
R_v : Volumetric runoff coefficient

Precipitation depth (*P*) for water quality volume

Annual sediment load is more a function of the **number of storms** than the magnitude of the storm. Therefore, storm percentiles are used to calculate the precipitation depth instead of the usual annual exceedance probabilities (e.g., a 100-year storm has a probability of occurrence of 1% per year).

The cumulative curve for storms is always used to compute the water quality volume. Usually a 90th percentile precipitation depth is used. For example, in the USA, the water quality volume equation in the state of Ohio uses $P = 22.86$ mm[8]. US EPA[9] (2016) provided a list where most states use P between 19.05 and 38.10 mm. While it is cost prohibitive to implement water quality storage facilities to capture all sediment in runoff from all storms, research shows that capturing 80% of sediment annually can be accomplished by capturing runoff from the 90th percentile storm.[10]

Volumetric runoff coefficient, R_v

The volumetric runoff coefficient is the key component of the formula for water quality volume computation. *The water quality volume requirement is a key parameter in structural SuDS design.* Define site imperviousness as *i*, which is a fraction of the total area of the site. Therefore, the equation for volumetric runoff coefficient is given as:

$$R_v = a + bi + ci^2 + di^2$$

The coefficients *a*, *b*, *c*, and *d* vary from place to place. There are two primary equations used to compute R_v, that is, the Driscoll equation and the Urbonas equation.[11]

The **Driscoll equation** uses only the coefficients *a* and *b* in the R_v equation. The **Urbonas equation** has values for all of the coefficients *a*, *b*, *c*, and *d*.[10] The water quality volume allows you to select whether the coefficients use i in decimal or percent. From Driscoll equation can be reduced to:

$$R_v = 0.05 + 0.009(i)$$

Where $a = 0.05$ and $b = 0.009$ based on i in percent, or

$$R_v = 0.05 + 0.9(i)$$

Where $a = 0.05$ and $b = 0.9$ based on i in decimals.

Worked Example 3.1: Water quality volume calculation
If $P = 22.86$ mm, $A = 20$ acres, and $i = 40\%$ use Driscoll equation to find R_v (using i in decimal), and calculate the water quality volume.
Solution

$$R_v = 0.05 + (0.9) \times (0.4) = 0.41$$

$$WQ_V = 4.047 P A R_v = 4.047 \times 22.86 \times 20 \times 0.41 = 758.62 \text{m}^3$$

Note: In the water quality volume calculator, P, A, and i must be entered as greater than or equal to zero. Depending on the values for the a, b, c, and d coefficients, R_v and/or WQ_v could be computed as negative which would be unrealistic. Coefficients for the volumetric runoff coefficient R_v equation vary from place to place.

Pretreatment surface area determination based on water quality volume (WQ_V)
The determination of pretreatment area is essential. Several stormwater regulations include special requirements for handling of the water quality volume WQ_v. Although the specific definition varies, WQ_v is commonly considered to be the runoff volume that includes 90% of all rainfall events in a given year. The Camp-Hazen Equation (3.3) is used to size pretreatment settling basin surface area. It was derived by the Washington State Department of Ecology.

$$A_s = -\left(\frac{Q_0}{w}\right)(\ln(1-E)) \tag{3.3}$$

Where:

A_s : Sedimentation basin surface area (m²)
E : Trap efficiency, which is the target removal efficiency of suspended solids (set equal to 90%)
w : Particle settling velocity; for target particle size (silt), use the following:
  • Settling velocity = 0.000122 m/s for I < 75%
  • Settling velocity = 0.0010 m/s for I > 75%
  where I is percentage of impervious area
Q_0 : Rate of outflow from the basin, which is equal to the water quality volume (WQ_v) divided by the detention time (t_d); usually 24 hours is used

The equations simplify to:

- $A_s = 0.218 W Q_v$ m² for I < 75%
- $A_s = 0.027 W Q_v$ m² for I ≥ 75%

Design volume determination

The depth and surface are obtained to calculate the volume. Then treatment volume is determined and account for runoff volumes from storms that exceed the design storm.

Infiltration trench design criteria

The depth of infiltration trench is given by the Equation (3.4) below:

$$d_{max} = \frac{T_s}{n} f \qquad (3.4)$$

Where:

d_{max} : Maximum depth of the trench
f : Final infiltration rate (mm/h)
T_s : Maximum allowable storage time (h)
n : Porosity

Surface area of the infiltration trench is given by the equation below:

$$A_t = \frac{V_w}{nd_t + fT} \qquad (3.5)$$

Where:

A_t : Surface area of the trench
V_w : Design volume entering trench, e.g., WQ_v
n : Porosity
T : Time to fill the trench (generally assumed to be less than 2 hours)
f : Infiltration rate of the trench
d_t : Trench depth based on the depth required above seasonal groundwater table or a depth less than d_{max} whichever is smaller

Infiltration basin design criteria

Maximum basin depth is calculated from the equation below:

$$d_{max} = fT_p \qquad (3.6)$$

d_{max} : Maximum depth
f : Final infiltration rate (mm/h)
T_p : Maximum allowable time (h)

Bottom surface area of the infiltration basin is calculated from the equation below:

$$A_b = \frac{(2V_w - A_t d_b)}{(d_b + 2P + 2fT)} \quad (3.7)$$

Where:

A_b : Bottom surface area of the basin
V_w : Design volume entering basin (e.g., WQ_v)
A_t : Top surface area of the basin
d_b : Basin depth based on the depth required above seasonal groundwater table or a depth less than d_{max}, whichever is smaller
P : Design storm rainfall depth
f : Infiltration rate of the basin (mm/h)
T : Time to fill the basin (generally assumed to be < 2 hours)

The top length of the basin (assuming a rectilinear shape) is determined by:

$$L_t = \frac{(V_w + Zd(W_t - 2Zd_b))}{(W_t(d_b - P) - Zd_b^2)} \quad (3.8)$$

Where:

L_t : Top length of the basin
V_w : Design volume entering basin (e.g., WQ_v)
Z : Side slope ratio of the basin (h: v)
d_b : Basin depth based on the depth required above seasonal groundwater table or a depth less than d_{max}, whichever is smaller
W_t : Top width of the basin
P : Design rainfall event

Note: the basin top length and width should be greater than $2Zd_b$.

f. Stage 6. Bypassing excess volumes or oversize the basin

The excess water must be bypassed and diverted to another SuDS so that the design infiltration occurs within 48 hours, or generally within 72 hours under specific local and watershed regulations. In no case should the bypassed volume be included in the pollutant removal calculation. Otherwise, the basin will be oversized.

g. Stage 7. Maintenance considerations

Maintenance considerations are key aspects of the infiltration practice. To ensure longevity of the infiltration practice, maintenance culture should be evaluated and reliability metrics incorporated throughout the project phases. The contributing drainage area should be properly landscaped and dense vegetation cover applied over the entire pervious drainage area. An observation well should be installed in infiltration trench. Compaction of subsoils leads to failure of the infiltration system and should not be allowed. Therefore, during the design stage, it must be clearly outlined and stated in the specifications and method of work to ensure that contractors know that excavations must commence and progress without compacting subsoils. An underdrain pipe is recommended for dewatering especially where there are marginal soils. Infiltration basins, in particular, should be developed with dense vegetation on the side slopes and floor so that erosion and sloughing are prevented.

3.8.1.5 Probable cost of infiltration systems

The cost of infiltration systems varies widely from place to place. Indicative figures are listed below:

a. **Dry well (infiltration trench).** The construction costs (capital costs) range from US$44 to 96.77/m^3, and the maintenance costs are about 5% to 10% of capital costs.
b. **Infiltration trench.** The construction costs (capital costs) range from US$215 to 322.58/m^3, and the maintenance costs are about 5% to 10% of capital costs.
c. **Infiltration basin.** The construction costs vary depending on the excavation, the type of plantings, and the pipe configuration.
d. **Subsurface infiltration bed.** The Construction costs are about US$139.78/m^3.

3.8.2 Bioretention gardens

3.8.2.1 Description and purpose

Bioretention gardens or rain gardens were first conceived in 1990 by stormwater specialists in the state of Maryland in the USA. The goal was to design bioretention to mimic naturally occurring functions that existed in nature before humans began to alter the earth's surface features.

A bioretention garden or rain garden is a type of filtration basin with landscaped shrubs and other various or native plants, a filter media, and a mulch cover to enhance pollutant removal. A bioretention gets its name from the ability of the biomass to retain nutrients and other pollutants. It is a terrestrial-based (upland as opposed to wetland) water quality and water quantity control practice using the chemical, biological, and physical properties of plants, microbes, and soils for removal of pollutants from stormwater runoff. Native plants are usually planted in rain gardens in localized depressions allowing runoff water from impervious surfaces to collect in the gardens and infiltrate into the subsurface soil. The use of native plants reduces the need for irrigation since they are accustomed to the rainfall cycle of the local environment. Carefully selected native vegetation is used to treat and capture runoff and sometimes it is underlain by sand or gravel storage/infiltration bed.

In traditional landscaping practice, institutional and commercial landscapes are often designed as areas built up from the surrounding ground in raised islands, generally to bring them into a higher visual plane. Then, irrigation water is applied, taken up by the plants, and soaks into the ground. Excess water flows off the landscape and sometimes over the pavement or sidewalk, and ultimately into the drainage system. The bioretention twists the traditional landscaping practice and when designed and developed properly can be more beneficial. The simple inversion of the rain garden is a tremendous advantage in managing stormwater runoff. Instead of having direct runoff from pavement to pipe or drain, the rain garden **slows the runoff**, thus increasing **groundwater recharge** and decreasing **downstream runoff volumes**. Therefore, the bioretention can significantly reduce the risk of flooding.

3.8.2.2 How does a bioretention work?

Bioretention gardens capture rainwater runoff to be filtered through a specially prepared soil medium. When the pore space capacity of the medium is exceeded, stormwater begins to pond at the surface of the planting soil. This ponding water is then dewatered either through infiltration into the subsoil (infiltration design), or by means of an underdrain (filter design), or by a combination of the two methods. Bioretention gardens require little maintenance once established and often replace areas that were intensively landscaped and required high maintenance. An illustration of a bioretention garden is shown in Figure 3.5.

A bioretention garden employs a site integrated and terrestrial-based design that provides opportunity for runoff infiltration, filtration, and storage and water uptake by vegetation. In addition to managing and controlling runoff volume and mitigating peak discharge rates, bioretention garden process filters suspended solids and associated pollutants from stormwater runoff. Bioretention can be designed into a landscape as a garden feature that helps to improve water quality while reducing runoff quantity.

Bioretention can take a variety of forms in addition to the landscaped depression in a parking lot. They can be integrated into a site with a high degree of flexibility and can balance well with other SuDS including porous pavement parking lots, infiltration trenches, and other nonstructural stormwater SuDS. Bioretention gardens can be installed in a series and have visible water during wet periods. Tree trenches along sidewalks and roads can become stormwater runoff receptors with the appropriate consideration of elevation and flow paths.

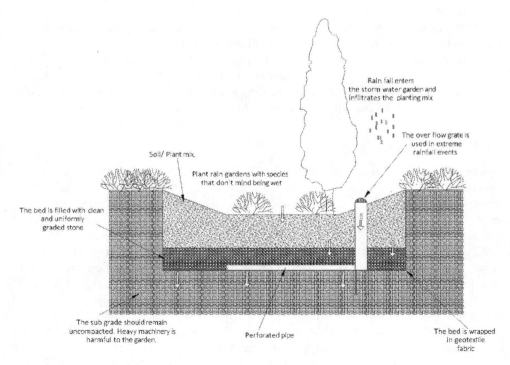

Figure 3.5 A conceptual bioretention garden.

Additionally, runoff water that does not infiltrate into the subsurface of the bioretention is nevertheless slowed down significantly.

The main elements of a bioretention garden include the following:

a. **Nonwoven geotextile**—this is a permeable geosynthetic made of nonwoven materials used with soil, rock, or other geotechnical-related material as an integral part of a civil engineering project, structure, or system. They are used in bioretention systems for two purposes: soil layers separation and permeability.
b. **Underdrain perforated pipe**—this comprises a length of perforated pipe embedded into a layer of aggregate. They are an optional component of bioretention systems. Their design varies according to the drainage requirements of the installation, and the available maintenance access.
  Underdrain pipe is commonly made with uPVC material. Alternative pipe material may include HDPE. Underdrain pipe must be perforated. The total opening area must exceed the expected flow capacity of the underdrain. During construction, the best solution is to have holes pointing downward closest to the pipe invert to achieve maximum potential for draining the bioretention garden. Perforated uPVC pipe commonly has 6- and 12 mm perforations, 150-mm center-to-center perforation along two or three longitudinal lines.
c. **Observation/clean-out stand pipe**—this is a pipe with a cap that provides access to the bottom of the bioretention system, especially the underdrain is usually connected to the clean-out pipe, so that blockages can be removed.
d. **Filter materials**—these generally consist of clean washed sand, gravel, or crushed rock. Manufactured aggregates are also occasionally used, and these often include blast furnace slags. Filter gravel is an extremely effective filter media because of its ability to hold back precipitates containing impurities/pollutants.
e. **Native plants**—a plant is considered native if it has occurred naturally in a particular region, ecosystem, or habitat without human introduction. Bioretention systems require native plants to perform better because native plants are already used to the regional climate and their longevity and sustainability is guaranteed.

3.8.2.3 Bioretention design criteria

Several regions have unique design approaches for a bioretention garden. The design criteria provided in this section is nonexhaustive. It is intended to provide an outline on how to approach the bioretention garden design. Design approaches may vary from one area to another. However, the key components required in the design of a bioretention are discussed and incorporated in the criteria. Bioretention design and development require experienced designers who can produce workable designs and specifications that can enable the bioretention when constructed, to stand a test of time through the operation and maintenance phase, effectively performing the intended purpose, reliably and safely. Experienced designers always benchmark on statutory and regulatory frameworks, especially regional manuals that specify codes of practice and standards for bioretention facilities.

A bioretention garden is flexible in design, providing several opportunities for the designer to be creative.[12] Site suitability and location, integration and site distribution of bioretention areas, and site grading considerations are the main issues for the designer to consider. The bioretention site integrating criteria include the following:

a. County right of way
b. Wellheads
c. Septic fields
d. Basements
e. Building foundations
f. Property lines
g. Outlet drainage
h. Underdrains,[13]
i. Soil restrictions
j. Cross-lot drainage
k. Groundwater
l. Minimum depth criteria
m. Slopes and existing grades
n. Wooded areas
o. Median and traffic island considerations and utilities

The effectiveness and reliability of a bioretention garden is a function of the design and the construction techniques employed. Construction is far more important in achieving quality results. Poor construction technique may cause the best-designed bioretention facility to fail prematurely, usually from sedimentation, clogging, or both. To counter this problem, adequate and proper inspection is vital.

A bioretention garden can be designed to filtrate runoff or infiltrate runoff. Once the soil pore space capacity of the medium is exceeded, stormwater begins to pool at the surface of the planting soil. This pooled water can then be dewatered either through infiltration into the subsoil (infiltration design), by means of an underdrain (filter design), or by a combination of both the methods. Unlike end-of-pipe SuDS, bioretention facilities are typically shallow depressions located in upland areas. The strategic, uniform distribution of bioretention facilities across a development site results in smaller, more manageable sub-watersheds, which control runoff close to the source where it is generated. The sizing of a bioretention garden follows computation methods and procedures for sizing and design of bioretention facilities.

The bioretention design has two types:

a. The filtration design type
This particular design emphasizes practices that capture and temporarily store the water quality volume WQ_v and pass it through **a filter bed of sand**, organic matter, and soil or other media that are considered to be filtering practices. The filtered runoff may be collected and returned to the conveyance system through the perforated underdrains. The design variants include the following:

- Surface sand filter
- Underground sand filter
- Perimeter sand filter
- Organic filter
- Pocket sand filter
- Bioretention*

*May also be used for infiltration.

b. Infiltration design type

This particular design emphasizes practices that capture and temporarily store the water quality volume WQ_v before allowing it to infiltrate into the soil over a two-day period include the following:

- Infiltration trench
- Infiltration basin

3.8.2.3.1 Bioretention design goals and objectives

Bioretention is flexible in design and can be used as either an infiltration SuDS or a filter SuDS. Bioretention can often be designed to meet one or more of the stormwater criteria, which include the following:

a. Water quality volume, WQ_v
b. Recharge volume, Re_v
c. Channel protection storage volume, Cp_v
d. Overbank flood protection volume, Q_p
e. Extreme flood volume, Q_f

3.8.2.3.2 General filtration design process for sizing the bioretention garden

A surface depression is created to receive the runoff (see Figure 3.6). This 230- to 305-mm depth will receive up to 150 mm of runoff from the impermeable area. Since the depth of runoff expected from 25 mm of rain falling on an impermeable surface (CN = 98) is 20 mm, the 150-mm storage in the bioretention garden represents the runoff from an impermeable area with an area 7.5 times larger than the bioretention garden. Thus, if the impermeable area is 3 acres in size, the surface area of the bioretention cell will need to be 3.0/7.5 = 0.4 acres in

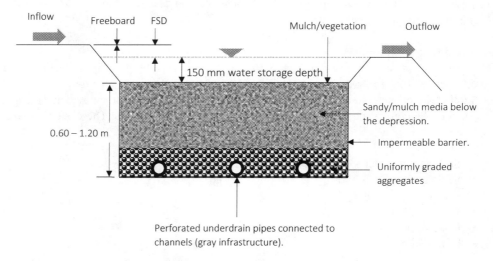

Figure 3.6 Schematic of a bioretention cell with underdrains (filter design).

size. This will permit the 20-mm first flush from the 3-acre impermeable area to fill the 0.4-acre bioretention cell to a depth of 150 mm. If the depth of storage in the bioretention area is increased to greater than 6 in or decreased to less than 150 mm, the ratio of impermeable area to first flush can be adjusted accordingly.

A bioretention garden should be located where it can receive water from impermeable area(s) especially in slightly depressed areas, provided the depressions do not have a seasonally high water table that reaches within 1.2 m of the soil surface and not designated as a wetland. Flat areas are always engineered.

When the natural soils are permeable enough to infiltrate the captured water (permeability > 50 mm/h), a small downstream depression should be created in a flat area to receive the impermeable area runoff. When the natural soil permeability is too slow to infiltrate captured runoff water (permeability < 50 mm/h), it may be necessary to excavate the area under the bioretention garden to a depth of 600 to 1200 mm. Then, an underdrain perforated pipe system is then placed at the bottom of the excavated area before the hole is filled with a mixture of sandy soil and mulch with a permeability of at least 50 mm/h (see Figure 3.6). And well-graded aggregates are placed at the bottom. A surface depression is created to receive the runoff (see Figure 3.6).

3.8.2.3.3 Sizing the bioretention, North American standard12

Note: Bioretention gardens were first conceived in 1990 by stormwater specialists in the state of Maryland in the USA. The goal was to design bioretention to mimic naturally occurring functions that existed in nature before humans began to alter the Earth's surface features. According to Maryland department of environment the following **steps** are considered:

1. **Step 1: Delineate the development site drainage area** in the pre- and postdevelopment condition. Delineate subdrainage divides for the postdevelopment condition, identifying strategic locations for possible bioretention facilities.
2. **Step 2: Using Technical Release #55 (TR-55) methodology**, determine the pre- and postdevelopment Curve Numbers (CN) for the proposed development site. Adjust the CN value by measuring the actual impervious versus pervious areas. Remember to incorporate other LID site design techniques to help reduce the postdevelopment CN value. Methods such as increasing the percentage of disconnected impervious areas, preserving wooded areas, and reducing impervious surfaces will minimize the amount of control required.
3. **Step 3: Select the required design storm and design depth** for the bioretention facility or facilities. The stormwater management concept application submission must identify the intent to use an LID approach through the use of bioretention. The design storm used depends on the objectives of the LID approach, which can vary from stream and ecosystem protection to load reduction for total maximum daily load (TMDL) requirements. **Note**: the design storm can vary significantly depending on the stormwater management objective.
4. **Step 4: Determine the storage volume required to maintain runoff volume or CN.**
5. **Step 5: Determine the storage volume required to maintain the predevelopment peak runoff volume using 100% retention.** Calculate the percentage of site area required to maintain the predevelopment peak runoff rate using 100% retention.

80 Integrated Drainage Systems Planning and Design for Municipal Engineers

6. **Step 6:** Determine the percentage of the site needed to maintain both the predevelopment peak runoff and the runoff volume.
7. **Step 7:** Determine the appropriate percentage of the site available for retention practices. If the percentage of the site available for retention practices is less than the percent determined in step 5, recalculate the amount of SuDS required to maintain the peak runoff rate while attenuating some volume.

3.8.2.3.4 Summary of the bioretention design criteria, North American standard[12]

Note: Bioretention can often be designed to meet one or more of the stormwater sizing criteria. The designer can adopt this process in consultation with local standards and manuals to customize the design to regional standards.

Sizing criteria	Description of stormwater sizing criteria
Water quality volume, WQ_v determination	$WQ_V = 4.047 PAR_v$ WQ_V : Water quality volume (m³) P : Precipitation depth (mm) A : Watershed area (acres) R_v : Volumetric runoff coefficient
Volumetric runoff coefficient, R_v determination	The volumetric runoff coefficient is the key component of the formula for water quality volume computation. Define site imperviousness as i, which is a fraction of the total area of the site. Therefore, the equation for *volumetric runoff coefficient is given as:* $R_v = a + bi + ci^2 + di^2$ The coefficients a, b, c, and d vary from place to place. There are two primary equations used to compute R_v, i.e., the Driscoll equation and the Urbonas equation. The Driscoll equation uses the coefficients a and b in the R_v equation. The Urbonas equation has values for all of the coefficients a, b, c, and d.[10] From Driscoll equation
Recharge volume, Re_v	This is the fraction of WQ_v which depends on predevelopment soil hydrologic group. The recharge volume is considered part of the total WQ_v that must be provided at a site and can be achieved either by a structural practice (e.g., infiltration or bioretention), a nonstructural practice (e.g., buffers, disconnection of rooftops), or a combination of both. $Re_v = 4.047 SR_v A$ S : Soil specific recharge factor (mm) A : Watershed area (acres) R_v : Volumetric runoff coefficient

Channel protection storage volume, Cp_v	Channel protection storage volume is computed to protect channels from erosion. Channel protection storage volume Cp_v means the volume used to design structural management practices to control stream channel erosion. Methods for calculating the channel protection storage volume are specified in the 2000 Maryland Stormwater Design Manual.[14] According to 2000 Maryland Stormwater Design Manual, Cp_v = 24 hour (12 hour in USE II and IV USA watersheds) extended detention of postdeveloped one-year, 24-hour storm event. The Cp_v requirement does not apply to direct discharges to tidal water unless specified by an appropriate review authority on a case-by-case basis. Local governments may wish to use alternative methods to provide equivalent stream channel protection such as the distributed runoff control method.
Overbank flood protection volume, Q_v	Overbank Flood Protection Volume (Q_v) means the volume controlled by structural stormwater management practices computed to prevent an increase in the frequency of out-of-bank flooding generated by development, for which the method of calculation is specified in the 2000 Maryland Stormwater Design Manual, Volumes I and II (Maryland Department of the Environment, April 2000).[14]
Extreme flood volume, Q_f	Extreme flood volume Q_f means the storage volume computed to control those infrequent but large storm events in which overbank flows reach or exceed the boundaries of the 100-year floodplain. Normally, no control is needed if development is excluded from 100-year floodplain and downstream conveyance is adequate.

3.8.2.3.5 Probable cost

Every site is unique, requiring specific cost estimation to account for the variability and this creates huge differences in cost. Cost will vary depending on the garden size and the types of vegetation used; typical costs are about US$108 to 185 per m². In estimating the cost of using bioretention, a number of factors need to be considered which include the following:

- Site restrictions—both physical and regulatory
- Availability of materials, equipment, and labor
- Scheduling tasks for efficiency

3.8.3 Ponds/basins

There are two kinds of ponds: wet and dry. The wet pods are generally referred to as retention ponds, and dry ponds are referred to as detention ponds. Retention ponds aim to limit the outlet flow into existing sewers or channels, while allowing some stored runoff to recharge groundwater levels. On the other hand, detention uses a watertight system to redirect flow to only the regulated discharge outlets. These systems have two categories: subsurface and surface retention/detention systems. The materials used to develop these ponds vary widely from concrete to UPVC pipes and HDPE.

Local environmental regulations or the groundwater situation influence the decision to select a retention system or a detention system. An underground retention system typically uses **perforated pipe** so that the stored runoff can recharge groundwater, and a discharge outlet is designed to limit the flow rate into the receiving sewers or channels. A detention system typically uses **non-perforated pipe** and watertight joints so that the stored runoff exits only through the regulated discharge outlet.

In Greenfield development, detention/retention systems are sized to maintain the runoff rate prior to development so that existing storm sewers, channels, and other waterways are not flooded. Underground pipe networks allow the land above the system to be used for recreation, parking, and other purposes.[15]

3.8.4 Detention ponds/basins (dry ponds/basins)

3.8.4.1 Description and purpose of detention ponds/basins

Detention ponds (dry ponds) or dry detention basins and retention ponds are developed to collect and store stormwater. The difference between the two is that a detention pond releases stormwater immediately after the storm typically not exceeding 24 hours while a retention pond always has a permanent pool of water. All the water runs out of the detention pond between storms and it usually remains dry. The water rarely remains in dry detention basins for longer than 24 hours, but the longer it stays in the basin, the more the water quality improves. The water level is controlled by a low-flow orifice. A detention pond has an orifice level at the bottom of the basin and does not have a permanent pool of water. In most cases, the orifice is part of a metal or concrete structure called a riser. Figure 3.7 shows a schematic diagram of a typical dry detention pond.

Detention ponds are a type of SuDS used primarily for controlling water quantity versus quality, as they only confine stormwater for a short period and later release it to flow. Note that detention ponds do not reduce stormwater quantity but reduce the peak flow thereby minimizing the risk of flooding downstream. The detention pond reduces the peak discharge through attenuation, that is, prolonging the flow time of runoff to reduce the peak discharge.

Detention ponds are of two types: subsurface and surface systems. Due to the land requirements for a pond, planners and developers choose to locate detention below ground, enabling the dual use of the same area for both surface access and drainage. In some areas, the below ground type is known as **detention tanks**.

Generally, detention ponds provide flood control measures. A detention pond is also called a dry pond because it is always empty before it rains. The design of these ponds follows statutory and local guidelines. The most important requirement to design and develop these ponds is **space**. In many cases, they consume large spaces. Also, they come with considerable initial capital investments. To be successful, detention ponds do not require any specific climate conditions that is why they are commonly applied to control stormwater. They fit in almost all climates although with minor design alterations.

The primary purpose of detention and retention ponds is to collect, store, treat, and later release water to natural receiving water bodies. Detention ponds are usually designed to:

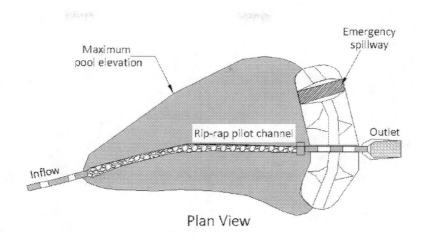

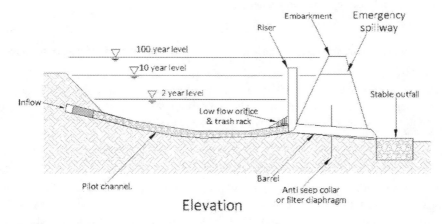

Figure 3.7 Schematic diagram of a typical dry detention pond.

a. Minimize peak flows from stormwater runoff.
b. Accommodate existing and future upstream development.
c. Control erosion downstream.

The detention and retention systems can both be designed and developed as subsurface systems or surface water systems. The subsurface detention system is shown in Figure 3.8.

3.8.4.2 *Features and benefits of surface detention system*

Detention basins are surface storage basins or facilities that provide **flow control through attenuation of stormwater runoff**. They are located on the surface and facilitate some settling of particulate pollutants. Detention basins are normally dry and in certain situations they may also function as a recreational facility.

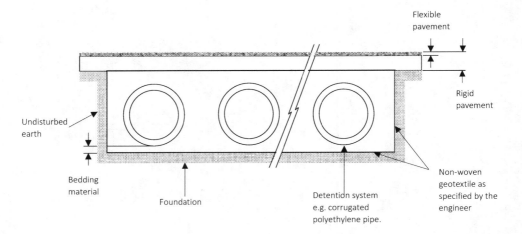

Figure 3.8 Typical subsurface detention system.

3.8.4.3 Features and benefits of below surface (subsurface) detention system

When aboveground space is at a premium, underground stormwater detention systems are an efficient way to store, detain, or infiltrate stormwater runoff while the ground above is used for parking, parks, or other features. These detention systems also may include filters to remove sediment and debris before the water enters the system.

3.8.4.4 Potential limitations for detention systems

In some areas, detention basins may require relatively large spaces in case a surface detention pond is found more feasible than below ground detention system. Enough impervious land area is required to develop the detention basins because if the area is small, the outlet system could become ineffective due to clogging after raining in case surface detention system is to be developed. Where the area is highly contaminated, detention basins may not be applicable and may not work in isolation especially in flooding hotspots. Detention ponds have a risk of turning into mosquito breeding sites especially in tropical countries. They may also lower the property value.

3.8.4.5 Environmental benefits of detention ponds

a. Detention ponds can improve water quality by removing sediment loads and uptake of hydrocarbon pollutants.
b. Detention ponds promote the amphibian habitat and fish habitat.
c. Detention ponds can be a source of nutrients to creek system.
d. Detention ponds provide aesthetic purposes with varying water levels, diverse plantings, and walking trails.

3.8.4.6 How does a detention pond work?

The pond/basin fills with stormwater runoff during and immediately following storm events and particles and pollutants settle to the bottom. An outlet designed to throttle flow rate drives water into nearby streams or storm sewer systems, leaving the dry pond empty whenever the water reaches a certain level. Dry ponds are designed with an emergency spillway to guide water away from the basin during major storm events. Both the outlet and the emergency spillway must be designed and maintained to eliminate the risk of erosion.

3.8.4.7 Detention ponds' design criteria

Local ordinances and statutory requirements dictate the approach to design and develop detention systems. Retention/detention systems vary widely in design, from open ponds to subsurface piping systems and underground vaults to gravel pits. Stormwater detention systems vary widely in design, from open ponds to subsurface piping systems and underground vaults to gravel pits. Checking statutory and local regulations for surface and subsurface retention/detention systems is the first step to designing the systems.

Several SuDS are designed to cope with the most frequent storms, for example, 2-year or 5-year storms. Particularly, the detention and retention basins are designed to withstand storms of such frequency. The modified rational method is the primary method used to determine the flow in detention pond design (Section 4.4.3.1.3).

EPA SMM mode provides explicit steps to cater for infiltration practices and detention systems within the model basins. Integrated drainage systems usually prioritize detention systems to collect and delay the flooding upstream. In gray infrastructure drainage networks, detention ponds may be fitted in the system as nodes (see Figure 3.9). The figure shows nodes A and B. A is a detention pond and B is a manhole. They are both fed by interconnecting links. The design of these junctions requires the estimation of flow quantities, and safe routing of the flows. The famous flow quantity determination of such a system is the modified rational method. The primary assumption of the modified rational method is that when the storm duration exceeds the time of concentration, the rate of runoff would rise to the *peak value obtained from the rational formula and then stay constant until the net rain stops.* Therefore, the detention pond volume can be obtained accordingly.

3.8.4.8 Probable detention pond cost

The cost provided here is indicative. It is important to note that costs vary from one location to another. According to the US Environmental Protection Agency (EPA),[16] typical costs for wet

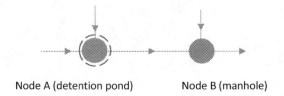

Node A (detention pond)   Node B (manhole)

Figure 3.9 Illustration of interconnected SuDS (detention pond) and gray infrastructure.

detention ponds range from US$17.50 to 35.00 per m³ of storage area. Dry detention basins typically cost around $10 per m² for smaller basins and $5 per m² for larger basins. However, the total cost for a pond or detention basin needs to include allowances for permitting, design and construction, and maintenance costs. The permitting costs may vary depending on state and local regulations.

3.8.5 Retention ponds/basins

3.8.5.1 Description and purpose of retention ponds

A retention basin or pond is designed and developed in the same way as detention ponds (Section 3.8.4). On surface retention pond has a riser and orifice at a higher point and therefore retains a permanent pool of water. It plays an important role in controlling stormwater runoff.

The depth of the permanent pool of water in a retention pond is generally designed considering water quality volumes and hence a retention pond also acts as a water treatment device. Additional storage capacity is provided above the permanent pool for temporary runoff storage (see Figure 3.10). Retention ponds always have a permanent pool of water. The primary purpose of retention system is to control stormwater runoff volumes. A retention basin or pond has a riser and orifice at a higher point and therefore retains a permanent pool of water.

3.8.5.2 Integrated on surface detention and retention system

This is a system which serves as both detention and retention system. This is typically surface system as shown in Figure 3.11. The design or an integrated surface detention/retention system follows similar principles for detention/retention systems.

3.8.6 Vegetated/bio-swales

3.8.6.1 Description and purpose of swales

These are shallow channels usually densely planted with a variety of grasses, shrubs, and/or trees providing a sustainable and competitive **alternative to traditional kerb and gutter**

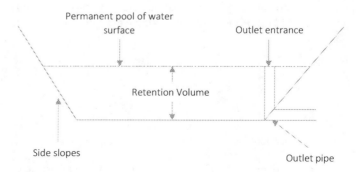

Figure 3.10 Typical retention system.

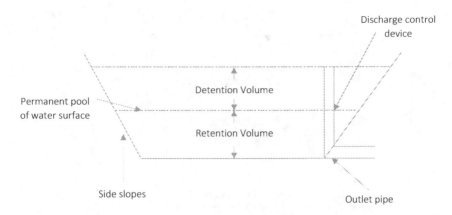

Figure 3.11 Integrated surface detention and retention system.

conveyance systems specifically through pretreatment and stormwater distribution to other SuDS. They are purposely designed to slow, filter, and infiltrate stormwater runoff.

Bio-swales are generally at least 30-m long, 0.6-m wide, range in longitudinal slope from 0.5% to 6%, and located in series with detention ponds, which store runoff and reduce peak discharges. Although they are designed to convey runoff from the 100-year 24-hour storm event, they are only intended to treat runoff effectively from much smaller and more frequent storms, typically up to the 2-year 24-hour storm event.[17, 18]

Vegetated swales can be applied on steeply sloping grounds with check dams installed to control sediment load, drop velocity, and provide additional storage. That means swales can be easily incorporated into plans and layouts with unique topography because the longitudinal slopes can vary between 0.5% and 6% while the side slopes vary from 3:1 to 5:1 (H: V). Swales provide several other benefits such as water quality improvement through filtration, peak, and volume control mechanism and are cost-effective because of replacing the costly standard gray kerb and gutter gray systems.

Bio-swales provide some benefits, for example, typical grassed channel swales and dry swales provide a considerable amount of groundwater recharge considering a high degree of infiltration is designed. However, wet swales do not contribute to groundwater recharge, because accumulation of organic debris on the bottom of the swale lowers infiltration. Bio-swales are generally applied along streets, roads, and around parking lots, where substantial automotive pollution settles on the pavement and is flushed by the first instance of rain, well-known as the first flush.

3.8.6.2 How does a vegetated swale work?

Vegetated swales are shallow stormwater channels commonly used as pretreatment devices capturing runoff especially from roadway. They substitute for kerb and gutter gray drainage structures. The grass, trees, or shrubs planted in the swale can slow, filter, and infiltrate stormwater runoff. In many cases, swales effectively treat stormwater runoff coming from highly impervious surfaces. The performance of swales is usually enhanced through additional pretreatment measures. See Figure 3.12 for a typical cross-section of a swale.

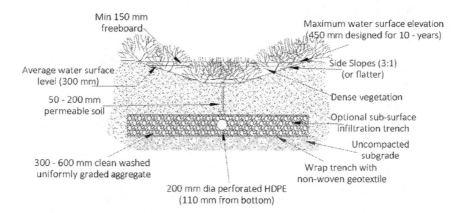

Figure 3.12 Vegetated swale.

3.8.6.3 Potential limitations of swales

They have limited removal of fine sediment and dissolved pollutants. When designed improperly, bio-swales have very little pollutant removal potential. They also do not seem to be effective at reducing bacteria levels in stormwater runoff. For most bio-swales, flow velocity and hydraulic loading during storm events appear too large to permit sedimentation of silt and clay particles, even with dense vegetation and abundant organic litter.[19] Thus, herbaceous vegetation abundance may not provide a good indication of bio-swale treatment performance, and actual stormwater treatment may be much poorer than is generally anticipated.

Bio-swales have a limited application in areas where space is a concern. They use more land area than kerb and gutter system and restrict certain activities like car parking. Bio-swales have a limited peak and volume control unless designed for infiltration. Bio-swales require a sunny aspect for plant growth, which limits their application in shaded areas. Heavily shaded areas overwhelm all other environmental factors. Where light is adequate, vegetation and organic litter biomass is strongly and inversely related to the proportion of time bio-swales are inundated above 25-mm depth during the driest time of year, for example, summer.

3.8.6.4 Variations in bio-swale designs

Bio-swales specifically designed to **attenuate and treat stormwater runoff** for a defined water volume. There are a number of design variations in bio-swales. These include vegetated swales with infiltration trench, grass swales, wet swales, and dry swales.

a. Wet swales

Wet swales behave almost like a bio-filtration or linear wetland cell treatment system intersecting the groundwater. Wet swales typically occur when the water table is located very close to the surface or water does not readily drain out of the swale. The wet swale design incorporates a shallow permanent pool and wetland vegetation to provide stormwater treatment. Wet swales do not provide volume reduction and have limited treatment capability.

The shallow standing water poses a health hazard; it can become a breeding site for mosquitoes and other insects. For that reason, wet swales are unpopular in residential areas.

b. Grass swales

A grass swale is a stable turf, parabolic or trapezoidal channel used for water quality or to convey stormwater runoff, which does not rely on the permeability of the soil as a pollutant removal mechanism. The total suspended solid (TSS) removal rate for a grass swale designed according to this chapter is 50%.

c. Dry swales

The dry swale is a soil filter system that temporarily stores and then filters the desired treatment volume. Dry swales rely on a premixed soil media filter below the channel that is the same as that used for bioretention. Dry swales are essentially bioretention cells that are shallower, configured as linear channels, and covered with turf or other surface material (other than mulch and ornamental plants).[20]

3.8.6.5 Probable cost

Indicative figures range between US$16 and 65 per linear meter depending on extent of grading and infrastructure required, as well as the vegetation used.

3.8.7 Reforestation and afforestation

3.8.7.1 Description and purpose

Afforestation is the practice of planting trees in an area where they never existed while reforestation is the practice of planting trees in an area which was previously deforested. **Afforestation** is the establishment of a forest or stand of trees (forestation) in an area where there was no previous tree cover. During site clearing for project development, naturally occurring trees may need to be cut down to create a way for developing structures. That means well-planted trees are planted after the project is completed to bring back the developed area to a condition close to the original predevelopment conditions. Some sites never have trees. In that case, planting trees is a stormwater BMP that is found relevant when integrated in project developments. The purpose is to restore predevelopment hydrology to the extent possible or to afforest an area to tap into trees benefits. In several countries, vertical foresting has become a trending norm because of space constraints and the desire to design and develop ecologically adaptive and resilient built environments.

Afforestation and reforestation activities must be considered systematically and integrally. More and more studies have shown that making afforestation plans from the perspective of forest ecosystems is the future trend. Using multifactor methods to analyze forest site characteristics will become the primary site evaluation and classification method.[21]

Reforested and afforested areas are well-known for providing a carbon sink and are at the forefront of fighting climate change than any other BMP because they mimic nature forests acting as 'lungs' that absorb carbon and create oxygen. Trees stabilize unstable soils and can be planted on steep slopes; they do not have slope constraints unlike other types of SuDS. The area becomes aesthetically appealing with trees planted across the estate providing shades for people and a cool temperature range. They provide energy benefits (cooing/heating) and a reduction in urban heat island. They act as erosion control BMP solving huge sediment loads that would otherwise end in storm sewers, ponds, and receiving waters. Trees promote the ecological system acting as a shelter for a number of animal and bird species.

3.8.7.2 How does reforestation and afforestation work?

Reforestation and afforestation practices work in a similar way natural forests work. However, they need to be planted following local guidelines and regulations. Because the lifecycle of forests is long and the complexity of changing the forests once afforested, many factors must be carefully evaluated so that the success of afforestation is achieved. These factors include forest site, selection of afforestation material, and site preparation.

3.8.7.3 Potential afforestation and reforestation limitations

Trees absorb water from the soil which can lead to consolidation and settlements especially in fine-grained soils. The effect depends on three factors:

a. Plasticity index of the soil
b. The proximity of the tree from structures, for example, houses with shallow foundations
c. The tree water demand

The implication is that houses with shallow foundations built near afforested or reforested areas would face the effect of consolidation and settlement. It is therefore relevant to space trees quite far from houses with shallow foundations depending on the number and type of trees planted. The plasticity index is determined through laboratory testing of soils and it is used to calculate the consolidation effect. Where the plasticity index of the soil is not known, it is a good practice to assume high plasticity. The following suggested minimum foundation depths are based on the assumption that low water demand trees are located 0.2 times the mature height from the building, moderate water demand trees at 0.5 times the mature height, and high water demand trees at 1.25 times the mature height of the tree.

3.8.7.4 Afforestation and reforestation design criteria

Reforestation and afforestation practices work in a similar way natural forests work. However, they need to be planted following local guidelines and regulations. Because the lifecycle of forests is long and the complexity of changing the forests once afforested, many factors must be carefully evaluated so that the success of afforestation is achieved. These factors include forest site, selection of afforestation material, and site preparation.

3.8.8 Planter boxes

3.8.8.1 Description and purpose

These are carefully constructed containers usually enclosed features either below or above the ground boxes planted with vegetation that capture stormwater within the box, (Figure 3.13). Planter boxes are usually located close to downspouts to receive stormwater runoff from the rooftops. Just like other SuDS, they play an important role in reducing the volume and peak flow. The planter boxes work very similar to bioretention gardens, as they reduce stormwater

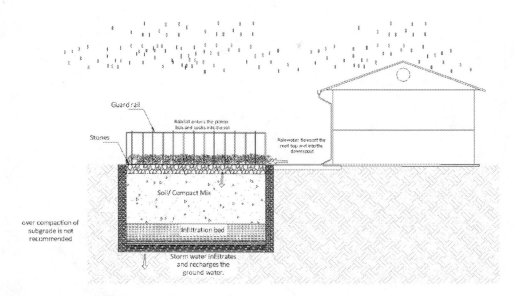

Figure 3.13 Planter box.

through evapotranspiration, capture pollutant loads, and creating a greener and aesthetic environment. They are restricted due to their relatively high cost due to structural components for some variations. There are limited stormwater quantity/quality benefits. The commonest materials used to make planter boxes are usually wood and concrete. However, they require proper waterproofing when constructed near buildings to protect foundations. Some of the benefits of planter boxes include the following:

a. They enhance site aesthetics and habitat.
b. They provide air quality and climate benefits.
c. They have a potential to reduce runoff and combined sewer overflow.
d. They have a wide applicability including ultra-urban areas.

3.8.8.2 Planter box probable cost

The cost varies based on type, size, plant selection, etc., but indicative figures range approximately between US$85 and 160 per m^2.

3.8.9 Tree trenches

3.8.9.1 Description and purpose

Tree trenches perform similar functions that other infiltration practices perform, which include infiltration, storage, evapotranspiration, and pollutant removal. In addition, tree trenches provide an increased tree canopy. Figure 3.14 shows a tree trench integrated with permeable pavers to infiltrate stormwater. Figure 3.15 provides an example of a tree trench with permeable pavers and subsurface infiltration bed.

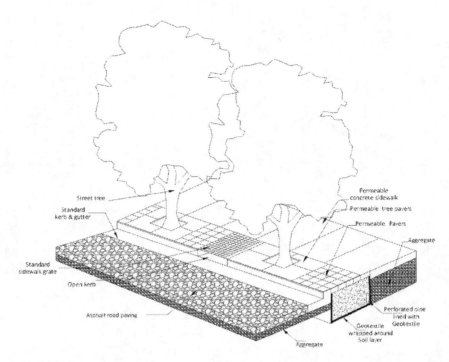

Figure 3.14 Example of a street tree trench with structural soil and adjacent infiltration trench.

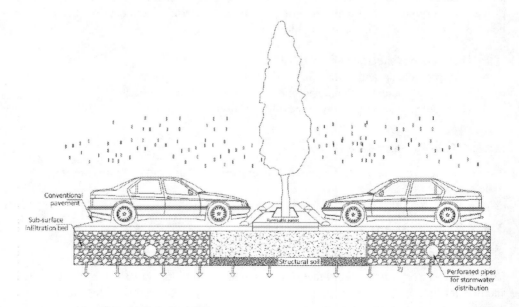

Figure 3.15 Tree trench with porous pavers and subsurface infiltration bed.

Increased canopy cover is one of the primary benefits associated with tree trenches. As a result of tree trenches the site becomes aesthetically appealing. Several climate benefits are realized. In urban areas, tree trenches significantly reduce the runoff quantities, lead to improved air and water quality, have a high fast-track potential and enhanced tree longevity.

3.8.9.2 Tree trench design criteria

The tree trenches vary in terms of the paving technology used, structural soil or alternative such as Silva cell. Some tree trenches are open vegetated strips planted with ground cover like grass while others are covered with porous pavers. Others have tree grates, alternate storage media (modular storage units). In some places, tree pits are usually prefabricated. The design criteria follow a number of site factors which include the following:

a. Overhead clearance is needed so that trees don't conflict with the natural and built environment such as utility lines (electricity, water, and telecoms). Any obstruction above or below the surface of the tree trench must be evaluated and removed or mitigated.
b. The water table must be relatively low to enable the infiltration of water into the subsoil.
c. The root zone must be considered so that hazards resulting from roots as trees grow are minimized or removed.
d. Soil permeability must be evaluated so that the rate of water absorption is calculated and thus the efficiency of soils for water to permeate through is known.

The key design features include the following:

a. The system is usually designed with native plants/trees that are already used to the environmental conditions.
b. It is designed with flexibility in size and infiltration capacity.
c. It allows for kerb cuts, new inlets, or similar approaches to introduce runoff into the trench.
d. Tree trenches require linear infiltration and quick drawdown.
e. Sufficient tree species selection and spacing are vital.

3.8.9.3 Potential limitations of tree trenches

The following are the possible limitations:

a. Required careful selection of tree species.
b. Required appropriate root zone area.
c. Utility conflicts, including overhead electric wires, posts, signs, etc.
d. Conflicts with other structures (basements, foundations, etc.).

3.8.9.4 Probable cost for tree trenches

The cost of tree trenches varies from place to place.

- US$850 per tree.
- US$ 110 to 165 per m^2.

- US$8000 to 10,000 to purchase one prefabricated tree pit system including filter material, plants, and some maintenance.
- US$1500 to 6000 for installation.

3.8.10 Rainwater harvesting systems

3.8.10.1 Description and purpose

RWH is the simple process or technology used to conserve rainwater by collecting, storing, conveying, and purifying rainwater that runs off from rooftops, parks, roads, open grounds, etc. for later use. RWH is the commonest SuDS practiced worldwide. It is a process of harvesting rainwater for storage and use in cleaning, feeding animals, flushing toilets, irrigation, washing, etc. RWH systems can be designed to be located above or below ground. It involves collection and storage of rainwater using artificially designed systems that collect runoff from natural or manmade catchment areas like rooftops, compounds, rock surface, hill slopes, and artificially repaired impervious or semi-pervious land surface.

The variation in RWH systems includes rain barrels, jars, cisterns, pots, and tanks. They can be made in various sizes, materials, and shapes ranging from a few liters to thousands of liters depending on the community served. They are usually made of stainless steel, uPVC, HDPE, ferro-cement tanks, concrete tanks, etc. Figure 3.16 shows rainwater harvested from a residential house using HDPE tank.

The advantages of HDPE water tanks include the following: they are made of food-grade polyethylene, seamless therefore leak-proof, ultraviolet (UV) stabilizers, and usually have about 30 years of life span.

The advantages of stainless steel water tanks include rust free, fire-resistant, and solid body, free from fungus, algae, insect, and water contamination, base discharge point makes it easy to drain sludge and residue, lightweight, easy to install, customized tank stand, and usually have about 50 years of life span.

RWH systems generally consist of the following components:

a. **Catchment.** This is used to collect and store the collected rainwater.
b. **Conveyance system.** This is used to transport the harvested water from the catchment to the recharge zone.

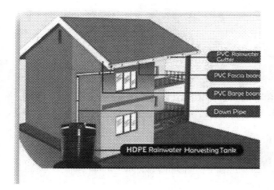

Figure 3.16 HDPE rainwater harvesting system and stainless steel tanks.

c. **Flush.** This is used to flush out the first spell of rain.
d. **Filter.** This is used for filtering the collected rainwater and remove pollutants.
e. **Tanks and the recharge structures.** This is used to store the filtered water which is ready to use.

Several factors play a vital role in the amount of water harvested. Some of these factors are as follows:

a. Types of the roof, its slope and its materials
b. The quantity of runoff
c. The frequency, quantity, and the quality of the rainfall
d. Features of the catchments
e. Impact on the environment
f. Availability of the technology
g. The capacity of the storage tanks
h. The speed and ease with which the rainwater penetrates through the subsoil to recharge the groundwater

RWH can be done in different approaches, that is, on surface storage and below surface storage. In that case, the underground structures usually store relatively large volumes of stormwater. There are two types of rainwater catchment systems: the cistern and the rain barrel.

Cisterns are typically large underground containers or tanks although on surface tanks are not uncommon, with a larger storage capacity than a rain barrel, typically used to supplement gray water needs (i.e., toilet flushing) in buildings, as well as irrigation. This requires pumping from an underground cistern to a gravity tank. Cisterns are larger storage containers that can store 750 up to 75,000 liters of water in residential settings.

Rain barrels are rooftop downspouts directed to an aboveground (typically) container that collects rainwater and stores it until needed for a specific use, such as landscape irrigation, feeding animals, cleaning, etc. Rain barrels typically store 180 liters and 750 liters. They require very little space and can be connected or 'daisy chained' to increase total storage capacity.

Cisterns and rain barrels are structures designed to intercept and store runoff from rooftops to allow for its reuse, reducing volume and overall water quality. They come in various sizes, materials, and shapes. Stormwater is contained in the cistern or rain barrel and reused for irrigation or other water needs. This type of SuDS technology reduces potable water needs and peak discharges. It also provides supplemental water supply, environmental benefits and reduces costs and runoff impacts.

3.8.10.2 Benefits of rainwater harvesting

RWH has several benefits which include the following:

a. It is associated with relatively less cost. This technology is relatively simple, easy to install and operate.
b. It helps in reducing the water bill as it decreases the demand for water and the need for imported water.

c. It is an excellent source of water for landscape irrigation with no chemicals and dissolved salts and free from all minerals.
d. It does not require a filtration system for landscape irrigation.
e. It promotes both water and energy conservation and improves the quality and quantity of groundwater.
f. It reduces soil erosion, stormwater runoff, flooding, and pollution of surface water with fertilizers, pesticides, metals, and other sediments.

3.8.10.3 Limitations of potential rainwater harvesting systems

The primary limitation of RWH systems is that it requires the developer to have an immediate use for the harvested/stored water because RWH systems are designed to manage stormwater for minor storm event. The water has to be discharged from the tank or rain barrel between successive rainfall events.

In addition, RWH systems might require additional management of stormwater runoff as appropriate.

Unpredictable rainfall sometimes may create shortages of water and the system becomes unreliable in case it was designed as the primary source of water. Therefore, limited and no rainfall can limit the supply of Rainwater.

Unavailability of the proper storage system in some areas may limit effective and constant supply of rainwater to the end users and this is one of the significant drawbacks of the RWH system, that is, storage limits.

Regular maintenance may be required for some RWH techniques.

In some cases, RWH requires some technical skills for installation and in some areas this may not be readily available especially underground RWH systems.

If not installed correctly, it may attract mosquitoes and other waterborne diseases. Rainwater systems need to be properly installed in tropical climates.

3.8.10.4 How does a rainwater harvesting system work?

RWH systems collect stormwater from the roof, that is, roof gutters. It also has a downspout pipe that connects to the storage facility, that is, cistern or rain barrel. Aboveground RWH systems have overflow and outlet pipes, which are used to remove excess stormwater and discharge collected/stored stormwater, respectively.

Ferro-cement rainwater harvesting tanks are good for sustainability because when designed and constructed properly, they can meet the social, economic, and environmental aspects. They can be constructed with local materials such as sand, cement, aggregates, rock fill, and steel reinforcement. In areas with limited financial envelopes, ferro-cement RWH tanks can be very helpful. Because they are fabricated on site, women and youth can learn many skills, which is a plus point on sustainability going forward. Figure 3.17 is a cross-section of a ferro-cement tank.

3.8.10.5 Rainwater harvesting systems design criteria

RWH systems include rain barrels, cisterns (both underground and aboveground), and tanks. They come in various sizes, materials, and shapes. The design of the aboveground RWH

Planning for sustainability 97

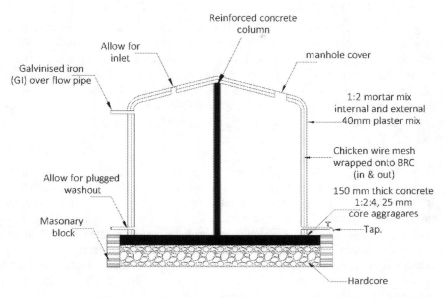

Figure 3.17 Cross-section of a ferro-cement rainwater harvesting tank.

systems leverages the concept of hydrostatic pressure distribution. In that case, the pressure acting on the walls of the RWH tank is hydrostatically distributed and increases with the depth. This depends on the volume of water to be stored in the tank, which influences the level of reinforcement in concrete tanks and the required tensile and flexural strength in HDPE tanks. The below ground systems necessitate adequate ground investigations to determine the water table, chemical composition of the soils, possibility of ground contamination, and susceptibility of excavated walls collapsing due to weak soils.

The hydrologic methods used for overland drainage are also applicable to RWH systems through estimating the quantity/volume received per unit area for a given rainfall intensity of a given frequency. Usually, RWH systems are designed to withstand frequently occurring storms. They are usually emptied before another storm falls.

The key design features of RWH system include the following:

a. Most structures consider the most frequent storm events, that is, **2-year and 5-year** annual recurrence intervals (ARI).
b. Water must be discharged/used before the next storm fills the tank/rain barrel.
c. To eliminate pumping needs during irrigation or flushing, the tank/cistern is placed upland where possible; the site topography is an essential consideration.

3.8.10.6 Probable cost

The cost varies from place to place. However, indicative figures for rain barrels range from US$110 to 330 and cisterns typically range from US$550 to 5500.

3.8.11 Open spaces

According to the U.S. Environmental Protection Agency (EPA), open space is any open piece of land that is undeveloped (has no buildings or other built structures) and is accessible to the public.[22] Open space provides recreational areas for residents and helps to enhance the beauty and environmental quality of neighborhoods. But with this broad range of recreational sites comes an equally broad range of environmental issues. Just as in any other land uses, the way parks are managed can have good or bad environmental impacts, from pesticide runoff, siltation from overused hiking and logging trails, and destruction of habitat. Open spaces include the following:

a. **Green spaces.** These are partly or completely covered with grass, trees, shrubs, or other vegetation. Green spaces include parks, community gardens, and cemeteries. Cities' green spaces help preserve and restore natural areas, such as cultural sites, forests, stream buffers, and wetlands. One most important aspect of green infrastructure as far as sustainability is concerned is its potential to absorb emissions, especially carbon dioxide, and hence reduce the harmful effect these could present to the urban population.
b. **Schoolyards.** The schoolyard is the large open area with a hard surface just outside a school building, where the schoolchildren/students can typically play games or sports and do other activities.
c. **Playgrounds.** A playground, play park, or play area is a place designed to provide an environment for children that facilitates play, typically outdoors. The social skills that children develop on the playground often become lifelong skill sets that are carried forward into their adulthood.
d. **Public seating areas.** These provide different types of seating options such as ledges, steps, benches, moveable chairs, as well as different places or locations within the same area, such as in the sun, in the shade, in groups, alone, close to activity, or somewhat removed from activity. Seating gives people a stopping point, whether it be for lunch, for a rest, a place to wait and meet others, or simply a place to sit and take in the scenery.
e. **Public plazas.** A public plaza is a community amenity that serves a variety of users including building tenants, visitors, and members of the public—brings people together. This space type may function as pedestrian site arrival points, homes for public art, settings for recreation and relaxation, and inconspicuous security features for high-profile buildings. Public plazas offer an excellent opportunity to improve sustainability in downtown while also educating the public about its importance.[23] This has become particularly important with the increase in awareness of global climate change and resilience. Public plazas should be designed to support and enhance the vitality of the adjacent businesses, restaurants, offices, and other uses. When you think of plazas and community growth, you have to consider the revenue the area will bring, either through the arts, markets, or other events.
f. **Vacant lots.** This refers to a piece of land that is not being used.

3.8.12 Porous/permeable pavers

3.8.12.1 Description and purpose

Permeable pavement is a type of SuDS that combines stormwater infiltration, storage, and structural pavement consisting of a permeable surface underlain by a storage/infiltration bed.

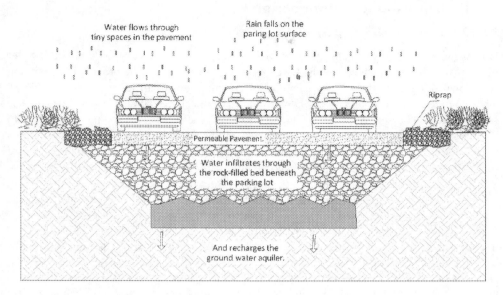

Figure 3.18 Permeable pavers in parking lot.

Permeable paving surfaces are made of either a porous material enabling stormwater to flow through it or nonporous blocks spaced so that water can flow between the gaps. Permeable pavement is well suited for parking lots, walking paths, sidewalks, playgrounds, plazas, tennis courts, and other similar uses. Permeable paving can have a variety of surfacing techniques for roads, parking lots, and pedestrian walkways. Permeable pavement surfaces may be composed of pervious concrete, porous asphalt, paving stones, or interlocking pavers.

The beauty about permeable pavers is that they allow water to infiltrate into the ground and they are reusable. A pervious pavement system consists of a pervious surface course underlain by a storage bed placed on uncompacted subgrade to facilitate stormwater infiltration. Figure 3.18 is an example of a car parking lot with permeable pavers. The storage reservoir may consist of a stone bed of uniformly graded, clean and washed course aggregate with a void space of approximately 40% or other premanufactured structural storage units. The pervious pavement may consist of asphalt, concrete, permeable paver blocks, reinforced turf/gravel, or other emerging types of pavement.

3.8.12.2 Probable cost

   a. Varies by porous pavement type
   b. Local quarry needed for stone-filled infiltration bed
   c. $75 to 160 per m², including underground infiltration bed
   d. Generally, more than standard pavement, but saves on cost of other BMPs and traditional drainage infrastructure

3.8.13 Downspout disconnection

3.8.13.1 Description and purpose

Downspouts are usually directed onto a paved surface, such as a driveway, which sends water directly into the street when it rains. Disconnecting downspouts is the process of separating roof downspouts from the sewer system and redirecting roof runoff onto pervious surfaces, most commonly a lawn or garden. Downspout disconnection **reduces stormwater in the sewer system**. The disconnection is simple, inexpensive, effective, and easily integrated into the landscape design. Disconnecting downspouts allows for volume reductions through infiltration and evapotranspiration. Disconnected downspouts improve water quality because roof runoff contains deposited atmospheric pollutants, particles of roofing materials, nutrients, and Biochemical oxygen demand (BOD) loadings from bird droppings. You can install rain barrels on a disconnected downspout. The gardens can be watered using rainwater and thus help conserve water. See Figure 3.19 for a downspout disconnection.

Downspout disconnection is known for providing supplemental water supply when used in conjunction with capture/reuse systems such as RWH systems and detention systems. It has significantly wide applications because it can be directed to lawns, Greenfield, vegetated areas, open/green spaces, and gardens as appropriate. In some instances, downspout disconnection reduces potable water use and costs when used in conjunction with capture/reuse systems.

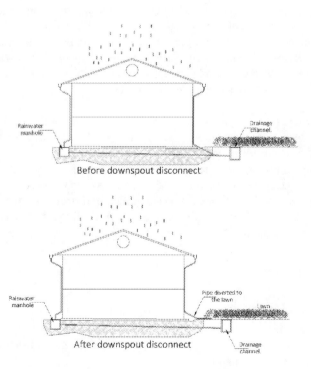

Figure 3.19 Downspout disconnection.

Downspout disconnection comes with environmental benefits such as crops' irrigation potential, provides opportunity for removal of debris from the roof, birds' droppings, and debris from the roof runoff hence improving water quality. Through reduced runoff volume, downspout disconnection reduces costs associated with maintenance of large sewers during combined sewer over flow peak events.

3.8.13.2 Potential limitations of downspout disconnection

Downspout disconnection may not be appropriate for all locations. Sometimes, it is not recommended that disconnected downspouts direct rainwater toward hard surfaces such as sidewalks and driveways where a slip and fall may occur. For buildings with internal drainage, disconnecting internal downspouts may be difficult or impractical. It is impractical to disconnect onto adjacent property owner. The designer needs to consider local regulations and the condition of the area where to direct water to avoid property damage, unsafe conditions, or other potential problems. Disconnected downspout requires enough receiving area.

3.8.13.3 How does the downspout disconnection work?

Downspout disconnection is the process of disconnecting the pipes that are directed toward the lawn. Instead of stormwater discharging to the standard storm sewer, it discharges on the lawn and stormwater infiltrates into the surrounding soil or vegetation.

3.8.13.4 Downspout disconnection design criteria: What are the factors to consider?

Downspout disconnection has a number of variations: scuppers, drip chains, and decorative gargoyles. The key design features include install splash block at the end of the extension to prevent erosion and the roof runoff must be discharged at least 1.5 m away from property lines including basements and porches.

3.8.13.5 Probable cost

It is affordable as materials are readily available at hardware store such as uPVC pipes.

3.8.14 Vegetated kerb extensions

Vegetated kerb extensions, also called stormwater kerb extensions, are landscaped areas within the parking zone of a street that capture stormwater runoff in a depressed planting bed.[24] The landscaped area can be designed similar to a bioretention garden or bio-wale, utilizing infiltration and evapotranspiration for stormwater management. The advantages of vegetated kerb extensions include traffic calming and pedestrian safety, enhanced site aesthetics, potential air quality and climate benefits, potential combined sewer overflow reductions, wide applicability, including in ultra-urban areas, reduced runoff, and improved water quality. Some potential limitations are as follows: It may require removal of on-street parking, may cause a conflict with bike lane, and utility and fire hydrant conflicts.

3.8.15 Green roofs

3.8.15.1 Description and purpose

A green roof is a low-maintenance vegetated roof system that stores rainwater in a lightweight engineered soil medium, where the water is taken up by plants and transpired into the air.[25] In the end, much less water runs off the roof, as compared to conventional rooftops. Green roofs, also called 'vegetated rooftops' can be optimized to achieve water quantity and water quality benefits. The green roof is composed of a veneer of vegetation grown on and covers an otherwise conventional flat or pitched roof with less than 30° slope, endowing the roof with hydrologic characteristics that more closely match surface vegetation than the roof. Figure 3.20 shows a typical green roof.

Air quality is improved on green roofs as a result of the vegetation grown on the roof. The same vegetation improves the aesthetics of the building and greatly reduces the urban 'heat island' effect. Green roofs provide an extra layer of insulation that reduces heating and cooling costs, and they are likely to last much longer than conventional roofs because the roofing material itself is shielded from ultraviolet light and thermal stress. The overall appearance of the building improves generally as a result of incorporating the green rooftop. Green roofs are easy to incorporate into new construction and can even be used on many existing buildings.

The overall thickness of the vegetation veneer typically ranges from 50 to 150 mm and may contain multiple layers, consisting of waterproofing, non-soil engineered growth media, synthetic insulation, fabrics, and synthetic components. Through the appropriate selection of

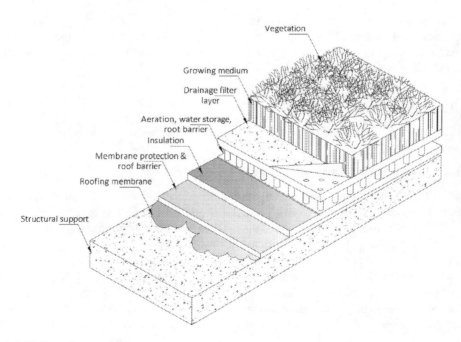

Figure 3.20 Typical green roof.

materials, even thin vegetated roof covers can provide significant rainfall retention and detention functions. Vegetated roofs have been in use in Europe for more than 30 years.

Depending on the plant material and planned usage for the roof area, modern vegetated roofs can be categorized as systems that are intensive, semi-intensive, or extensive. All types of green roofs aim to transform rooftops from 'wasted space' into a form of infrastructure that has environmental, economic, aesthetic, and social benefits.

   a. Intensive green roofs utilize a wide variety of plant species that may include small trees and shrubs planted in more than 150 mm of growing medium. They are often designed as accessible building amenities. They require deeper substrate layers and are generally limited to flat roofs. They require intense maintenance and are often park-like areas accessible to the general public.
   b. In contrast, extensive green roofs require less than 150 mm of soil medium and support mostly herbaceous plants. These utilitarian 'roof meadows' generally have no public access and require little maintenance. Extensive green roofs are limited to herbs, grasses, mosses, and drought-tolerant succulents such as sedum, can be sustained in a shallow substrate layer. They require minimal maintenance once established and are generally not designed for access by the public. These vegetated roofs are typically intended to achieve a specific environmental benefit, such as rainfall runoff mitigation. Extensive green roofs are well suited to rooftops with little load-bearing capacity and sites that are not meant to be used as roof gardens.
   c. Semi-intensive vegetated roofs fall between intensive and extensive vegetated roof systems. They call for more maintenance and higher costs.

3.8.15.2 Benefits and effectiveness of green roofs

   a. **High volume reduction (annual basis).** Green roofs effectively reduce stormwater runoff.
   b. **Moderate ecological value and habitat.** Green roofs can serve as a habitat for insects, and animals, and promote the conservation of the ecological system.
   c. **High aesthetic value.** Green roofs have demonstrated aesthetic benefits that can increase community acceptance of a high visibility project. They may also add value to the property if marketed effectively.
   d. **Energy benefits (heating/cooling).** Green roofs lower heating and cooling costs because the trapped air in the underdrain layer and in the root layer helps to insulate the roof of the building. During the summer, sunlight drives evaporation and plant growth, instead of heating the roof surface. During the winter, a green roof can reduce heat loss by 25% or more. Because green roofs shield roof membranes from intense heat and direct sunlight, the entire roofing system has a longer life span than conventional roofs.
   e. **Urban heat island reduction.** The presence of a green roof helps to reduce air temperatures around the building, reducing the 'heat island' effect and the production of smog and ozone, which forms in the intense heat (175°C) over large conventional roofs. The vegetation on green roofs also consumes carbon dioxide and increases the local levels of oxygen and humidity.
   f. Green roofs reduce peak discharge rates by retaining runoff and creating longer flow paths. Research indicates that peak flow rates are reduced by 50% to 90% compared to conventional roofs, and peak discharge is delayed by an hour or more.

3.8.15.3 Potential limitations of green roofs

a. The initial construction cost is higher than conventional roofs.
b. Higher maintenance needs until vegetation is established but after that maintenance costs reduce.
c. The main limitation for green roofs in retrofit applications is usually load restrictions. A professional engineer must assess the necessary load reserves and design a roof structure that meets state and local codes.
d. Slopes greater than **15%** require a wooden lath grid or other retention system to hold substrate in place until plants form a thick vegetation mat.
e. Green roofs should not be used where groundwater recharge is a priority, such as in aquifer recharge areas or watersheds experiencing low-flow stresses.
f. The need for adequate roof structure can be challenging on retrofit applications.

3.8.15.4 Key design features for green roofs

a. The variations in green roofs design include single media system, dual media system, dual media system with synthetic layer and intensive, extensive, and semi-intensive types of green roofs.
b. The engineered media should have a high mineral content. For example, the engineered media for extensive green roof covers is typically 85% to 97% nonorganic.
c. Access routes should be identified during the design phase, and access paths of gravel or other inert materials provided, as well as safety harness hooks for inspection and maintenance personnel.
d. Green roofs typically have between 50 and 150 mm of non-soil engineered media. Assemblies that are 100 mm and deeper may include more than one type of engineered media.
e. Waterproof membranes are made of various materials, such as modified asphalts (bitumens), synthetic rubber (EPDM), hypolan (CPSE), and reinforced PVC. The most common design used in Europe is 6080 mil PVC single ply roof systems. The waterproofing must be resistant to biological and root attack. In many instances, a supplemental roof-fast layer is installed to protect the primary waterproofing. Modified asphalts usually require a root barrier, while EPDM and reinforced PVC generally do not. Attention to seams is critical because some glues and cements are not always root impermeable.
f. Irrigation is generally not required (or even desirable) for optimal stormwater management using vegetated covers.
g. Vegetation should be low growing, spreading perennial or self-sowing annuals that are drought tolerant. Appropriate varieties include sedum, delospermum, sempervivium, creeping thyme, allium, phloxes, anntenaria, ameria, and abretia. Vegetation may be planted as vegetation mats, plugs or potted plants, sprigs (cuttings), or seeds. Vegetation mats are the most expensive but achieve immediate full coverage. Potted plants are also expensive and labor intensive to install. Sprigs are often the most cost-effective option, even considering that initial irrigation is necessary and repeat installations may be required due to mortality.
h. Internal building drainage, including provision to cover and protect deck drains or scuppers, must anticipate the need to manage large rainfall events without inundating the cover.

i. Vegetated roof covers intended to achieve water quality benefits should not be fertilized.
j. Assemblies planned for roofs with pitches steeper than 2:12 (9.5°) must incorporate supplemental measures to ensure stability against sliding.
k. The roof structure must be evaluated for compatibility with the maximum predicted dead and live loads. Typical dead loads for wet extensive vegetated covers range from 39.06 to 175.77 kg/m^2.
l. The underdrain layer may be constructed of perforated plastic sheets or a thin layer of gravel. Pitched roofs and small flat roofs may not require an underdrain.
m. Conventional sod should not be used because it requires irrigation, mowing, and maintenance.

3.8.15.5 Probable cost of green roofs

The costs for a green roof generally range from US$55 to 540 per m^2, including all structural components, soil, and plants. They generally cost more to install than conventional roofs, but are financially competitive on a lifecycle basis because of longer life spans (up to 40 years), increased energy efficiency, and reduced stormwater runoff. Green roofs are generally less expensive to install on new roof versus retrofit on existing roof. If the application is a retrofit, structural upgrades may increase the cost.

3.9 STORMWATER QUALITY IMPROVEMENT AND QUANTITY REDUCTION POTENTIAL FOR DIFFERENT SUDS TYPES

The overarching goal of SuDS is to capture pollutants and control stormwater quality and reduce stormwater quantities that convert into runoff. The potential of SuDS to meet these goals varies widely from one particular SuDS to another. Table 3.3 provides the stormwater quality indicators and lifecycle considerations used to determine the effectiveness of each particular SuDS. These are described as **low, medium,** and **high**.

Note: The stormwater quality indicators are as follows:

TSS : Total suspended solids
TP : Total phosphorus
TN : Total nitrogen

3.10 RIVER RESTORATION PROGRAMS

For sustainability purposes going forward, evaluating the possibility of river restoration techniques becomes necessary during the planning and design of major drainage systems that would discharge in streams and rivers especially in an urban watershed. The peak flows generated from the watershed would leave the stream and river banks eroded. For that matter, the drainage engineer or planner may find it appropriate to propose and design river restoration programs to be able to accommodate the peak flows generated from the urban watershed and at the same time contribute to the sustainability of the river. River restoration techniques would also support the control of the quality of the runoff that gets discharged into the river.

Table 3.3 Stormwater quality indicators and lifecycle considerations for SuDS

Planter box

Stormwater quality functions		Stormwater quantity functions		Lifecycle considerations	
TP	Medium	Flood protection	Low	Capital investment	Low/medium
TN	Medium	Groundwater recharge	Low	Maintenance	Medium
Temperature	Medium	Peak flow rate	Low	Fast-track potential	Low
TSS	Medium	Erosion reduction	Low	Aesthetics	High
		Volume	Low/medium		

Infiltration system (basin)

Stormwater quality functions		Stormwater quantity functions		Lifecycle considerations	
TP	Medium/high (85%)	Flood protection	High	Capital investment	Medium
TN	Medium (30%)	Groundwater recharge	High	Maintenance	Medium
Temperature	High	Peak flow rate	High	Fast-track potential	Medium
TSS	High (85%)	Erosion reduction	Medium	Aesthetics	Medium
		Volume	High		

Bioretention garden

Stormwater quality functions		Stormwater quantity functions		Lifecycle considerations	
TP	Medium (60%)	Flood protection	Low/medium	Capital investment	Medium
TN	Medium (40%–50%)	Groundwater recharge	Medium/high	Maintenance	Medium
Temperature	High	Peak flow rate	Medium	Fast-track potential	Medium
TSS	High (70%–90%)	Erosion reduction	Medium	Aesthetics	High
		Volume	Medium/high		

Downspout disconnection

Stormwater quality functions		Stormwater quantity functions		Lifecycle considerations	
TP	N/A	Flood protection	Low	Capital investment	Low
TN	N/A	Groundwater recharge	Medium/high	Maintenance	Low

Table 3.3 Cont.

Downspout disconnection

Stormwater quality functions		Stormwater quantity functions		Lifecycle considerations	
Temperature	Medium/high	Peak flow rate	Medium	Fast-track potential	Low/medium
TSS	Medium	Erosion reduction	Medium	Aesthetics	High
		Volume	Medium		

Rainwater harvesting system

Stormwater quality functions		*Stormwater quantity functions*		*Lifecycle considerations*	
TP	Medium	Flood protection	Low/medium	Capital investment	Low/medium
TN	Medium	Groundwater recharge	Low	Maintenance	Medium
Temperature	Medium	Peak flow rate	Low	Fast-track potential	Medium/high
TSS	Medium	Erosion reduction	Low	Aesthetics	Low/medium
		Volume	Low/medium		

Tree trench

Stormwater quality functions		*Stormwater quantity functions*		*Lifecycle considerations*	
TP	Medium (60%)	Flood protection	Low/medium	Capital investment	Medium
TN	Medium (40%–50%)	Groundwater recharge	Medium	Maintenance	Medium
Temperature	High	Peak flow rate	Medium	Fast-track potential	High
TSS	High (70%–90%)	Erosion reduction	Medium	Aesthetics	High
		Volume	Medium		

Permeable pavement

Stormwater quality functions		*Stormwater quantity functions*		*Lifecycle considerations*	
TP	Medium	Flood protection	Medium/high	Capital investment	Medium
TN	High	Groundwater recharge	High	Maintenance	Medium
Temperature	High	Peak flow rate	Medium/high	Fast-track potential	Low/medium
TSS	High	Erosion reduction	Medium/high	Aesthetics	Low/medium
		Volume	High		

(continued)

Table 3.3 Cont.

Green roof

Stormwater quality functions		Stormwater quantity functions		Lifecycle considerations	
TP	Medium	Flood protection	Low/medium	Capital investment	High
TN	Medium	Groundwater recharge	Low	Maintenance	Medium
Temperature	Medium	Peak flow rate	Medium	Fast-track potential	Low
TSS	Medium	Erosion reduction	Low/medium	Aesthetics	High
		Volume	Medium/high		

Green street

Stormwater quality functions		Stormwater quantity functions		Lifecycle considerations	
TP	Medium (60%)	Flood protection	Low/medium	Capital investment	Medium
TN	Medium (40%–50%)	Groundwater recharge	Medium	Maintenance	Medium/high
Temperature	High	Peak flow rate	Medium	Fast-track potential	Low/medium
TSS	High (70%–90%)	Erosion reduction	Medium	Aesthetics	High
		Volume	Medium		

Bio-swale

Stormwater quality functions		Stormwater quantity functions		Lifecycle considerations	
TP	Low/high	Flood protection	Low	Capital investment	Low/medium
TN	Medium	Groundwater recharge	Low/medium	Maintenance	Low/medium
Temperature	Medium/high	Peak flow rate	Low/medium	Fast-track potential	High
TSS	Medium/high (50%)	Erosion reduction	Medium	Aesthetics	Medium
		Volume	Low/medium		

The fact that most towns and cities are developed near rivers describes the importance of rivers to humans. Rivers and floodplains provide benefits to society. Such benefits include flood regulation, freshwater supply, recreation activities, tourism, water purification, carbon storage, and improved human health. Many of these benefits, together with biodiversity and habitat, become compromised when humans modify the rivers for land development purposes. Therefore, river restoration projects are aimed primarily to encourage local communities to engage in their local environment by raising awareness of environmental issues. Considering the root cause of river degradation, for example, natural processes can be reintroduced through a variety of river restoration mechanisms to reshape rivers to provide the diversity of habitats required for a healthy river ecosystem and ensure their long-term recovery.

By definition, river restoration is the reestablishment of natural physical processes (e.g., variation of flow and sediment movement), features (e.g., sediment sizes and river shape), and physical habitats of a river system (including submerged, bank, and floodplain areas).[26] Rivers are reinstated to their natural processes to restore biodiversity providing benefits to both the people and the wildlife. This is aimed at reintroducing natural processes to reshape rivers to provide the diversity of habitats required for a healthy river ecosystem. Several rivers worldwide have been extensively modified to accommodate societal needs for food production, flood protection, and economic activity so it is not always possible or desirable to restore to an original condition. In this case, improvement of river structure and habitats within the adjacent constraints can increase the overall biodiversity and mitigate adverse conditions.

The main techniques for restoring rivers include the following:

a. Managing catchments (catchment basis)
In this approach to restore rivers, communities are subdivided into groups and prioritized as part of the wider catchment. In doing so, individuals and organizations are constantly engaged from across society to improve freshwater environments. This approach is called the 'community-led based approach'.

On a catchment basis, the identified groups list the issues to address and then agree how to restore the river. The advantage that the community-led based approach to managing river restoration presents is that the river system is viewed as a whole which enables several stakeholders and communities involved to be consulted and work together across all the river reaches. By focusing on a specific reach, without a wider catchment understanding, can lead to detrimental effects elsewhere. Therefore, a **holistic approach** is always essential and beneficial to promote sustainability.

b. Restoring a more natural river course
In some urban areas, rivers are straightened, diverted, and over-deepened to develop infrastructure, enhance river navigation, reduce flooding, and improve land drainage, or for agricultural purposes. Rivers which have been straightened may lack flow and habitat diversity because of removing the natural features and tampering with their profiles. Therefore, restoring the river to a more natural course means re-meandering the river to reinstate a more natural course and river profile. River straightening may increase the risk of flooding downstream as water moves at higher velocities as it routes through the modified sections and hence increase the downstream discharges.

When the river is re-meandered, habitat diversity and biodiversity are improved as flow can be returned to the former river course. In some cases, it may not be possible to identify the old channel course or may not be accessible due to land developments and a new course is constructed.

c. Floodplain reconnection and wetland creation
For purposes of flood mitigation and control, floodplains are important. Floodplains provide benefits such as fish refuge, habitat diversity, and flood storage, and they are important features of a riverine landscape. However, they are constantly disconnected and drained to

protect housing from flooding and to create land for agriculture or development. The disconnection and reduction in storage may instead lead to a greater flood risk downstream as water moves through a catchment quicker.

In order to restore connectivity, flood banks can be breached or set back in properly judged locations, allowing water to spill out onto the floodplain again. Several benefits of reconnecting the floodplain can be realized which include recreation of wetland habitat, an increase in flood storage area, reintroduction of wetland species, and creating refuge for fish during high flows—working as a sustainable drainage system altogether.

d. In-stream enhancement

In-stream enhancement is conducted especially where space for large-scale river restoration is not possible due to space constraints or when floodplains have already been reclaimed and developed. In-stream enhancement include the provision of some form of roughness in-channel such as creating berms, woody material, and reworking gravels. In-stream enhancement may be the next best option especially where rivers have concrete banks or beds.

e. Removing or passing barriers

As part of river restoration programs, barriers such as dams, weirs, and other controls constructed or installed in rivers may need to be removed. Barriers such as weirs and dams are common features of the riverine environment. Sometimes, these barriers restrict the passage of wildlife within the river and also modify the river habitats through the creation of deeper and slower flows upstream. When the barrier becomes obsolete, unreliable, or unsustainable, then complete removal becomes the next best thing so that upstream and downstream connectivity is restored. Where complete removal of the barrier is not feasible, alternative methods can be adopted, for example, lowering weirs or creating bypass channels with fish passes. In some instances, when the barrier becomes unreliable, unsustainable, old, and obsolete, it may prevent efficient transportation sediment loads leading to downstream erosion and high maintenance costs.

3.11 FLOOD CONTROL STRUCTURES

Flood control structures can be designed and built in different materials such as concrete, gabion baskets, earthen embankment, among others. Different kinds of floods exist such as flash floods, river floods, and ocean floods. Flood control structures may be constructed from different materials and the designer always chooses the most economical, environmentally sustainable materials to build flood defense structures. While planning for sustainability, flood control structures may be evaluated to ascertain the practicability to develop them as a part of the drainage system. Modern flood defenses include the following:

- Dams and spillways
- Diversion canals
- Floodplains and groundwater replenishment
- River defenses, for example, levees, floodwalls, bunds, reservoirs, and weirs
- Flood defense walls built to safeguard flash floods in urban or residential areas
- Costal defenses, for example, groynes, seawalls, revetments, and gabions
- Retention and detention ponds

- Moveable gates
- Moveable barriers
- Channel improvements

Due to climate change, many incidences of flooding happen around the world and it is likely to get worse in the future unless mitigation measures are taken seriously to reduce climate change. Several kinds of dams are built to control floods. A dam is a barrier that restricts or stops the flow of water, helps suppress floods, as well as providing irrigation, industrial, and aquaculture uses. Flood control dams include the following:

- **Detention dam.** Detention dams are specifically constructed for flood control by retarding flow downstream, helping reduce flash floods (to some extent). The water is retained in a reservoir to be gradually released later.
- **Embankment dam.** An embankment dam is a large, artificial dam that is constructed with natural excavated materials or industrial waste materials, such as compacted plastics, and various compositions of soil, sand, rock, and clay.
- **Cofferdam.** A **cofferdam** is a temporary, portable dam used for a variety of projects including bridge repair, shoreline restoration, pipeline installation, and many other construction projects. A cofferdam is used to close off some or all of a construction area.[27]

To be more resilient, flood control structures should be built to higher levels and with more resilient materials and designed to withstand repeated and more extreme floods. Several practices such as structural changes (e.g., changes to embankment slopes) and policy changes (e.g., changes to zoning codes, relocation, and designing redundancy plans) must be considered by weighing options to mitigate worst scenarios. Climate stressors such as changes in precipitation, sea levels, extreme events, and resulting storm surges can significantly affect the efficacy and durability of flood defense structures.

Future climate scenarios must be well-studied to take a firm decision on the type of flood control structure, by studying several factors such as how sea levels or extreme event intensity would change in the future. Sea level rise projections, hydrology, and physiography of the watershed should be properly incorporated in the planning and design of flood control structures to minimize or avoid unintended adverse impacts. Using this reliability data and information collected, engineers, designers, and planners can identify the needed changes to the design, construction, and maintenance of structures. Experienced engineers and designers should be tasked to plan and design flood control structures, since untested flood control structures may pose a direct threat to human life.

3.12 CONCLUSION

Planning for sustainability while developing drainage structures takes on a holistic approach and considers a back-and-forth approach that is aimed at mimicking the natural environment as much as possible. The integration of SuDS in the drainage network that are specific to the area supports the reduction of flow volumes and peak flows. At the same time, it contributes to a reduction in GHGs because trees and vegetation planted in the SuDS absorb emissions such as carbon dioxide from the atmosphere. Therefore, SuDS contribute to a carbon sink.

The back-and-forth criteria of integrating SuDS with standard gray infrastructure allow the designer to obtain an optimal solution that strikes a balance between capital carbon emissions from the construction of gray infrastructure and the initial capital investments in SuDS. In gray infrastructure, reducing capital cost means reducing embedded carbon too. A reduction in storm sewer sizes is achieved when SuDS are integrated in the drainage networks especially infiltration systems and bioretention gardens. The reduction in storm sewer sizes implies a reduction in cost of construction and embedded carbon. However, the initial cost of constructing SuDS and the overall SuDS lifecycle costs must be competitive to make the entire integrated stormwater management program feasible.

In some cases, the operational carbon from SuDS as a result of the equipment, trucks, and machinery used to maintain the SuDS competes favorably with the operational carbon for storm sewers. The emissions in the operational phase basically originate from trucks and equipment used in the desilting of sewers, mowing, pruning, and pumping. All efforts must be geared toward reducing capital carbon in gray infrastructure especially reducing the amount of reinforced concrete used in construction of hydraulic structures. The reason is to scale down the use of cement and steel whose production is highly energy intensive and produces huge amounts of emissions. In addition to a reduction in storm sewer sizes, the use of ground granulated blast-furnace slag (GGBS) concrete as it produces lower carbon emissions should be encouraged (www.ukcsma.co.uk/ggbs-concrete/).

For sustainability purposes going forward, designing integrated drainage systems looks at the entire ecosystem and how the drainage network affects it. The receiving water bodies must be evaluated so that the stormwater that ends there does not have an adverse effect on the flora and fauna, cause erosion of banks, collapse of structures, or cause flooding in downstream areas. That means SuDS developed in the drainage networks must control the quality of runoff before excess flows are allowed to discharge in the storm sewers which finally discharge in the receiving water bodies. The water quality volumes for each specific SuDS must be properly evaluated. The cumulative curve for storms is always used to compute the water quality volume. Usually a 90th percentile precipitation depth is used. Therefore, river restoration programs and flood control structures must be evaluated based on the level of risk from the entire proposed integrated stormwater management program.

References

1. *Report of the World Commission on Environment and Development: Our Common Future (Brundtland Report)* (Oslo: World Commission on Environment, 20 March 1987), Article 3 (27).
2. Slater, T., Lawrence, I. R., Otosaka, I. N., Shepherd, A., Gourmelen, N., Jakob, L., Tepes, P., Gilbert, L., & Nienow P. (2021). Review article: Earth's ice imbalance. *The Cryosphere*, 15, 233–246. https://doi.org/10.5194/tc-15-233-2021
3. Vaughan, D. G., Comiso, J. C., Allison, I., Carrasco, J., Kaser, G., Kwok, R., Mote, P., Murray, T., Paul, F., Ren, J., Rignot, E., Solomina, O., Steffen, K., & Zhang, T. (2013). Observations: Cryosphere, in: Climate change (2013): The physical science basis. In Stocker, T. F., Qin, D., Plattner, G. K., Tignor, M., Allen, S. K., Boschung, J., Nauels, A., Xia, Y., Bex, V., & Midgley, P. M. (Eds.), *Contribution of Working Group I to the Fifth Assessment Report of the Intergovernmental Panel on Climate Change* (pp. 317–382). Cambridge: Cambridge University Press.

4. The IMBIE Team: Mass balance of the Antarctic Ice Sheet from 1992 to 2017, *Nature*, 558, 219–222. https://doi.org/10.1038/s41586-018-0179-y, 2018 (data available at: http://imbie.org/data-downloads/)
5. Six ways loss of Arctic ice impacts everyone | Pages | WWF (worldwildlife.org). www.worldwildlife.org/pages/six-ways-loss-of-arctic-ice-impacts-everyone. Accessed May 12, 2022.
6. Minnesota Stormwater Manual. Shallow soils and shallow depth to bedrock. https://stormwater.pca.state.mn.us/index.php?title=Shallow_soils_and_shallow_depth_to_bedrock. Accessed May 12, 2022.
7. Trimedia: Environmental & Engineering. 4 examples of stormwater best management practices (BMPs). (August 23, 2021) | Environmental. https://trimediaee.com/blog/environmental/4-examples-of-stormwater-best-management-practices-bmps/
8. Ohio Environmental Protection Agency Division of Surface Water. (October 2018). Post-construction storm water questions and answers: Water quality volume. NPDES Construction General Permit #OHC000005. www.epa.state.oh.us/portals/35/storm/CGP5-PC-QA%20WQV.pdf
9. U.S. Environmental Protection Agency. (July 2016). Summary of state post construction stormwater standards. www.epa.gov/sites/production/files/2016-08/documents/swstdsummary_7-13-16_508.pdf
10. Water Environment Federation and American Society of Civil Engineers. (2012). Design of Urban Stormwater Controls. WEF Manual of Practice No. 23, ASCE/EWRI Manuals and Reports on Engineering Practice No. 87. Prepared by the Design of Urban Stormwater Controls Task Force of the Water Environment Federation and the American Society of Civil Engineers/Environmental and Water Resources Institute. WEF Press.
11. Water quality volume, WQv (lmnoeng.com). www.lmnoeng.com/Hydrology/WaterQualityVolume.php. Accessed June 21, 2022.
12. Bio-retention Manual, Environmental Services Division, Department of Environmental Resources. The Prince George's County, Maryland, December 2007.
13. Design criteria for bio-retention – Minnesota stormwater manual design criteria for bio-retention, February 2021. https://stormwater.pca.state.mn.us/index.php/Design_criteria_for_bioretention
14. Maryland Stormwater Design Manual, Volumes I and II. (October 2000, Revised May 2009). https://mde.maryland.gov/programs/water/stormwatermanagementprogram/pages/stormwater_design.aspx
15. Plastic pipe institute, 'Stormwater Management'. https://plasticpipe.org/Drainage/Drainage/Applications/Stormwater-Management.aspx. Accessed May 14, 2022.
16. US EPA Stormwater technology factsheet: Wet detention ponds. United States Environmental Protection Agency. Office of Water. Washington, D.C. EPA 832-F-99-048 September, 1999.
17. Department of Ecology (Ecology). (1996). Selecting best management practices for stormwater management. Water Quality Program – Urban Nonpoint Unit. Olympia, WA.
18. King County. (1995). Evaluation of water quality ponds and swales in the Issaquah/East Lake Sammamish Basins. *Final Report for Task 5 of Grant Agreement cTAX90096-Issaquah/East Lake Sammamish Nonpoint Plans*. King County Surface Water Management Division, Seattle, WA.
19. Mazer, G., Booth, D., & Ewing, K. (2001). Limitations to vegetation establishment and growth in biofiltration swales. *Ecological Engineering*, 17, 429–443. https://doi.org/10.1016/S0925-8574(00)00173-7
20. VIRGINIA DCR stormwater design specification No. 10 dry swales. Version 2.0. January 1, 2013. https://swbmpvwrrc.wp.prod.es.cloud.vt.edu/wp-content/uploads/2018/07/BMP_Spec_No_10_DRY_SWALE.pdf
21. Duan, J., & Abduwali, D. (2021). Basic Theory and Methods of Afforestation. In (Ed.), *Silviculture*. IntechOpen. https://doi.org/10.5772/intechopen.96164

22. United States Environmental Protection Agency. www3.epa.gov/region1/eco/uep/openspace.html
23. Dave Lustberg. (2020). Public Plazas: 8 key considerations when planning a public plaza – Downtown New Jersey (downtownnj.com). www.downtownnj.com/public-plazas/
24. VegetatedCurbExtension.pdf (saveitlancaster.com). www.saveitlancaster.com/wp-content/uploads/2011/10/11_VegetatedCurbExtension.pdf
25. Massachusetts clean water Toolkit. (2022). Green roofs. https://megamanual.geosyntec.com/npsmanual/greenroofs.aspx. Accessed May 9, 2022.
26. River restoration guidance factsheet funded by the Esmée Fairbairn Foundation. The River Restoration Centre (RRC), UK. www.therrc.co.uk www.therrc.co.uk/sites/default/files/general/Training/esmee/what_is_river_restoration_final.pdf
27. GGBS Concrete. www.ukcsma.co.uk/ggbs-concrete Accessed July 14, 2022.

Further reading

Bio-retention Manual Environmental Services Division, Department of Environmental Resources, The Prince George's County, Maryland, December 2007.

2000 Maryland stormwater design manual volume I & II. Maryland department of the environment. Water management administration. 2500 Broening highway. Baltimore Maryland 21224. http://mde.state.md.us

SuDS Manual C753 Chapter List (ciria.org). www.ciria.org//Memberships/The_SuDs_Manual_C753_Chapters.aspx

Chapter 4

Useful hydrology

4.1 INTRODUCTION

The knowledge of hydrology is essential to support a drainage engineer and planner carrying out the planning and design of integrated sustainable drainage systems to control and manage stormwater. Hydrology is the science of water that embraces the occurrence, distribution, movement, and properties of the waters of the Earth. As a drainage systems planner, it is imperative to have a basic understanding of calculus, statistics, and probability density functions to derive relationships between observed variables, forecast probable storms or droughts, or find trends in collected data. This allows stormwater management crews to make informed decisions. Probability and statistical measurements are typically based on historic data over an extended period and are used for risk analysis. The application of the laws of probability underlies most studies of the statistical nature of repeated observations or trials.

Knowledge of a spectrum of probability density functions and statistics is essential while developing hydrologic models which entirely depend on it. Several helpful applications are widely harnessed in hydrology to define climate patterns, such as rainfall analysis, groundwater recharge, river stream flows, flood warning systems, and infiltration models, developing intensity-duration-frequency (IDF) curves, and forecasting climate trends of an area, etc. Supporting sustainable drainage practices requires the designers and planners to forecast and predict the hydrology of the area after new developments are in place, how average temperatures will change to affect rainfall patterns, the amount of stormwater to infiltrate into the ground, and how flows and water quality in receiving water bodies will vary.

Some of the extensively used probability distribution functions include the Weibull distribution, the poison distribution, normal distribution, Gumbel extreme value distribution, log-Pearson type III distribution, and the log-normal distribution, among others.

Extreme value analysis (EVA) is a statistical technique used to estimate the likelihood of the occurrence of extreme values based on a few basic assumptions and observed/measured data. EVA applies several distributions which include the log-normal distribution, log-Pearson type III distribution, Gumbel extreme value distribution, and normal distribution. In drainage evaluation, planning, and design, the planner or engineer may be tasked to forecast the future flows in receiving waters such as rivers and streams when the watershed experiences rapid changes over the years. As a result of climate change, several receiving water bodies like rivers and steams' banks are constantly eroded due to the enormous peak flows generated upstream from impervious surfaces. The extreme value analysis is especially relevant where municipalities and towns have rivers and streams that need sound restoration programs. The development of cities and municipalities comes with increased imperviousness which is scaled down through the integration of appropriate sustainable drainage systems (SuDS) in the drainage networks.

As a drainage planner, knowing the range of probability distribution functions (PDFs) applicable to the task at hand is pertinent to solving drainage problems. For example, the Weibull distribution is applied in developing IDF curves for use in the rational method and predicting the likelihood of receiving more or less rainfall in the future. The chapter explicitly outlines one case study where EVA was used and where Weibull distribution was applied. These case studies demonstrate how PDFs can be used to communicate helpful information to public authorities' leadership and management teams about the likely future impacts of the climate trend and possibly employ measures to control adverse impacts as appropriate. In today's flood risk management approaches, and the need to make decisions that consider sustainability as an overarching goal, holistic recommendations are made for leaders to scrutinize.

4.2 HYDROLOGIC CYCLE

The hydrologic cycle, also known as the water cycle or the hydrological cycle, is a biogeochemical cycle that describes the continuous movement of water on, above, and below the surface of the Earth. The exchange of energy across leading to temperature fluctuations influences the climate. For example, when the water evaporates from water bodies, energy from its surroundings is used which cools the environment and when condenses in the atmosphere, it releases energy and warms the environment.

When the sun heats lakes, ponds, rivers, plants, forests, and land, water rises in the form of water vapor through different physical processes, that is, evaporation, evapotranspiration, and transpiration. Evapotranspiration is when water is transpired from plants and evaporated from the soil. Evaporation is a process in which water changes from a liquid state to a gaseous state; as part of the hydrologic cycle, it takes place in the water bodies. In natural environments, much of the rainfall soaks into the ground as infiltration. Some water infiltrates deep into the ground and recharges the groundwater resources and replenishes aquifers, which can store freshwater for long periods of time.

The water vapor in the atmosphere condenses to form clouds which are later heated by the sun's rays causing rainfall or precipitation to fall on the Earth. Part of the rainfall infiltrates or percolates into the ground while the other converts into runoff when it falls on impermeable surfaces as shown in Figure 4.1.

4.3 HYDROLOGIC MODELS, TOOLS, AND TECHNIQUES

A technique is usually a way of carrying out a particular task, especially the execution or performance procedure, for example, an engineering or scientific procedure. For example, suppose the famous rational method is used to estimate peak flow from a catchment. In that case, the rational method is the technique used to obtain the peak flow. Again, suppose a customized software such as Autodesk Storm and Sanitary Analysis is used for planning and designing integrated drainage systems. In that case, the software is a tool which usually operates on algorithms inbuilt following theoretical techniques such as the rational method mentioned earlier.

Useful hydrology 117

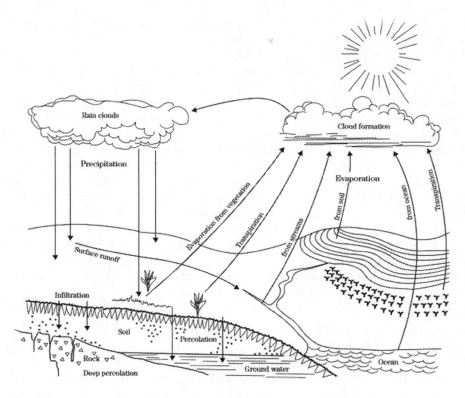

Figure 4.1 The hydrologic cycle.

Developing hydrologic and sewer hydraulic networks within the software, for example, is termed modeling. Hydrologic models are based on a spatially lumped form of the continuity equation, often called water budget or balance, and a flux relation expressing storage as a function of inflow and outflow.[1] Hydrologic models are mathematical models which can either be empirical or statistical or founded on physical laws. Mathematical models describe how a system's elements respond to inputs (stimulus).

Several computer simulation techniques are employed to study the hydrology of an area—in computation of runoff flow peaks or volumes and routing of the runoff. Most of these techniques involve solution of the one-dimensional equations of fluid motion called the Saint Venant equations which consist of the continuity equation and the momentum equation. Hydrologic computer simulation techniques are used worldwide to predict the rainfall-runoff models. Examples of computer software tools used in stormwater and runoff modeling include HEC-1, U.S. Environmental Protection Agency Storm Water Management Model (EPA SWMM),[2] ILLUDAS, Autodesk Storm and Sanitary Analysis, and STORM.

The ILLUDAS is the U.S. version of the Transport Road Research Laboratory (TRRL) method developed in the UK. The model breaks down the catchment area into subcatchments and the subcatchments are further subdivided into paved areas, grassed areas, and supplemental grassed areas.

HEC-1 was developed by the U.S. Army Corps of Engineers. The program has several storage routing techniques and has the ability to model pumping from one basin to another. It has several hydrologic and hydraulic computer models.

EPA SWMM developed a model used worldwide. The SMM software has the capability to model quality and quantity of flow. It can also be used to simulate and analyzes combined and partially combined systems. This model was developed by the EPA. The most significant attribute about SWMM is its ability to keep track of flows in excess of conduit capacity.

STORM was designed by the U.S. Army Corps of Engineers to assist in estimating pollutant loadings resulting from urban runoff. It requires less data as compared to HEC-1. The primary purpose of STORM is to analyze water pollution and detailed modeling of hydraulic features is out of scope. However, it permits analysis of multiple storm events.

Autodesk Storm and Sanitary Analysis was developed by Autodesk Inc. It is a subbasin-node-link–based model that performs hydrology, hydraulics, and water quality analysis of stormwater and wastewater drainage systems, including sewage treatment plants and water quality control devices.

The above computer software tools employ several techniques to estimate and route runoff such as Soil Conservation Service curve number (SCS-CN) method, statistical and probability techniques, Manning's equation, rational methods, continuity and momentum equations, the UK TRRL methods, hydrograph methods, and many others.

4.4 RAINFALL ANALYSIS

4.4.1 Average recurrence interval

The annual average recurrence interval (ARI) and annual exceedance probability (AEP) are used synonymously. ARI is the average time period between floods of a certain size. For example, a 25-year ARI flow will occur on average once in every 25 years. Alternatively, AEP is the probability of a certain size of flood flow occurring in a single year. In that case, 25-year ARI has an AEP of 0.04. The likelihood of a 25-year flood occurring in any given year is 0.04. Therefore, it is synonymous to speak of a 4% storm as a 25-year storm. Also, the likelihood of a 25-year ARI flood occurring at least once in the next 25 years is 0.64 while that for 10 years is 0.34. The 25-year ARI translates into a design life span of 25 years for the hydraulic structure.

The design flood, such as a 25-year flood is selected based on the perceived acceptable risks. Several standard manuals usually recommend 25-year ARI for storm sewers and 100-year ARI for streams and rivers' planning and restoration programs. A 100-year flood is huge and rare that it is estimated to occur once in every 100 years. In some cases, individual or residential drainage systems are generally designed to cope with more frequent storms—those with a 20% (5-year ARI) chance of occurring in a year. Any excess water travels along planned overland flow paths that carry water away from residential properties. This prevents them from flooding in the majority of storms—up to those with a 1% (100-year ARI) chance of occurring in a year.

In integrated drainage systems planning for sustainability, the planner/designer evaluates the AEP for residential sewers/drains, municipal council drains, and rivers/streams. It is not uncommon to consider between 5- and 25-year storm for residential drainage structures, 25-year storm for council drainage structures, and 100-year storm for rivers and streams hydraulic structures. However, since this is a probabilistic assessment that does not highly guarantee that no heavy storms within years will occur to cause serious damage to the drainage systems, designers are shifting to nonconventional approaches reinforcing drainage systems with appropriate engineering controls that derive volumes and peak flows from a spectrum of contributing factors beyond what is perceived normal.

4.4.2 What does a design year flood such as a 100-year flood mean?

A 100-year flood is quite big and rare that it will normally occur only once in every 100 years. That does not mean that a 100-year flood cannot occur immediately in the year after the occurrence of a 100-year flood! Every year has exactly the same probability—one in 100—of producing a 100-year flood, even in the area that experienced a 100-year flood in the year before.

4.4.3 Runoff estimation

Several methods (techniques) are used to forecast stormwater runoff, infiltration, evaporation, and evapotranspiration for use in drainage planning and design projects. Some methods generate the peak flows while others generate the flow volumes. It is entirely at the discretion of planner or designer following local regulations and standards to apply whichever technique is applicable. There is no consensus on which method is more appropriate for a specific area. That means it entirely depends on local area characteristics, data requirement of the method, and the holistic engineering judgment.

The SCS technique, the unit hydrograph method, or continuous simulation models may be used for high-level design and evaluation. Whatever method is employed, the engineer or planner must be certain that it has been properly calibrated for the area being studied. The estimation of peak flows or flow volumes for a given return period (ARI) is of utmost importance for planning and design of hydraulic structures.

Several methods used to estimate runoff quantities from relatively small urban watershed are classified as physical-process methods and empirical methods. Empirical methods result from relationships derived from observation of rainfall-produced runoff. Physical-process methods follow and manipulate the laws of physics and equations of motion. Rainfall data need to be specified for the catchments or subcatchments/subbasins under study.

In some literature, methods to calculate runoff quantities are termed as either conventional or nonconventional methods. **Conventional methods** are traditional approaches, which can be empirical or physical-process methods, that are well-known almost worldwide and these include the rational method, U.K. modified rational method, EPA SWMM, Hydrologic Engineering Center (HEC1-1) method, Soil Conservation Center, that is, SCS TR-20 and SCS TR-55, and unit hydrograph method. These are collectively termed **hydrologic methods**. For the **rational methods**, the user generates **IDF curves** to be used. All other hydrologic methods use a **rain gauge** to assign a rainfall time series. While using a specialized software, the IDF curve data are used based on standard equations as well as direct entry of a single intensity, or a table of intensity values.

Nonconventional methods were developed primarily to improve the interaction between drainage systems, the environment, and society to address the shortfalls in the conventional management methods which previously ignored these issues. Conventional rainfall-runoff modeling has been challenged by the evergrowing urbanization and climate change.

4.4.3.1 Hydrologic methods

4.4.3.1.1 EPA SWMM

The EPA SWMM is used throughout the world for planning, analysis, and design related to stormwater runoff, combined and sanitary sewers, and other drainage systems. It was designed to simulate the runoff of a drainage basin for any given rainfall pattern. The SWMM model breaks down the total watershed into a finite number of smaller units or subbasins describing them by their hydraulic or geometric properties. It can be used to evaluate gray infrastructure stormwater control strategies, such as pipes and storm drains, and is a useful tool for creating cost-effective SuDS/gray hybrid stormwater control solutions. SWMM was developed to help support local, state, and national stormwater management objectives to reduce runoff through infiltration and retention and help to reduce discharges that cause impairment of water bodies.

The SWMM model can be used to determine the locations and magnitudes of localized floods for short-duration storms, and the quantity and quality of stormwater runoff at several locations both in the system and in the receiving water bodies. A flowchart for the SWMM model process is shown in Figure 4.2. Detailed information about the EPA SWWM model can be obtained from hydrology textbooks.

4.4.3.1.2 Rational method

The rational method expresses a relationship between rainfall intensity and catchment area as independent variables and the peak flood discharge resulting from the rainfall as the dependent variable. It has been used for over 150 years and known as the rational method for nearly 100 years. The rational method expresses flow basing on intensity and area as independent variables. The rational method is given by Equation (4.1):

$$Q = 0.278CiA \tag{4.1}$$

Where:

Q : Peak discharge (m³/s)
C : Runoff coefficient
i : Rainfall intensity (mm/h)
A : Catchment area (km²)

Rational method primary assumptions

- Area in size does not exceed **80 hectares.**
- Time of concentration equals the rainfall duration.

4.4.3.1.3 UK modified rational method

Unlike the rational method where the peak flow rate is estimated, in the modified rational method, a full hydrograph is developed. The primary assumption is that when the storm duration exceeds the time of concentration, the rate of runoff would rise to the peak value obtained from the rational formula and then stay constant until the net rain stops. The modified rational method takes the standard rational to a different level in order to yield a hydrograph

Useful hydrology 121

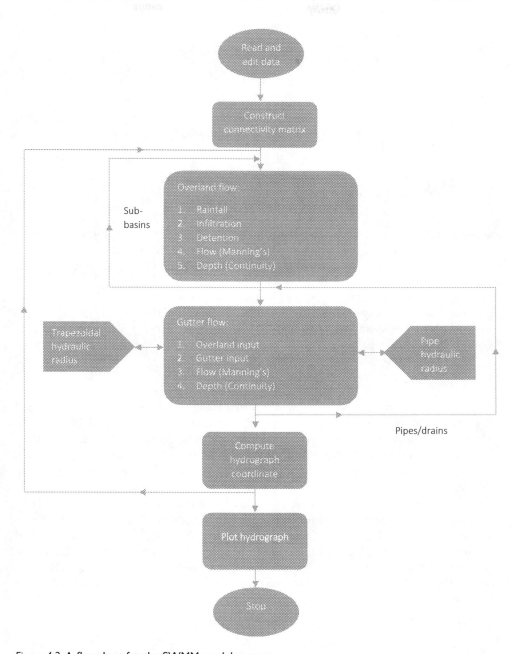

Figure 4.2 A flowchart for the SWMM model process.

mainly for use in **detention pond design**. According to the rational method, the highest discharge Q occurs when the rainfall duration equals time of concentration t_c.

When the rainfall stops, runoff rates start decreasing to zero as excess rain gets released from the basin, that is, **detention pond**. In case the rainfall excess time is equal to the time of

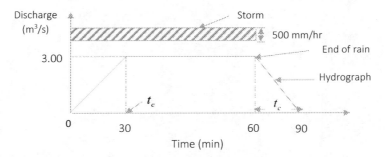

Figure 4.3 Modified rational method hydrograph.

concentration, the hydrograph would approximate to a trapezoidal shape (Figure 4.3) rising to peak at $t = t_c$ and remaining flat until t = the duration of the rain d and then falling along a straight line until $t = d + t_c$. The modified rational method is applied to watershed up to 20 hectares in size.

In modern storm sewer design, except for small areas, more sophisticated procedures are used other than the rational method and modified rational method. For example, SCS-CN, the East African Transport Road Research Laboratory (EA TRRL) flood model, and EPA SWMM are particularly helpful.

4.4.3.1.4 The soil conservation service curve number method

The U.S. Soil Conservation Service developed methods for estimating runoff volume and peak rates of discharge from urban areas, applicable for small watersheds, in the 1950s. These procedures revised in 1975 are the **Technical Release 55** (TR-55) graphical method, the chart method, and the tabular method. The first two procedures are used for estimating peak flows, and the last one is used for synthesizing complete hydrographs. All techniques are used with 24-hour storms.

TR-55 presents simplified procedures to calculate storm runoff volume, peak rate of discharge, hydrographs, and storage volumes required for floodwater reservoirs. These procedures are applicable in small watersheds, especially urbanizing watersheds. The method takes into account initial abstraction, consisting of interception, infiltration, and depression storage, and retention capacity of the catchment, which refers to continuing infiltration following the runoff initiation.[3] The governing equation for the SCS-CN method is as shown in Equation (4.2).

$$P_e = \frac{(P - I_a)^2}{(P - I_a) + S}; \text{ which is valid for } P > I_a; P = 0 \text{ otherwise} \tag{4.2}$$

$$S = \frac{1000}{CN} - 10 \tag{4.3}$$

Where:

P_e	:	Depth of effective precipitation (runoff in mm)
S	:	Depth of effective precipitation (runoff in mm)
P	:	Total rainfall depth in storm (mm)
CN	:	[1]Runoff curve number
I_a	:	Equivalent depth of initial abstraction (mm)

A broad graphical and tabular solution to this method was presented in Technical Release 55.[4] Curve number tables are used to derive CN which is based on antecedent moisture condition (for preceding 5 days),[5] soil group and cover type, or land use. The CN parameter reflects human activities and urbanization impacts and it should be revised and updated with respect to the urban development plan, reflecting the change in development density. Therefore, the area of physical development plan (PDP) is usually referenced during the design and planning of drainage assets.

SCS method assumptions

- Runoff cannot begin until the initial abstraction has been met, Equation (4.6).
- The graphical method, that is, SCS TR-55 was developed for homogenous watershed up to 20 mi² in size.
- The chart method is used for watersheds up to 2000 acres.
- All three procedures were developed for use with 24-hour storm.

Comparison of the SCS method and rational method

a. The SCS-CN method became popular due to its simplicity and easy application. It also accounts for most of the watershed characteristics associated with runoff production and incorporates them in the CN parameter.
b. The SCS-CN method could be utilized for surface runoff estimation while the rational method could estimate the peak runoff rate only.[6]
c. The SCS-CN accounts for the antecedent conditions, which are neglected in the rational method. Therefore, the CNvalues need to be accurately assessed for each locale.[5,7] It could depend on rainfall depth in some locations.[8]
d. The SCS-CN does not consider the spatial and temporal variability of the antecedent condition and continuing losses, which could make it a less reliable predictive method.[9]
e. The SCS method provides better details in describing physical characteristics of the watershed, which led to its applicability to bigger watersheds than those in the rational method.[10]

4.4.3.1.5 The SCS unit hydrograph

The SCS proposed a parametric unit hydrograph (UH) model; this model is included in the program. The model is based on averages of UH derived from gaged rainfall and runoff for a large number of small agricultural watersheds throughout the USA. SCS TR-55[4] (1986) and the National Engineering Handbook (1971) describe the UH in detail.

The SCS dimensionless UH procedure is one of the well-known methods for deriving synthetic unit hydrographs in use today. References for this method can be found in most hydrology textbooks or handbooks.

Hydrograph techniques

This is defined as the hydrograph of surface runoff resulting from an effective rain falling for a unit of time. The unit of time may be any value less than the time of concentration.

Primary assumptions of the unit hydrograph technique

a. All single storms in a given watershed which have equal duration produce runoff during equal lengths of time. For example, all storms on a watershed with a duration of 10 hours may result in runoff extending over 4 days.
b. The ordinates of the runoff hydrograph are proportional to the rainfall excess net rainfall minus abstractions).

A UH can be constructed from existing records of rainfall and stream flow. In Figure 4.4(a), A to C represents the hydrograph of a storm of unit duration, assumed equal to 10 hours. The time intervals in the figure are 10 hours.

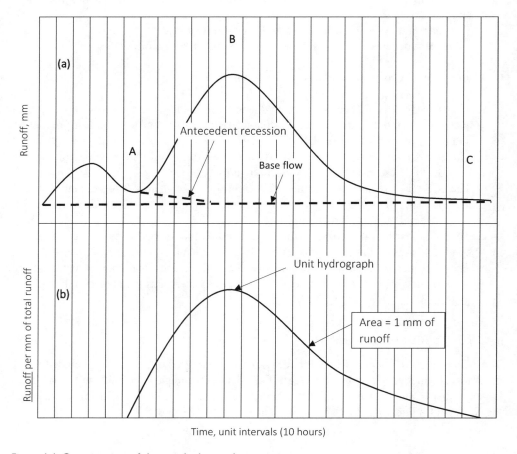

Figure 4.4 Construction of the unit hydrograph.

The runoff curve before point A results from antecedent precipitation, and the recession from point C is a continuation of the base flow. Figure 4.4(b) is obtained by subtracting the effects of antecedent rainfall and the base flow and dividing the ordinates by the total runoff. It represents a 10-hour unit hydrograph for the drainage basin.

The concept of the UH is applied to small urban watersheds (0.04–38 km²). **Parametric equations** are developed defining the hydrograph shape based on physical characteristics of the catchment. For example, parametric equations based on a 10-minute unit time are as shown below:

$$T_R = 4.1 L^{0.23} S^{-0.25} I^{-0.18} \Phi^{1.57} \tag{4.4}$$

$$Q = 13.27 A^{0.96} T_R^{-1.07} \tag{4.5}$$

$$T_B = 71.21 A Q^{-0.95} \tag{4.6}$$

$$W_{50} = 12.08 A^{0.93} Q^{-0.92} \tag{4.7}$$

$$W_{75} = 7.21 A^{0.79} Q^{-0.78} \tag{4.8}$$

Where:

T_R : Rise time of the hydrograph, minutes
Q : Peak discharge, m³/s per mm net rain
T_B : Time base of the hydrograph, minutes.
W_{50} : Width of the hydrograph, minutes, at 50% Q.
W_{75} : Width of hydrograph, minutes, at 75% Q
L : Total distance along main channel, m
S : Main channel slope (generally neglecting flatter 20% of upstream length)
I : Imperviousness, %
Φ : Dimensionless conveyance factor (0.6–1.3)
A : Watershed area, km²

4.4.3.1.6 Transport Road Research Laboratory method

The TRRL method is an urban runoff model which utilizes the time-area runoff routing method. It was developed in the UK specifically for the analysis of urban runoff and ignores all pervious areas that are not directly connected to the storm drain system. Therefore, the estimates of peak flow rates and runoff volumes are likely to be low for systems that have these features. The TRRL model could be used for continuous stream flow simulation but tends to be used as an event-simulation model.

4.4.3.1.7 The East African TRRL flood model

The EA TRRL model was developed by the UK Transport and Road Research Laboratory on the basis of rainfall/runoff studies for a range of selected East African catchments. The method is commonly applied to catchments up to 200 km². The model requires inputs of 24-hour annual maximum rainfall and catchment characteristics.[11]

The model is made up of two parts: a linear reservoir model and a flood model. The linear reservoir part of the model describes the land phase of the flood cycle. This is the time between the rainfall reaching the ground and the water entering the stream system. The flood routing part of the model routes the flood down the water course to the catchment outfall. The model assumes that a storm rainfall of a given return period results in a peak flood of equal return period.

The most widely used dimensionless unit hydrograph is that of the U.S. Soil Conservation Service. The similar results for the ratio of time to peak to base time for other catchments is not satisfactorily applicable in East African catchments studied by TRRL. In the EA TRRL flood model the base time is assumed to be the time from 1% of peak flow on the rising limb to 10% of peak flow on the falling limb of the hydrograph. When defined this way, the ratio of base time to time to peak is approximately 3.0 for U.S. hydrographs. For East African catchments it varied between 2.7 and 11.0. The use of a single hydrograph base on time to peak was therefore not appropriate.

A much more stable ratio was found to be the *peak flow Q* divided by the *average flow measured over the base time* $\overline{Q}$ (peak flow factor).

$$F = \frac{Q}{\overline{Q}} \tag{4.9}$$

The peak flow can therefore be simply estimated if the average flow during the base time of the hydrograph can be calculated. The total volume of runoff R_V in m^3 is given by:

$$R_V = 1000(P-Y)AC_A \tag{4.10}$$

Where:

P : Rainfall (mm) during time period equal to the base time
Y : Initial retention
A : Catchment area (km²)
C_A : Contributing are coefficient

If the hydrograph base time is measured to a point on the recession curve at which the flow is **one-tenth** of the peak flow, then the volume under the hydrograph is approximately 7% less than the total runoff given by Equation (4.10). The average flow $\overline{Q}$ is therefore given by Equation (4.11):

$$\overline{Q} = \frac{0.93 * R_V}{3600 * T_B} \tag{4.11}$$

Where:

T_B : Hydrograph base time (hours)

The estimates of Y and C_A are required to calculate R_V and lag time K to calculate T_B. Parameters relied on in the EA TRRL model are as follows:

1. Initial retention (*Y*)

In arid and semi-arid zones an initial retention of 5 mm could be considered. Elsewhere zero initial retention could be assumed.

2. Contributing area coefficient (C_A)

The standard contributing area coefficient (C_A) varies with soil type and average catchment slope. The average catchment slopes have been classified according to the TRRL classification as follows: moderate (1%–4%), rolling (4%–10%), and hilly (10%–20%). The soil characteristics have been classified in accordance with the TRRL classification as presented in Table 4.1.

Contributing area coefficient is a coefficient that reflects the effects of the catchment wetness and the land use. A grassed catchment at field capacity is taken as a standard value of contributing area coefficient. The design value of the contributing area coefficient could be estimated from the following Equation (4.12).

$$C_A = C_S * C_W * C_L \tag{4.12}$$

Where:

C_S : The standard value of contributing area coefficient for a grassed catchment at field capacity
C_W : The catchment wetness factor
C_L : The land use factor

The three factors are given in Tables 4.2, 4.3, and 4.4.

Table 4.1 TRRL classification of soil characteristics

Soil charactreisitcs	Description
Impeded drainage	• Very low permeability • Clay soils with high swelling potential • Shallow soils over largely impermeable layer, very high water table
Slightly impeded drainage	• Low permeability • Draiange slightly impeded when soil is fully wetted
Well drained	• Very permeable • Soil with very high infiltration rates such as sands, gravel, and aggregated clays

Source: TRRL Labaratory Report 706, Transport and Road Research Labaratory, Department of Environment, UK, 1976.

Table 4.2 Standard contributing area coefficient (wet zone catchment, short grass cover)

Catchment slope	Soil type		
	Well drained	Slightly impeded drainage	Impeded drainage
Very flat < 1.0%		0.15	0.30
Moderate 1%–4%	0.09	0.38	0.40
Rolling 4%–10%	0.10	0.45	0.50
Hilly 10%–20%	0.11	0.50	
Mountainous > 20%	0.12		

Note: The soil types are based on the soil map contained in the Hand Book of Natural Resources of East Africa.

Table 4.3 Catchment wetness factor

Rainfall zone	Catchment wetness factor	
	Perennial streams	Ephemera streams
Wet zone	1.0	1.0
Semiarid zone	1.0	1.0
Dry zones	0.75	0.50
Dry zones (except Western Uganda)	0.60	0.30

Table 4.4 Land use factor (base assumes short grass cover)

Land use	Land use factor
Large bare soil	1.50
Intense cultivation (particularly in valleys)	1.50
Grass cover	1.00
Dense vegetation (particularly in valleys)	0.50
Ephemeral steam. Sand filled valley	0.50
Swamp filled valley	0.33
Forest	0.33

Table 4.5 Catchment lag time

Catchment type	Lag time (K) in hours
Arid	0.1
Very steep small catchments (slope > 20%)	0.1
Semiarid scrub (large bare soil patches)	0.3
Poor pasture	0.5
Good pasture	1.5
Cultivated land (down to river bank)	3.0
Forest, overgrown valley bottom	8.0
Papyrus swamp in valley bottom	20.0

The catchment wetness factor (C_W) is a measure of the antecedent wetness of the catchment.

The land use factor (C_L) is the factor that adjusts the runoff factor according to land usage relative to a catchment with short grass cover.

3. Catchment lag time (*K*)
This is the time for the recession curve of outflow from a linear reservoir to fall to one-third of its initial value. This parameter is strongly dependent on the vegetation cover. The appropriate value of lag time can be estimated from Table 4.5. In assessing which category to place a given catchment, it should be remembered that generally only small areas on either side of the stream are contributing to the flood hydrograph. These areas, therefore, must be assessed.

Table 4.6 Rainfall time (T_P) for East African 10-year storm

Zone	Index 'c'	Rainfall time (T_P) (hours)
Inland zone	0.96	0.75
Coastal zone	0.76	4.0
Kenya Aberdare Uluguru zone	0.85	2.0

4. Base time

The rainfall time T_p is the time during which the rainfall intensity remains at high level. This can be approximated by the time during which 60% of the total rainfall occurs. Using the general IDF equation,

$$i = \frac{a}{(0.33 + t_d)^c} \tag{4.13}$$

The time to give 60% of the total rainfall is given by solving Equation (4.13).

$$0.6 = \frac{t_d}{24}\left(\frac{24.33}{t_d + 0.33}\right)^c \tag{4.14}$$

The values for the various rainfall zones of East Africa are given in Table 4.6.
  The flood wave attenuation T_A could be estimated from Equation (4.15).

$$T_A = \frac{0.028 L}{\overline{Q}^{1/4} S^{1/2}} \tag{4.15}$$

Where:

L : Length of main stream (km)
Q : Average flow during base time (m³/s)
S : Average slope along main stream

The base time T_B is, therefore, estimated from Equation (4.16):

$$T_B = T_P + 1.23K + T_A \tag{4.16}$$

Note that $\overline{Q}$ appears in Equation (4.15). Therefore, an iterative or trial-and-error solution is required. If initially T_A is assumed zero, two iterations could be adequate. Knowing Q and F, the peak flow is calculated using Equation (4.9).

4.4.3.2 Nonconventional design methods

The world is increasingly adopting nonconventional design methods to estimate stormwater volumes and peak discharges specifically due to climate change which has triggered severe floods. User-defined runoff methods are increasingly adopted worldwide. This is because of the variations in climatic zones, land cover, soils, and topography which necessitate a shift from rainfall-runoff models to a more flexible method that provides an interaction between drainage systems, the environment, and society. The conventional rainfall-runoff modeling has been challenged by the evergrowing urbanization and climate change. In several cases, the changing climate is largely taken into consideration neglecting the impact of urbanization on the runoff. Recent studies have shown that climate change and urbanization have to be modeled simultaneously. Global climate change could affect the drainage systems in many ways. It places a strong barrier to the provision of reliable hydrological input data for the design process.[12]

Climate change and urbanization impacts are woven together and could not be considered in isolation from one another. These combined effects could result in reduction in groundwater level and consequently land subsidence in the long term.[13] Without considering both factors together, the planning, designing, and developing a framework for a reliable drainage system, able to adapt to future changes, would fail. In this case, both climate scenarios and urbanization storylines need to be tailored to the future changes in each location.[12]

4.5 STATISTICAL MODELS AND PROBABILITY DISTRIBUTION FUNCTIONS

Statistical models are a type of mathematical models that are commonly used in hydrology to describe data, as well as relationships between data. Using statistical methods, hydrologists develop empirical relationships between observed variables, find trends in historical data, or forecast probable storm or drought events.

Several probability distributions are used to describe hydrologic processes. The theoretical probability distribution functions are not actual representations of the natural processes but provide seemingly comparable descriptions that approximate the underlying phenomenon.

Both discrete and continuous distributions are usually employed. However, continuous probability distributions are the mostly employed techniques in hydrologic processes because the random events such as rainfall receive in a particular area is a continuous event. Hydrologic process deals with random events that demonstrate variability which is not usually adequately explained by analytical measures of physical processes.

Statistical analysis in hydrology is essential because it is used to communicate helpful information and to take well-informed decisions. Collected data are usually summarized to present information precisely and in a meaningful way to come up with characteristics of the observed phenomenon and to predict the future behaviors. In distribution statistics, the measure of central tendency is of great significance, that is, mean, mode, and median. This is used to communicate useful information out of the processed data. The range, variability around the central tendency, degree of uncertainty, and frequency of occurrence of values are key parameters of concern in studying random events for which inference is made. Three characteristics of probability density functions are usually considered in the analysis of random events, that is, **the measure of central tendency, variability**, and **skewness**.

The measure of central tendency is of sound interest in hydrologic studies. The mean is the mostly used measure of central tendency which is estimated by the first moment about the origin for the sample data. The **mode** and **median** are the additional measures of central tendency. The mode is the value occurring the most frequently in discrete random variables while in continuous variables it is the peak value of the probability density function. The median is the middle value of the observed data and divides the distribution into equal areas.

Variability, which is the deviation about the mean, is of significant importance in hydrologic studies. The mean squared deviation measured by the second moment about the mean is of statistical importance which is termed as **variance**. In probability theory and statistics, **skewness** is a measure of the asymmetry of the probability distribution of a real-valued random variable about its mean. A distribution can be symmetrical, skewed to the right, or skewed to the left. The detailed discussion of the mean, median, mode, skewness, and variance of probability distribution functions can be obtained from standard statistics books.

Most hydrologic random variables are assumed to follow continuous probability distribution functions. Reliability studies entirely depend on the manipulation of continuous probability distribution functions. In coming up with sustainable drainage systems, the study of reliability is important. Hydrologic random variables such as **rainfall, temperature**, and **relative humidity** are studied and results analyzed to process meaningful information for designing reliable and sustainable drainage systems. In most cases, continuous probability distributions are employed to evaluate, analyze, design, and develop drainage solutions—coming up with characteristics of the observed phenomenon and to predict the future behaviors. The commonest distributions applied in hydrologic studies concerning drainage schemes is tabulated in Table 4.7. These are discussed henceforth, and detailed discussions of these distributions and other standard probability distributions can be obtained from specialized statistics books.

Table 4.7 gives the PDFs that are commonly applied in drainage analysis and design specifically applied to process hydrologic information used to forecast behaviors of observed phenomena. The forecasted results are used to model drainage systems reliability so that the systems perform satisfactorily. The PDFs are also used to plan for safeguards in drainage management so that the observed phenomena are dealt with appropriately.

4.5.1 Applications of continuous probability distribution functions

4.5.1.1 Frequency and return period

The theoretical return period is the inverse of the probability that the event will be exceeded in any one year (or more accurately the inverse of the expected number of occurrences in a year). Imagine the flooding event to be denoted with letter F and $P(F) = 0.04$. This implies that 4% is the chance that each year the flood will 'occur' which means that the flood will be exceeded. In theory, the probability of any exact value of any single event, of a continuous random variable is 0. This means that the flood to occur, the flood level may either be reached or exceeded. Therefore, the flood level would be reached or exceeded on average once in every 25 years. Thus, theoretically the average return period or average recurrence interval T in years is defined as follows:

$$T = \frac{1}{P(F)} \qquad (4.17)$$

Also, the probability that F will be equaled or exceeded in any year is the inverse of the return period.

Table 4.7 Common continuous probability distributions used in integrated drainage systems analysis and design

Distribution of a random variable X	PDF and CDF	Range	Mean $\bar{x}$ or $\propto$	Variance s^2 or σ^2
Exponential	$f(x) = \dfrac{1}{a} e^{-(x/a)}$	$0 \leq x \leq \infty$	a	a^2
Weibull	$f(x) = \dfrac{\beta}{\alpha}\left(\dfrac{x}{\alpha}\right)^{\beta-1} e^{-(x/\alpha)^\beta}$ $F(x) = 1 - e^{-\left[\frac{x}{\alpha}\right]^\beta}$	$x \geq 0; \alpha, \beta \geq 0$	$\mu = \alpha\Gamma\left(1+\dfrac{1}{\beta}\right)$	$\sigma^2 = \alpha^2\left\{\Gamma\left[\left(1+\dfrac{2}{\beta}\right)\right] - \left[\Gamma\left[\left(1+\dfrac{1}{\beta}\right)\right]\right]^2\right\}$
Normal	$f(x) = \dfrac{1}{\sigma\sqrt{2\pi}} e^{-(x-\mu)^2/2\sigma^2}$	$-\infty \leq x \leq \infty$	μ	σ^2
Log-normal ($y = \ln x$)	$f(y) = \dfrac{1}{x\sigma_y\sqrt{2\pi}} e^{-(y-\mu_y)^2/2\sigma_y^2}$	$-\infty \leq x \leq \infty$	μ_y	σ_y^2
Gamma	$f(x) = \dfrac{x^\alpha e^{-x/\beta}}{\beta^{\alpha+1}\Gamma(\alpha+1)}$	$0 \leq x \leq \infty$	$\beta(\alpha+1)$	$\beta^2(\alpha+1)$
Gumbel	$f(x) = \dfrac{1}{\alpha} e^{-\left[\frac{x-\xi}{\alpha}\right] - e^{-\left[\frac{x-\xi}{\alpha}\right]}}$ $F(x) = e^{-e^{-\left[\frac{x-\xi}{\alpha}\right]}}$	$-\infty \leq x \leq \infty$	$\mu = \xi + 0.5772\alpha$	$\sigma^2 = \dfrac{\pi^2\alpha^2}{6} \approx 1.645\alpha^2$
Extreme value	$f(x) = \alpha e^{-\alpha(x-u) - e^{-\alpha(x-u)}}$	$-\infty \leq x \leq \infty$	$u + \dfrac{0.5772}{\alpha}$	$\dfrac{\pi^2}{6\alpha^2}$
Log-Pearson III ($y = \ln x$)	$f(x) = \dfrac{(y-\gamma)^\alpha}{\beta^2 x\Gamma(\alpha+1)} e^{-\left[\frac{(y-\gamma)}{\beta}\right]}$	$-\infty \leq x \leq \infty$ $(0 \leq x \leq \infty)$	$\mu_y = \gamma + \beta(\alpha+1)$	$\sigma_y^2 = \beta^2(\alpha+1)$

$$P(F) = \frac{1}{T} \qquad (4.18)$$

The probability that F will not be exceeded in any given year is obtained as follows:

$$P(\overline{F}) = 1 - P(F) = 1 - \frac{1}{T} \qquad (4.19)$$

The probability that F will not be equaled or exceeded in any of n successive years is as follows:

$$P_1(\overline{F}) \times P_2(\overline{F}) \times ... \times P_n(\overline{F}) = P(\overline{F})^n = \left(1 - \frac{1}{T}\right)^n \qquad (4.20)$$

The probability that R, called risk that F, will be equaled or exceeded at least once in n successive years is as follows:

$$R = 1 - \left(1 - \frac{1}{T}\right)^n = 1 - \left[P(\overline{F})\right]^n \qquad (4.21)$$

Case study 4.1 (Section 4.6.1), provides a worked example of generating IDF **curves** used for the famous rational method.

4.5.1.2 Forecasting behaviors

This is used to obtain the characteristics of the observed phenomenon and to predict the future behaviors. Case study 4.2 (Section 4.6.2) provides a worked example where the rainfall trend is perceived to move for the years ahead. The rainfall received in the area is one of the primary random variables of great concern for drainage planners and engineers. The amount of storm received in an area varies from time to time. This variability can be modeled to forecast how the future is likely to be; how much rainfall would be received.

The changing climate is primarily a result of temperature increase. The increased rainfall must be due to increased temperatures which heat up the water bodies, land, and plants thereby increasing the rates of evaporation, evapotranspiration, and transpiration, respectively. The water holding capacity of the air increases by about 7% per 1°C warming, which leads to increased water vapor in the atmosphere which leads to increased rainfall. Hence, storms, whether individual thunderstorms, extratropical rain or snow storms, or tropical cyclones, supplied with increased moisture, produce more intense precipitation events. Such events are observed to be widely occurring, even where total precipitation is decreasing: **'it never rains but it pours!'** This increases the risk of flooding.[14]

The two-parameter Weibull distribution can be used to forecast rainfall trends which is a very good technique to check for the likelihood of high storms that could lead to flooding. It is on of the commonest techniques used to conduct reliability studies. This is an important technique engineers use to determine the extent of climate change for a given area specifically focusing on the rainfall received in the area. Case study 4.2 provides an example where the two-parameter Weibull distribution was applied to forecast the rainfall trend.

4.5.1.3 Extreme value analysis

EVA is used to model streams and rivers' storage capacity. It deals with the extreme deviations from the median of probability distributions. It seeks to assess, from a given ordered sample of a given random variable, the probability of events that are more extreme than any previously observed. It has wide applications in flooding hydrology, river restoration programs, and climate engineering. Although infrequent, extremes usually have large impact on the environment.

In sustainable drainage planning and design, **EVA** is constantly applied to receiving water bodies such as rivers and streams to determine the impact caused to the receiving water bodies as a result of the of the changes in the watershed. Land use and water use changes in watersheds can alter the frequency of high flows in receiving water bodies such as streams and rivers. These changes, which are predominantly caused by humans, include:

a. Urbanization
b. Construction sites
c. Reservoir construction, with the resulting attenuation and evaporation
d. Stream diversions
e. Construction of transportation corridors that increase drainage density
f. Deforestation from logging, infestation, and high-intensity fire
g. Reforestation

Flow frequency analysis is usually conducted to process the information about future stream or river performance. In case some or all the above changes occur in a watershed, stream or river restoration is important so as to improve and restore the physical, chemical, and biological functions of a stream or river system. Physical functions that might be improved are reduction in bank erosion and a self-sustaining water and sediment movement that does not require human intervention such as dredging. Chemical function improvements may include higher water quality and greater removal of impurities as the water flows through the river. Augmentations to biological function might be an expansion of habitat for diverse species, such as fish, aquatic insects, and other wildlife. However, given the large-scale degradation of landscape as a result of human activities, permanent return of the landscape to a predisturbed level is rarely achieved. Four distributions are most common in frequency analyses of hydrologic data, and these are as follows:

a. Normal distribution.
b. Log-normal distribution.
c. Gumbel extreme value distribution.
d. Log-Pearson type III distribution. The log-Pearson distribution is the most recommended in several countries.

a. Normal distribution
The normal or Gaussian distribution is one of the most popular distributions in statistics. It is also the basis for the Log-normal distribution, which is often used in hydrologic applications. The distribution, as used in **frequency analysis computations**, is provided:

$$X_{N,T} = \overline{X} + K_{N,T} S \tag{4.22}$$

Where:

$X_{N,T}$: Predicted discharge, at return period T
$\overline{X}$: Average annual peak discharge
$K_{N,T}$: Normal deviate (Z) for the standard normal curve, where area $= 0.5 - (1/T)$
S : Standard deviation, of annual peak discharge

b. Log-normal distribution

The annual maximum flow series is usually not well approximated by the normal distribution; it is skewed to the right, since flows are only positive in magnitude, while the normal distribution includes negative values. When a data series is left-bounded and positively skewed, a logarithmic transformation of the data may allow the use of normal distribution concepts through the use of the log-normal distribution. This transformation can correct this problem through the conversion of all flow values to logarithms. This is the method used in the log-normal distribution:

$$X_{LN,T} = \overline{X_1} + K_{LN,T} S_1 \qquad (4.23)$$

Where:

$X_{LN,T}$: Logarithm of predicted discharge, at return period T
$\overline{X_1}$: Average annual peak discharge logarithms
$K_{LN,T}$: Normal deviate (Z) of logarithms for the standard normal curve, where area $= 0.5 - (1/T)$
S_1 : Standard deviation, of logarithms of annual peak discharge

c. Log-Pearson type III distribution

The log-Pearson type III distribution applies to nearly all series of natural floods. It is similar to the normal distribution, except that the log-Pearson distribution accounts for the skew, instead of the two parameters, standard deviation and mean. When the skew is small, the log-Pearson distribution approximates to a normal distribution. The basic distribution is:

$$X_{LP,T} = \overline{X_1} + K_{LP,T} S_1 \qquad (4.25)$$

Where:

$X_{LP,T}$: Logarithm of predicted discharge, at return period T
$\overline{X_1}$: Average annual peak discharge logarithms
$K_{LP,T}$: A function of return period and skew coefficient provided in tables for K-values for the log-Pearson type III distribution
S_1 : Standard deviation, of logarithms of annual peak discharge

The mean in a log-Pearson type III distribution is approximately equal to the logarithm of the 2-year peak discharge. The standard deviation is the slope of the line, and the skew is shown by the curvature of the line.

d. Gumbel extreme value distribution

Peak discharges commonly have a positive skew, because one or more high values in the record result in the distribution not being log-normally distributed. Hence, the Gumbel extreme value distribution was developed.

$$X_{G,T} = \overline{X}_{G,T} + K_{G,T} S_1 \quad (4.24)$$

Where:

$X_{G,T}$: Predicted discharge, at return period T
$\overline{X}_{G,T}$: Average annual peak discharge
$K_{G,T}$: A function of return period and sample size provided in tables for K-values of the Gumbel distribution
S_1 : Standard deviation of annual peak discharge

4.6 CASE STUDIES AND WORKED EXAMPLES

4.6.1 Case study 4.1: Generating IDFs for the rational method, Kabale Municipal Council, Western Uganda (2016)

The rational method has been the commonest method used worldwide until when nonconventional methods became prominent following changing climates. The user generates IDF curves to be used to determine the design intensity used in the rational formula (4.1). A worked example for how IDFs are generated using the Watkins and Fiddes method in East African tropical climate is provided henceforth.

This worked example is taken from Kabale Municipal Council in Uganda which were used in designing sewers and drains for the drainage master plan, 2016. It has been observed in practice that the greater the intensity of rainfall, the shorter the duration of rainfall. In other words, intense storms occur for short durations. The intensity duration curves can be obtained by plotting the rainfall intensity against duration of the storm. IDF curves have been found helpful in the estimation of the design flow of hydraulic structures like drains and sewers. The IDF curves show the intensity of rainfall for a given duration and expected frequency. From the curves, intensities of given duration at required return periods are obtained and used in estimation of peak flows.

The IDF curves show the intensity of rainfall for a given duration and expected frequency. From the curves, intensities of given duration at required return periods are obtained and used in estimation of peak flows. The two methods of developing intensity duration curves are as follows:

a. The Watkins and Fiddes method
b. The Bells method

However, in a study[15] on data for different stations in Uganda observes that there is a good correlation between return period and the daily maximum rainfall, which is easier to measure accurately, whereas a poor correlation exists between the return period and hourly rainfall intensities on which Bells method is based. The Bells method was based on the North American catchments which have less thunderstorms days, hence the method of Watkins and Fiddes is the most appropriate for Uganda.

The Watkins and Fiddes method provides a good correlation between return period and the daily maximum rainfall, which is easier to measure accurately.[16] During design, the rainfall data for Kabale district were obtained from the Uganda National Meteorological Authority

(UNMA). The data were in depth of millimeters (mm). From the data the maximum monthly rainfall in every year for 10 years (2006–2015) was obtained. Table 4.8 provides the maximum rainfall per year.

The values of daily maximum rainfall together with the assumed duration of 1.5 hours were then ranked from 1 to n in decreasing order as shown in Table 4.9.

From Equation (4.26), the theoretical return period is the inverse of the probability that the event will be exceeded in any one year (or more accurately the inverse of the expected number of occurrences in a year). For example, a 10-year flood has a 1/10 = 0.1 {\displaystyle 1/10=0.1} or 10% chance of being exceeded in any one year and a 50-year flood has a 0.02 or 2% chance of being exceeded in any one year. The corresponding return period was estimated for each data set, **using Weibull's plotting position formula** (Table 4.10);

$$T = \frac{(n+1)}{r} \quad \text{For} \quad n = 10. \tag{4.26}$$

Table 4.8 Maximum rainfall received by Kabale per year

Year	Maximum monthly rainfall (mm)	Month of occurrence for maximum rainfall	Monthly totals (2006–2015) (mm)
2006	71.5	May	383.9
2007	72.3	July	159.1
2008	48.1	March	364.4
2009	44.7	November	396.1
2010	55.5	October	355.2
2011	53.8	March	364.4
2012	40.0	September	425.5
2013	54.5	September	425.5
2014	79.7	September	425.5
2015	54.3	February	380.4

Source: Uganda National Meteorological Authority (UNMA).[17]

Table 4.9 Ranking maximum rainfall received by Kabale for 10 years

Year	Maximum daily rainfall (mm)	Rank, r
2014	79.7	1
2007	72.3	2
2006	71.5	3
2010	55.5	4
2013	54.5	5
2015	54.3	6
2011	53.8	7
2008	48.1	8
2009	44.7	9
2012	40.0	10

Plotting position formulas

Plotting position formulas are used to obtain the frequency of an event. In many hydrologic study cases, annual maximum rainfalls are used to analyze rainfall trend for a specific location. The recurrence interval (return period) is usually approximated as the mean time in years, with N future trials, for the rth largest value to be exceeded once on average. The average number of exceedances for this condition is as shown below:

$$\bar{x} = N \frac{r}{n+1}$$

Where:
$\bar{x}$ = the mean number of exceedances
N = the number of future trials
n = the number of values
r = the rank of descending values with largest equal to 1.
In case the average number of exceedances $\bar{x} = 1$ and $N = T$, then,

$$T = \frac{n+1}{r}$$

This indicates that the return period (recurrence interval) is equal to the number of years of record plus 1, divided by the rank of the event. Table 4.5 gives seven plotting formulas commonly used.

Where:
n = the number of years of record
r = the rank

Therefore, adopting the Weibull plotting position formula, the theoretical return period, T was obtained as shown in Table 4.11.

The maximum rainfall depths were then plotted against return periods on a semi-log paper and a line of best fit plotted through the points as shown in Figure 4.5.

From the graph, maximum rainfall depths at desired return periods were then read off. The maximum 24-hour intensity and coefficient a^T were then computed for each selected return period using the formulae below:

$$i^T_{24} = \frac{R^T_{24}}{24} \quad \text{and} \quad a^T = i^T_{24} x (b+24)^n \tag{4.27}$$

$$n = \frac{\ln\left(\dfrac{14.4}{t_{eff}}\right)}{\ln\left(\dfrac{b+24}{b+t_{eff}}\right)} \tag{4.28}$$

Table 4.10 Plotting position formulas

#	Technique	Solve for $P(X > x)$	For $r=1$ and $n=10$. P	T
a	Tukey	$\dfrac{3r-1}{3n+1}$	0.065	15.5
b	Weibull	$\dfrac{r}{n+1}$	0.091	11
c	Chegadayev	$\dfrac{r-0.3}{n+0.4}$	0.067	14.9
d	Beard	$1-(0.5)^{1/n}$	0.067	14.9
e	California	$\dfrac{r}{n}$	0.10	10
f	Hazen	$\dfrac{2m-1}{2n}$	0.05	20
g	Blom	$\dfrac{r-\tfrac{3}{8}}{n+\tfrac{1}{4}}$	0.061	16.4

Table 4.11 Estimates of return period following Weibull plotting position formula (Kabale)

Year	Duration (hours)	Maximum daily rainfall (mm)	Rank, r	Return period, T
2014	1.5	79.7	1	11.00
2007	1.5	72.3	2	5.50
2006	1.5	71.5	3	3.67
2010	1.5	55.5	4	2.75
2013	1.5	54.5	5	2.20
2015	1.5	54.3	6	1.83
2011	1.5	53.8	7	1.57
2008	1.5	48.1	8	1.38
2009	1.5	44.7	9	1.22
2012	1.5	40.0	10	1.10

Where:

R^T_{24} : Maximum 24-hour rainfall
i^T_{24} : maximum 24-hour intensity
b : 1/3 (as developed from other studies in East Africa)[16] (Rugumayo, 2000)
t_{eff} : 1.5 hours

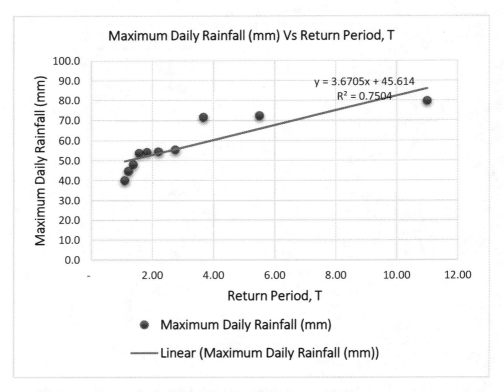

Figure 4.5 Graph of maximum daily rainfall (mm) against return period, t.

Therefore,

$$n = \frac{\ln\left(\dfrac{14.4}{t_{eff}}\right)}{\ln\left(\dfrac{b+24}{b+t_{eff}}\right)} = \frac{\ln\left(\dfrac{14.4}{1.5}\right)}{\ln\left(\dfrac{1/3+24}{1/3+1.5}\right)} = 0.8747$$

The value of n was calculated from the original equation of **Watkins and Fiddes**. These were then recorded in the Table 4.12.

The desired sequence of durations (10, 20, 30 minutes) for each data set was then chosen and used to calculate the rainfall intensity using the basic mathematical form of the IDF curve represented by the **rectangular hyperbola** shown below:

$$i = \frac{a}{(t+b)^n} \qquad (4.29)$$

Where i is the intensity in mm/h, t is the duration in hours, and a, b and n are constants developed for each IDF. From Table 4.13, the IDF curves were then developed by plotting values of intensity against duration for each return period as shown in Figure 4.6.

Table 4.12 Calculating a^T values for return periods

Return period, T (years)	Maximum 24-hour rainfall (mm)	Maximum 24-hour intensity (mm/h)	a^T
1	49.2845	2.053521	33.49729
2	52.955	2.206458	35.99202
5	63.9665	2.665271	43.47623
10	82.319	3.429958	55.94991
20	119.024	4.959333	80.89727
25	137.3765	5.724021	93.37094
50	229.139	9.547458	155.7393
100	412.664	17.19433	280.4761

Table 4.13 Intensity-duration-frequency table generated using a rectangular hyperbola, Equation (4.29)

Duration		Average recurrence interval (ARI), years							
t (minutes)	t (hours)	1	2	3	5	10	20	25	50
0	0.00	87.57	94.09	113.66	146.27	211.50	244.09	407.14	733.23
10	0.17	61.42	66.00	79.72	102.59	148.34	171.21	285.57	514.29
20	0.33	47.76	51.31	61.98	79.77	115.34	133.12	222.04	399.88
30	0.50	39.29	42.22	50.99	65.62	94.89	109.51	182.67	328.97
40	0.67	33.50	35.99	43.48	55.95	80.90	93.37	155.74	280.48
50	0.83	29.27	31.45	37.99	48.89	70.69	81.59	136.09	245.10
60	1.00	26.05	27.98	33.80	43.50	62.90	72.60	121.09	218.08
70	1.17	23.50	25.25	30.49	39.24	56.74	65.49	109.24	196.73
80	1.33	21.43	23.02	27.81	35.79	51.75	59.73	99.62	179.41
90	1.50	19.71	21.18	25.59	32.93	47.61	54.95	91.65	165.06
100	1.67	18.27	19.63	23.71	30.51	44.12	50.92	84.94	152.96

This was then exported to the design software to quicken the design process.

Using the time of concentration and considering a return period of 25 years as provided by the National Standard Manuals for Drainage Sewers, the design intensity for drains and storm sewers in the delineated catchments was determined from the IDF curve. According to Uganda's manual, the council drains/sewers were designed to meet ARI of 25-year storm and checked for a 50-year storm.

4.6.2 Case study 4.2: Kabale municipality rainfall forecasting with use of two-parameter Weibull distribution, Western Uganda (2016)

In this case study, the two-parameter Weibull distribution was used to forecast the likelihood of receiving more rainfall in future.

Guiding question: Is the municipal to receive more rainfall in future?

From Table 4.7, the two-parameter Weibull distribution is given by the formula below:

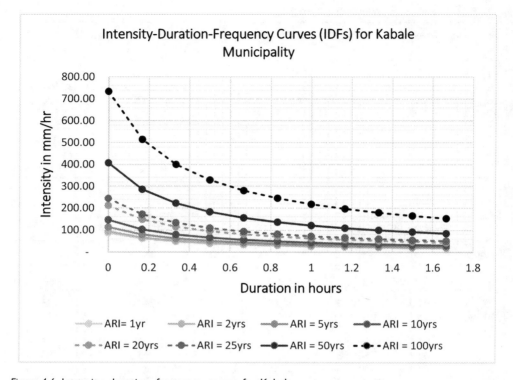

Figure 4.6 Intensity–duration–frequency curves for Kabale.

$$f(x;\alpha,\beta) = \begin{cases} \dfrac{\beta}{\alpha}\left(\dfrac{x}{\alpha}\right)^{\beta-1} e^{-(x/\alpha)^{\beta}} & x \geq 0, \\ 0 & x < 0, \end{cases} \quad (4.30)$$

Where:

x : is the random variable, in this case the monthly maximum rainfall received in a year
α : is the scale parameter, it scales the x variable. The scale parameter determines the range of the distribution
β : Shape parameter

The Weibull cumulative distribution $F(x)$ appears as follows:

$$\begin{aligned} F(x) &= 1 - e^{-\left[\frac{x}{\alpha}\right]^{\beta}} \\ 1 - F(x) &= e^{-\left[\frac{x}{\alpha}\right]^{\beta}} \\ \ln\left(\dfrac{1}{1-F(x)}\right) &= \left(\dfrac{x}{\alpha}\right)^{\beta} \\ \ln\left(\ln\left[\dfrac{1}{1-F(x)}\right]\right) &= \beta\ln x - \beta\ln\alpha \end{aligned} \quad (4.31)$$

Useful hydrology 143

For sample sizes less than 100, in the absence of median rank tables the true median rank values can be adequately approximated using Chegadayev's approximation[2]:

$$f(x) = \frac{(r-0.3)}{(n+0.4)} \tag{4.32}$$

Where:

- x : is the random variable, in this case the maximum rainfall in mm received in a year
- r : is the rank
- n : is the sample size

Table 4.14 gives the maximum annual rainfall for Kabale municipality ranked from 1 to 10 for years 2006 up to 2015.

Median rank values give the best estimate for the primary Weibull parameter and are best suitable for reliable confidence limits. Table 4.15 illustrates the median ranks. A plot of ln(ln(1/(1-median rank))) versus ln(maximum Daily Rainfall (mm)) is shown in Figure 4.7.

From Weibull cumulative distribution (4.31),

$$\beta = -4.87$$

Hence,

$$-\beta \ln \alpha = 19.108$$
$$\alpha = 50.46 \text{ mm}$$

Table 4.14 Maximum annual rainfall ranked

Year	2014	2007	2006	2010	2013	2015	2011	2008	2009	2012
Rainfall (mm)	79.7	72.3	71.5	55.5	54.5	54.3	53.8	48.1	44.7	40.0
Rank	1	2	3	4	5	6	7	8	9	10

Table 4.15 Median rank table

Year	Maximum rainfall (mm)	Rank	Median rank	1/(1-median rank)	ln(1/(1-median rank))	ln(ln(1/(1-median rank)))	ln(Maximum Daily Rainfall (mm))
2014	79.7	1	0.067308	1.072165	0.06968	−2.66384	4.37827
2007	72.3	2	0.163462	1.195402	0.178483	−1.72326	4.280824
2006	71.5	3	0.259615	1.350649	0.300585	−1.20202	4.269697
2010	55.5	4	0.355769	1.552239	0.439698	−0.82167	4.016383
2013	54.5	5	0.451923	1.824561	0.60134	−0.5086	3.998201
2015	54.3	6	0.548077	2.212766	0.794243	−0.23037	3.994524
2011	53.8	7	0.644231	2.810811	1.033473	0.032925	3.985273
2008	48.1	8	0.740385	3.851852	1.348554	0.299033	3.873282
2009	44.7	9	0.836538	6.117647	1.811178	0.593977	3.799974
2012	40.0	10	0.932692	14.85714	2.698481	0.992689	3.688879

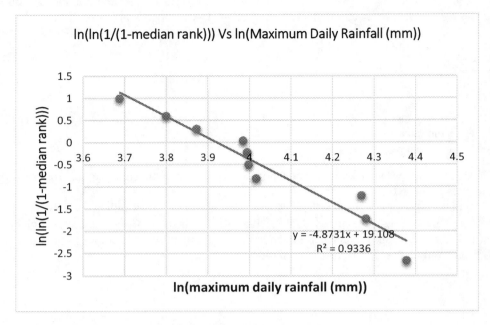

Figure 4.7 ln(ln(1/(1-median rank))) versus ln(maximum daily rainfall (mm)).

4.6.2.1 Findings

Yes, the municipality was found to have increased annual rainfalls in future—the shape parameter $\beta = -4.87 < 1$ indicated there is a more likely event for an increase in maximum monthly annual rainfalls received in Kabale municipality for years ahead. The maximum monthly annual rainfalls would be in the range of ±50.46 mm. Therefore, heavy rainfall would lead to floods that could destroy farmlands, roads, buildings, houses, and also impact on service delivery.

4.6.2.2 Analysis implications—decisions that followed

Specifying green practices, that is, SuDS—rainwater harvesting, infiltration systems, bioretention gardens, tree planting, etc., followed next. The primary goals of SuDS is to collect water, infiltrate runoff into soil, capture pollutants, provide green spaces, and groundwater recharge, reduce runoff velocity, and absorb greenhouse gases (GHGs). Therefore, recommendations for sustainable practices in order to reduce stormwater volume entering drainage system (minimize runoff hence reducing flooding risk) were done. A sustainable system/process/product is economical, socially beneficial, and environmentally friendly. A $660, 10,000-L rainwater harvesting (RWH)—HDPE tank, 30-year life span would be adequate for 464.52 m^2 plot, waiving about $1444 in 30 years, that is, $ 0.96/m^3. A typical tree could absorb around 21 kg of CO_2/year. The designer specified more than 90 trees to be planted on wider channels' banks. These would help to control temperatures, capture pollutants and CO_2.

4.6.3 Case study 4.3: Extreme value analysis for river Kiruruma in Kabale, Uganda (2016)

Figure 4.8 shows river Kiruruma, the main receiving water body for Kabale municipality. The municipality is located in Western Uganda. The current flow (dry weather) falls in the range of 2.53 to 2.68 m^3/s as determined by the leaf method. The river is relatively small. For small rivers it is appropriate to approximate water flow using the leaf method as the technique. Special equipment may not be needed to determine a quick estimate. From the river dimensions taken, full capacity was estimated in the range of 6.72 to 7.13 m^3/s.

In the analysis, data used were for one station, that is, R. Kiruruma North at Kabale-Kisoro road (81249). The assumption was made to consider 1998 as the base year and make a forecast of up to 120 years ARI to cater for the period between 1998 and 2016. Table 4.16 gives the station information, Table 4.17 gives peak flow per year, and Weibull plotting positions. Table 4.18 gives the peak flow for River Kiruruma sorted from 1955-1997. Table 4.19 gives four Statistical Distributions forecasting ARI Peak Discharges.

Note that this statistical analysis does not take into consideration the changes in the watershed. Some of these data were taken when the majority of areas in Kabale

Figure 4.8 River Kiruruma in Kabale municipality, Western Uganda.

Table 4.16 Data for Gage Station No. 81249 (R. Kiruruma North at Kabale Kisoro Road)

Station number	Latitude	Longitude	Elevation (meters)	Area (km^2)	Name
81249	1:12:51 S	29:56:58 E	1824.0	161.0	R. Kiruruma North at Kabale Kisoro Road

Table 4.17 Peak discharge per year for R. Kiruruma ranked from highest to lowest

Year	Peak discharge Q (m³/s)	Rank	LN (peak discharge)	Weibull plotting position (year)
1984	4.916	1	1.5924952	41
1988	4.567	2	1.5188565	20.5
1990	4.462	3	1.4955971	13.66667
1985	4.437	4	1.4899785	10.25
1989	4.246	5	1.4459774	8.2
1982	4.232	6	1.4426747	6.833333
1976	3.845	7	1.3467736	5.857143
1987	3.839	8	1.3452119	5.125
1978	3.657	9	1.2966431	4.555556
1986	3.438	10	1.2348899	4.1
1983	3.432	11	1.2331432	3.727273
1967	3.22	12	1.1693814	3.416667
1977	2.929	13	1.0746611	3.153846
1956	2.708	14	0.9962104	2.928571
1975	2.684	15	0.9873082	2.733333
1958	2.63	16	0.9669838	2.5625
1957	2.613	17	0.960499	2.411765
1966	2.544	18	0.9337376	2.277778
1991	2.362	19	0.8595087	2.157895
1971	2.211	20	0.7934449	2.05
1969	2.096	21	0.7400308	1.952381
1992	2.02	22	0.7030975	1.863636
1996	2.017	23	0.7016113	1.782609
1993	2.015	24	0.7006192	1.708333
1970	1.957	25	0.6714127	1.64
1997	1.921	26	0.6528459	1.576923
1968	1.839	27	0.6092219	1.518519
1962	1.704	28	0.5329784	1.464286
1995	1.679	29	0.5181984	1.413793
1963	1.58	30	0.4574248	1.366667
1974	1.519	31	0.4180522	1.322581
1955	1.513	32	0.4140944	1.28125
1965	1.488	33	0.3974329	1.242424
1964	1.309	34	0.2692635	1.205882
1972	1.302	35	0.2639015	1.171429
1994	1.29	36	0.2546422	1.138889
1960	1.139	37	0.1301507	1.108108
1959	1.045	38	0.0440169	1.078947
1961	0.934	39	−0.068279	1.051282
1973	0.887	40	−0.11991	1.025

Source: Directorate of Water Resources Management (DWRM), Uganda.[18]

municipality watershed and the environs consisted of forests and grass. The situation today is different and the future predictions give a highly disturbed environment. This analysis illustrates the analysis of the gaged flow data with the four common distributions outlined in Section 4.5.1.

These frequency curves in Figure 4.9 give a prediction of the peak flow discharges. However, they don't take into consideration the changes in the watershed. Such changes in

Table 4.18 Peak flow for River Kiruruma sorted from 1955 to 1997

Year	Peak flow (m³/s)	Year	Peak flow (m³/s)	Year	Peak flow (m³/s)	Year	Peak flow (m³/s)	Year	Peak flow (m³/s)
1955	1.513	1965	1.488	1975	2.684	1985	4.437	1995	1.679
1956	2.708	1966	2.544	1976	3.845	1986	3.438	1996	2.017
1957	2.613	1967	3.22	1977	2.929	1987	3.839	1997	1.921
1958	2.63	1968	1.839	1978	3.657	1988	4.567		
1959	1.045	1969	2.096	1979		1989	4.246		
1960	1.139	1970	1.957	1980		1990	4.462		
1961	0.934	1971	2.211	1981		1991	2.362		
1962	1.704	1972	1.302	1982	4.232	1992	2.02		
1963	1.58	1973	0.887	1983	3.432	1993	2.015		
1964	1.309	1974	1.519	1984	4.916	1994	1.29		

For Q

Mean	=	2.506
Standard deviation	=	1.152
Skew coefficient	=	0.551

For lnQ

Mean	=	0.812
Standard deviation	=	0.476
Skew Coefficient	=	−0.111

Table 4.19 Four statistical distributions forecasting ARI peak discharges

Method						Return period					
		1.25	2	5	10	19	20	25	50	100	120
Normal distribution	K_N	-	0.00	0.84	1.28	1.62	1.64	1.75	2.05	2.33	2.39
	Q_N(m³/s)	-	2.51	3.47	3.98	4.37	4.40	4.52	4.87	5.19	5.26
Log-normal distribution	K_{LN}	-	0.00	0.84	1.28	1.62	1.64	1.75	2.05	2.33	2.39
	Q_{LN}	-	0.812	1.212	1.421	1.583	1.593	1.645	1.788	1.921	1.950
	Q_{LN}(m³/s)		2.252	3.360	4.142	4.870	4.917	5.181	5.976	6.828	7.026
Gumbel distribution	K_G	−0.9	−0.16	0.84	1.5	1.95	2.01	2.33	2.94	3.55	4.2
	Q_G(m³/s)	1.469	2.322	3.474	4.234	4.752	4.822	5.190	5.893	6.596	7.344
Log-Pearson distribution	K_{LP}	−0.836	0.017	0.846	1.27	1.55	1.58	1.716	2.000	2.252	2.452
	Q_{LP}	0.414	0.820	1.215	1.417	1.550	1.564	1.629	1.764	1.884	1.979
	Q_{LP}(m³/s)	1.513	2.271	3.369	4.123	4.711	4.778	5.098	5.836	6.579	7.237

148 Integrated Drainage Systems Planning and Design for Municipal Engineers

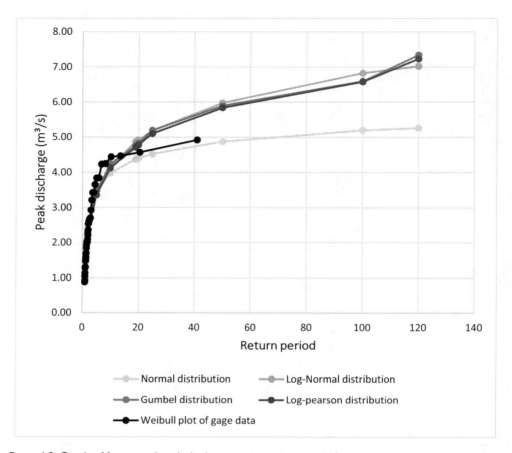

Figure 4.9 Graph of forecasted peak discharge against return period.

the watershed are outlined in Section 4.5.1.3. Discharge estimates from precipitation can be a helpful complement to gage data to obtain a more reliable 100-year peak flow.

4.6.3.1 Conclusion

The 100-year peak discharge would leave the river banks eroded. This should make the municipal council to plan restoration programs in advance. SuDS were highly recomended to reduce stormwater runoff quantities ending in river Kiruruma.

4.6.3.2 Recommendation from the analysis

a. The frequency curves give predicted flow discharges for the river at stream gage 81249. Considering the changes in the watershed as listed in Section 4.5.1.3, the peak flows might be higher than these forecasted by statistical methods. Therefore, the municipal should prepare for river conservation practices such as the following:

- SuDS throughout the municipality and along the river banks
- Reduction in bank erosion
- Sediment movement control (e.g., dredging)
- Installing bridges across the river

b. The design team prepared a basic statistical analysis to highlight the current performance and future performance of R. Kiruruma basing on only one gage station on the river but it would be better to analyze all the three gage stations that once existed on the river, that is, R. Kiruruma North at Kabale—Kisoro Road (81249), R. Kiruruma South at Kitumba (81250), and R. Kiruruma South at Rwanda Border (81253). Due to the removal of these gages from the river, the river missed data for 19 years (1998–2016). It was thus recommended for the municipal to use data from a nearby gaged R. Nyakizumba at Maziba station (81248) to estimate the current flow and future flow of the R. Kiruruma not forgetting the limitations of the approximate methods to be used. Some of the approximate methods include the following:

i. If gauge data are not available at the design location, discharge values can be estimated by transposition in case if a peak flow-frequency curve is available at a nearby gauged location. This method is appropriate for hydrologically similar watersheds that differ in area by less than 50%, with outlet locations less than 100 mi (160.93 km) apart. From the research of Asquith and Thompson (2008),[19] an estimate of the desired AEP peak flow at the ungauged site is provided by the equation (4.33):

$$Q_1 = Q_2 \sqrt{\frac{A_1}{A_2}} \qquad (4.33)$$

Where:

Q_1 : Estimated AEP discharge at ungauged watershed 1
Q_2 : Known AEP discharge at gauged watershed 2
A_1 : Area of watershed 1
A_2 : Area of watershed 2

ii. Regression equations to estimate the ARI discharges.

Note: These methods would be assessed for the appropriateness and applicability before using them to estimate flows since they were developed outside Uganda. If river restoration program projects were to be implemented, the municipal would first seek expert guidance by sourcing for a senior expert highly experienced to spearhead the implementation program.

4.7 HOW TO DEAL WITH RUNOFF QUALITY

Runoff quality is improved through a number of processes that capture pollutants commonly referred to as **'source control systems'**. The application of SuDS in stormwater drainage networks helps to remove pollutants through chemical, biological, and physical processes. SuDS good at pollutant removal include infiltration systems, bioretention gardens, detention/retention systems, swales, green roofs, etc. When stormwater passes through SuDS, its quality is significantly improved.

The fact that runoff is always so great, an integration of SuDS in storm sewer networks proves the most economical option to improve the quality of runoff. It is always quite expensive to collect, contain, and treat huge runoffs through the traditional approaches. The advancement of SuDS technology is proving the best approach in areas susceptible to heavy contamination and pollution.

4.8 SUDS HYDROLOGY

The development of SuDS requires an appreciation of hydrologic techniques so that they perform satisfactorily. The hydrological processes of SuDS are often modeled using process-based models.[20] The primary purpose of SuDS is to control pollutant loads in the runoff, reduce runoff volumes by infiltrating stormwater into the ground, reducing the peak flows by decelerating the flow at a particular time and more importantly to contribute to climate change mitigation or adaptation measures/techniques. SuDS mitigate the effect of human development on the natural water cycle, particularly surface runoff and water pollution trends.

SCS-CN method; TR-55 methodology is extensively used to obtain the design storm for several SuDS especially infiltration systems and bioretention gardens.

The EPA SWMM model provides explicit steps to cater for infiltration practices and detention systems within the model subbasins. SuDS make urban drainage systems compatible with components of the natural water cycle such as soil percolation, storm surge overflows, and bio-filtration. The movement of water through SuDS is particularly of great significance and this calls for an appreciation of different types of soils particularly the soil infiltration rates and percolation rates.

For SuDS such as infiltration systems to perform effectively, infiltration tests must be performed on soils that will be used as subsoil layers. Several SuDS are designed to cope with the most frequent storms, for example, 2-year or 5-year storms. Particularly, the detention and retention basins are designed to withstand storms of such frequency. The modified rational method is mainly used in detention pond design (Section 4.4.3.1.3).

Since SuDS are designed to mimic the natural environment, the effect of SuDS on the water cycle is evaluated. That means evaporation, percolation, infiltration, interception, groundwater recharging, transpiration, and evapotranspiration are topics evaluated in the design and development of several types of SuDS. SuDS such as infiltration systems and bioretention systems deal with removal of pollutant loads from the runoff, reducing runoff volumes through infiltration. Therefore, determining the infiltration rates of native soils or soils to be used as subsoil layers is paramount.

Various forms of vegetation developed in SuDS have the potential to intercept large quantities of stormwater runoff and therefore the type of precipitation, rainfall intensity and duration, and wind and atmospheric conditions affecting evaporation are some of the factors determining interception losses. Much as part of the stormwater is intercepted and it is returned to the atmosphere through evaporation; SuDS such as infiltration systems and bioretention gardens are designed to infiltrate the remaining part and the excess stormwater converts into runoff that ends in storm sewers. It is important to recognize that vegetation cover such as tree leaves or grass blades retain precipitation striking it as it flows down the stems of plants becoming stemflow, or fall off tree leaves becoming part of the throughfall. This behavior can be modeled practically in watershed management and SuDS development.

Understanding the native species that can adapt to the climate of the area is paramount in the design and development of SuDS. The type and amount of precipitation received in the area informs the most appropriate type of SuDS. Hydrology topics of interest in the design and development of SuDS include infiltration, evaporation, transpiration, evapotranspiration groundwater recharge, and pollution removal.

Detailed analysis of SuDS hydrology is out of the scope of this book.

4.9 CONCLUSION

The drainage design engineer and planner are tasked to evaluate hydrologic methods applicable in the circumstances. EVA is employed to evaluate extremes especially where flooding events are anticipated and where watershed changes are rapid. In that regard, the planners evaluate the volumes and quality of runoff ending in the receiving water bodies to plan for sustainable management of water resources.

A shift from conventional to nonconventional methods while estimating runoff quantities has been rapidly precipitated by the climate change as a result of which unprecedented flooding is inevitable across the world. In the interest of sustainable development, adaption strategies are fast becoming better techniques to safeguard communities from the effects of flooding. Therefore, integrated drainage systems planning that prioritizes SuDS is the way to go—and this requires an investigation into the climate of the area to design and specify the most appropriate techniques that can adapt to the area climate.

Probability density functions are numerous and are at the heart of hydrologic modeling. Therefore, deciding on what technique to use solely depends on the nature of model and understanding the project at hand. The case studies provided in this chapter brought out a few probability distributions used but there are many other scenarios that require the use of other PDFs differently.

Notes

1 The runoff curve number (CN) is an empirical parameter used in hydrology for predicting direct runoff or infiltration from rainfall excess. It has a range from 30 to 100. Lower numbers indicate low runoff potential while larger numbers are for increasing runoff potential. The lower the CN, the more permeable the soil is. **As can be seen in the curve number equation, runoff cannot begin until the initial abstraction has been met.**
2 See Table 4.2.

References

1. Singh, V.P. (1988). *Hydrologic systems: Rainfall-runoff modeling*. Prentice Hall, Englewood Cliffs, New Jersey.
2. United States Environmental Protection Agency. Storm water management model (SWMM) helps predict runoff quantity and quality from drainage systems. www.epa.gov/water-research/storm-water-management-model-swmm. Accessed July 22, 2022.
3. Methods, H., & Durrans, S. R. (2003). *Stormwater conveyance modeling and design*, 1st ed. Waterbury, CT: Haestad Press.
4. SCS. (1986) Urban hydrology for small watersheds. US Soil Conservation Service. *Technical Release 55*, 13.

5. Liu, X., & Li, J. (2008). Application of SCS model in estimation of runoff from small watershed in Loess Plateau of China. *Chinese Geographical Science, 18*(3), 235–241.
6. Varanou, E., Gkouvatsou, E., Baltas, E., & Mimikou, M. (2002). Quantity and quality integrated catchment modeling under climate change with use of soil and water assessment tool model. *Journal of Hydrologic Engineering, 7*(3), 228–244.
7. Wood, R., Mullan, A. B., Smart, G., Rouse, H., Hollis, M., McKerchar, A., Ibbitt, R., Dean, S., & Collins, D. (NIWA) (2010). *Tools for estimating the effects of climate change on flood flow*. Prepared for Ministry for the Environment, New Zealand.
8. Soulis, K. X., Valiantzas, J. D., Dercas, N. & Londra, P. A. (2009). Analysis of the runoff generation mechanism for the investigation of the SCS-CN method applicability to a partial area experimental watershed. *Hydrology & Earth System Sciences Discussions* 6(1), 605–615.
9. Ponce, V. M., & Hawkins, R. H. (1996). Runoff curve number: Has it reached maturity? *Journal of Hydrologic Engineering, 1*(1), 11–19.
10. Maidment, D. R. (1992). *Handbook of hydrology*. New York, NY: McGraw-Hill.
11. Nobert, J. (2013). Evaluation of methods for design discharge estimation in ungauged catchments: A case of Tigithe river catchment in Mara River Basin Joel Nobert1 1 University of Dar es Salaam, Department of Water Resources Engineering, Box 35131, Dar es Salaam, Tanzania.
12. Yazdanfar, Z. & Sharma A. (2015). Urban drainage system planning and design – challenges with climate change and urbanization: A review. *Water Science and Technology, 72*(2), 165–179.
13. Gu, C., Hu, L., Zhang, X., Wang, X. & Guo, J. (2011). Climate change and urbanization in the Yangtze River Delta. *Habitat International* 35 (4), 544–552.
14. Trenberth, K. E. (2011). Changes in precipitation with climate change. *Climate Research, 47*, 123–138. https://doi.org/10.3354/cr00953
15. Rugumayo, N. K., & Shima, J. (2003). *Rainfall reliability for crop production: A case study in Uganda*. Diffuse Pollution Conference Dublin, 2003.
16. Rugumayo, A. I. (2000). Drought intensity duration and frequency analysis: A case study of western Uganda. *Journal of CIWEM, 16*, 111–115.
17. Information management system. Uganda National Meteorological Authority (UNMA), Ministry of Water & Environment. Kampala, Uganda, 2022.
18. Information management system. Directorate of Water Resources Management (DWRM). Ministry of Water & Environment. Kampala, Uganda, 2022.
19. Asquith, W. H., & Thompson, D. B. (2008). *Alternative regression equations for estimation of annual peak-streamflow frequency for undeveloped watersheds in Texas using PRESS minimization*. U.S. Geological Survey Scientific Investigations Report 2008–5084, p. 40.
20. Yang, Y., & Chui, T. F. M. (2021). Modeling and interpreting hydrological responses of sustainable urban drainage systems with explainable machine learning methods, Hydrol. *Journal of Earth System Science, 25*, 5839–5858. https://doi.org/10.5194/hess-25-5839-2021

Further reading

Viessman, W. Jr., & Lewis, G. L. (2013). *Introduction to hydrology*, 5th ed. New Delhi: Asoke K. Ghosh, PHI Learning Private Limited.

Chapter 5

Hydraulics design principles

NOTATION

g	:	Acceleration due to gravity
Q	:	Specific discharge
N	:	Manning's roughness factor
V	:	Flow velocity
S	:	Slope
S_c	:	Critical slope
S_f	:	Friction slope
S_0	:	Bed slope
y	:	Flow depth
y_n	:	Normal depth
y_c	:	Critical depth
A	:	Flow cross-sectional area
P	:	Wetted perimeter
R	:	Hydraulic radius
Q	:	Discharge
E	:	Specific energy
α	:	Kinetic energy velocity distribution coefficient
t	:	Time
H_w	:	Headwater depth
F_r	:	Froude number
F	:	Force
B	:	Water surface breadth
β	:	The Boussinesq velocity distribution coefficient
ρ	:	Water density
h	:	Depth
V_f	:	Full flow velocity
Q_f	:	Full discharge
b	:	Breadth of a rectangular channel
F_{r_1}	:	Froude number at point 1
F_{r_2}	:	Froude number at point 2

5.1 INTRODUCTION

A 2-to 4-year-old child fails conservation problems. In many cases they do not understand that same tea quantity just changes volumetric shape when poured from a big mug to a small mug! This means their cognitive development has not yet marked logical thought! The understanding of conservation problems is important in hydraulic system design, that is, conservation of energy and momentum. Where solutions to hydraulic problems fall short of **energy equations, momentum equations** are applied. Therefore, these equations are applied to route stormwater flow. Figure 5.1 gives a simplified explanation of how water in

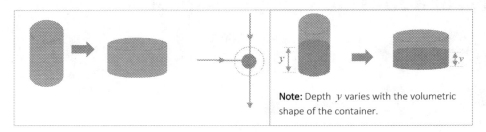

Figure 5.1 A conceptual diagram illustrating mass conservation.

high container changes volumetric shape when poured in a wide open container while conserving the quantity. The depth varies in both. Conservation of energy/momentum at a node junction, which could be a manhole or detention pond in a stormwater system, derives from the concepts of mass and energy conservation on which flow depths analysis, flow volumes, specific energy, and routing techniques are based to solve drainage problems.

5.2 HYDRAULIC DESIGN—OVERVIEW

Local, environmental, engineering, and construction regulations, standards and codes of practice must be relied on to design and develop **effective, efficient**, and **economical** drainage systems meeting value for money. The famous Manning's empirical equation, which expresses flow velocity with structure's geometry to size hydraulic structures, is widely adopted. Depending on evaluation process, the designer initially conducts as provided in Chapter 1, the Froude number (F_r) can be fixed and iterations for sizing hydraulic structures must obey assumptions. In several instances, but not always, uniform structures are designed, and no intermediate controls are proposed except where steep slopes risk high velocities, thereby proposing controls and sometimes engineering flow as largely critical. Engineering flow as critical answers a number of questions surrounding sustainability, especially in reducing and optimizing storm sewer capacities.

In the case of using a design software such as AutoCAD Storm and Sanitary Analysis, link output would be flooded or surcharged or OK. Therefore, iterations would be done accordingly. A choice would be made for a pipe to flow with inlet or outlet control. However, at large, pipe networks are designed to flow with outlet control—adopting partially full equations. This is because in outlet control the pipe can either flow full or partially full. In this case water is relatively deep and slower, and a disturbance propagates upstream. The rim elevation and invert elevations fixed for a node would influence available headwater at a node. When using a software program, the designer feeds[1] link length,[2] roughness factor,[3] invert elevations at inlet and outlet of link,[4] inlet and outlet types, pipe diameter (or channel width and height), and flow quantity to the model system. Self-cleansing velocities are considered in the range of 0.3 > v < 6 m/s avoiding erosion and pipe corrosion, and where velocities are considered high, controls must be proposed. In that case, exercising enough care while specifying construction slopes, inlet and outlet types for each drainage structure/pipeline is a must in order to ascertain available energy that would safely convey stormwater runoff downstream with sufficient velocities to move silt/sediment that may come with the runoff.

Hydraulics design principles 155

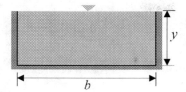

Figure 5.2 Demonstrating U-channel hydraulic radius.

Some of the common assumptions in drainage network design include the following:

- Neglecting lateral inflows.
- Proposing uniform conduits/drains.
- Largely normal flow conditions considered, and largely no intermediate controls.
- Engineering flow as critical.
- U-channels and circular pipes/sections are commonly considered.
- Considering concrete circular pipes, u-concrete drains (u-drains), and trapezoidal concrete/stone-pitched drains.

Engineering judgment is always exercised through taking enough care on specifying construction slopes by weighing on cost for a given hydraulic radius especially conduits; a large pipe on a flat slope can carry same flow as a smaller pipe on a steeper slope. To minimize earthworks/construction cost, easement cost, enough care is always exercised on optimal hydraulic radius, scaling b (Figure 5.2). This is a primary consideration meeting sustainability goals going forward.

The best hydraulic cross-section as far as sustainability goals are concerned going forward is the U-section and is evident as follows. The discharge will be a maximum for maximum hydraulic radius for a channel of given area and slope. Since maximum hydraulic radius for a channel of fixed area also requires minimum wetted perimeter, finding R_{max} will not only maximize discharge, but also minimizes the cost of construction hence reducing capital carbon. Imagine starting with a trapezoidal section shown in Figure 5.3, the proportions that will maximize R are obtained henceforth.

In Figure 5.3,

$$P = b + \frac{2y}{\cos(90-\alpha)}$$

$$A = by + \frac{y^2}{\tan \alpha}$$

Where,

$$b = \frac{A}{y} - \frac{y}{\tan \alpha}$$

$$R = \frac{A}{P} = \frac{A}{A/y - y/\tan \alpha + 2y/\cos(90-\alpha)}$$

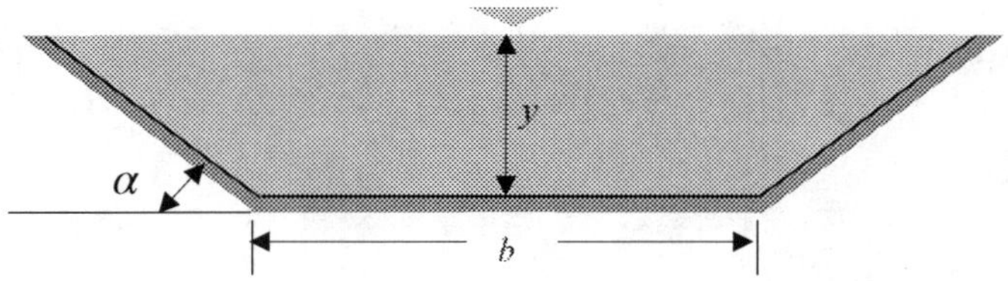

Figure 5.3 Trapezoidal section.

For R_{max}, $dR/dy = 0$. Then,

$$\frac{dR}{dy} = \frac{-A\left[-A/y^2 - 1/\tan\alpha + 2/\cos(90-\alpha)\right]}{\left[A/y - y/\tan\alpha + 2y/\cos(90-\alpha)\right]^2} = 0$$

$$A = y^2\left[\frac{-1}{\tan\alpha} + \frac{2}{\cos(90-\alpha)}\right]$$

$$R = \frac{y^2\left[-1/\tan\alpha + 2/\cos(90-\alpha)\right]}{2\left[-y/\tan\alpha + 2y/\cos(90-\alpha)\right]} = \frac{y}{2}$$

For maximum flow, a trapezoidal channel should be proportional so that the hydraulic radius equals one-half of its depth. As shown in Figure 5.2, a **rectangle is a special case of the trapezoidal** for which $\alpha = 90°$, hence $R = \frac{y}{2}$ for the best hydraulic section. R, for a rectangular section is also given by:

$$R = \frac{by}{b+2y}$$

$$\frac{y}{2} = \frac{by}{b+2y}$$

$$b = 2y$$

Therefore, the best and **economical** hydraulic section is a rectangular channel whose depth of flow is one-half of its width.

5.3 ENERGY EQUATIONS

As shown in Figure 5.1, energy equations are derived from theories of classical mechanics of fluid flow. The foundational axioms of fluid dynamics are the conservation laws, specifically, conservation of mass, conservation of linear momentum, and conservation of energy (also known as first law of thermodynamics). In case of using hydraulic design software platforms, proposed drainage networks are populated with survey data producing 1-D hydraulic models. Reinforced cement concrete (RCC) pipes and U-concrete channels are always the best choices going by sustainability goals, efficiency, and effectiveness. This is because U-channels' geometry provides a versatile situation in easing calculations (hydraulic radius) and also fitting within local site constraints, that is, available easements for stormwater sewer lines. For hydraulic performance, a circular pipe is the best geometric section among all.

The governing equations for nonuniform open channel flow commonly adopted are outlined from Equations (5.1) to (5.6):

$$\frac{dE}{dx} = S_0 - S_f \tag{5.1}$$

This can be expanded to,

$$\frac{\Delta E}{\Delta x} = \frac{\Delta\left(y + V^2/2g\right)}{\Delta x} = S_0 - \left(\frac{Vn}{R^{2/3}}\right)^2 \tag{5.2}$$

Also,

$$\frac{dE}{dy} = 1 - F_r^2 \tag{5.3}$$

Where,

$$F_r^2 = \frac{\alpha V^2}{g(A/B)} \tag{5.4}$$

Therefore,

$$\frac{dy}{dx} = \frac{S_0 - S_f}{dE/dy} = \frac{S_0 - S_f}{1 - F_r^2} \tag{5.5}$$

Manning's equation (Metric system)

$$V = \frac{1}{n} R^{2/3} S^{1/2} \tag{5.6}$$

5.3.1 Relevance of energy equations

Equations (5.1) and (5.2) are used to locate water flow profiles for open channel flow. In Equation (5.2), the friction slope keeps varying. This is because the ground slope keeps changing although in relatively small changes. The friction slope is usually taken as an average or approximation of the slope taken between two points. Figure 5.4 is a conceptual diagram showing a drain connecting through points A, B, C, D, and E which are relatively close to each other; the designer would take the average slope between points A and E. However, the slope between points A and E might slightly vary and not smooth. In theory, this variation in friction slope is responsible for creating nonuniform flow. Therefore, when locating the flow profile, small increments of the friction slope are usually taken as ΔS_f and the bed slope is in most cases considered a constant.

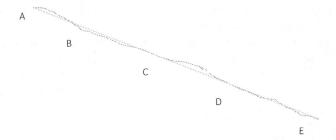

Figure 5.4 Drain passing through terrain of slopes with miniature changes.

- Equations (5.3) and (5.4) are used to forecast flow regime as critical, subcritical, or supercritical.
- Equation (5.5) provides the relationship between depth and distance from Equations (5.1) and (5.3).
- Equation (5.6) is an empirical equation that expresses flow velocity with channel geometry (applicable to both pipes and channels).

5.3.2 Derivation of key energy equations

From Figure 5.5, Total Head H,

$$H = Z + E \qquad (5.7)$$

Where,

$$E = y + \alpha \frac{V^2}{2g} \text{ (A simplification of Bernoulli's equation)}$$

Differentiating Equation (5.7) with respect to x

$$\frac{dH}{dx} = \frac{dZ}{dx} + \frac{dE}{dx}$$

The slope of the energy grade line and channel bed are downward in the direction of x as shown in Figure 5.6.

$$\frac{dH}{dx} = \text{Friction slope}, S_f$$

$$\frac{dZ}{dx} = \text{Bed slope}, S_0$$

$$\Rightarrow -S_f = -S_0 + \frac{dE}{dx}$$

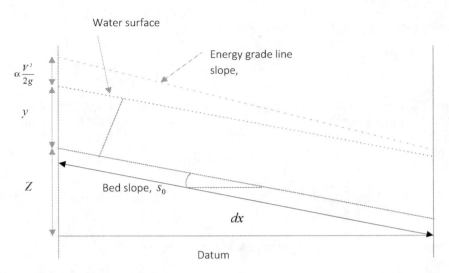

Figure 5.5 Derivation of energy in open channel flow.

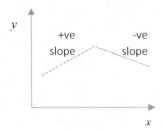

Figure 5.6 Slope of the energy grade line and channel bed.

$$\Rightarrow \frac{dE}{dx} = S_0 - S_f \qquad (5.8)$$

This is the equation for nonuniform flow which is used to locate the longitudinal surface water flow profile. The sign of $(S_0 - S_f)$ can vary and additional information can be obtained from books specialized on hydraulics.

Also,

$$\frac{dE}{dy} \cdot \frac{dy}{dx} = S_0 - S_f$$

$$\frac{dy}{dx} = \frac{S_0 - S_f}{(dE/dy)}$$

5.3.2.1 Specific energy

Specific energy is given by the formula below:

$$E = y + \frac{\alpha V^2}{2g}$$

Note: The coefficient on the velocity head terms arise from the integration of Euler's differential equation over the cross-sectional area of flow and reflect the nonuniform velocity distribution which exists in the flow of real fluids.

Therefore,

$$E = y + \frac{\alpha}{2g}\left(\frac{Q}{A}\right)^2 \tag{5.9}$$

Differentiate Equation (5.9) with respect to y

$$\frac{dE}{dy} = 1 - \frac{\alpha}{g}\frac{Q^2}{A^3}\frac{dA}{dy}$$

Note: Q is a constant and A depends on y
So,

$$\frac{dE}{dy} = 1 - \frac{\alpha Q^2}{g}\frac{B}{A^3}$$

B is the breadth of the water surface equaling to dA/dy.
For critical depth,

$$\frac{dE}{dy} = 1 - \frac{\alpha Q^2}{g}\frac{B}{A^3} = 0$$

$$\Rightarrow \frac{A_c^3}{B_c} = \frac{\alpha Q^2}{g}$$

Where, A_c and B_c are values when $y = y_c$.
So,

$$\frac{dE}{dy} = 1 - F_r^2$$

$$F_r^2 = \frac{\alpha V^2}{g(A/B)}$$

Where, F_r is the Froude number.

- For subcritical flow, $F_r < 1$
- For supercritical flow, $F_r > 1$
- For critical flow, $F_r \approx 1$ (here, flow velocity equals the wave celerity)

The Froude number is a measure of the ratio of inertial to gravitational forces, and arising from this, it gives the ratio of mean velocity of flow to the celerity of small surface waves. Consequently, while waves can travel upstream in subcritical flow, they cannot do so in supercritical flow.

Note:

- The depth of flow in a channel or partly full pipe is rarely uniform.
- y_n depends on S_0 but y_c does not.
- Bed slopes which make $y_n > y_c$ are mild, those which make $y_n < y_c$ are termed steep.
- Value of S_0 which makes $y_n = y_c$ is called the critical slope.

As shown in Figure 5.2, rectangular channels are the commonest structures fitting well enough in sustainability goals. For rectangular channels or U-channels,

$$A = by$$

$$V = \frac{1}{n} R^{2/3} S^{1/2} \text{ (Metric system)}$$

$$V = \frac{1}{n}\left(\frac{by}{2y+b}\right)^{2/3} S^{1/2}$$

For critical flow,
Since

$$\frac{A_c^3}{B_c} = \frac{\alpha Q^2}{g}$$

And

$$A_c = by_c$$
$$B_c = b$$

Taking $\alpha = 1$,

$$\frac{(by_c)^3}{b} = \frac{Q^2}{g}$$

$$y_c^3 = \frac{Q^2}{b^2 g}$$

$$y_c = \sqrt[3]{\frac{q^2}{g}}$$

Where,

$$q = \frac{Q}{b} \text{ (Specific discharge)}$$

In a similar manner, if one considers how q varies with y for constant E by rewriting Equation (5.9) as,

$$q^2 = 2gy^2(E-y), \text{ and taking } \alpha = 1.$$

Differentiating and equating to zero, and solving for y, we obtain;

$$q^2 = 2gy^2 E - 2gy^3$$
$$2\frac{dq}{dy} = 4Egy^3 - 6gy^4 = 0$$

Therefore,

$$2E - 3y = 0$$
$$y_c = \frac{2}{3}E$$

The graph of y versus q for constant E is shown in Figure 5.7. The two conditions illustrated in Figures 5.7 and 5.10 (minimum E and maximum q) are, in fact, different representations of the same circumstance. The point of minimum specific energy and maximum discharge is called critical flow. The velocity corresponding to this condition is called critical velocity, and for a rectangular channel is given by:

$$V_c = (gy_c)^{0.5}$$

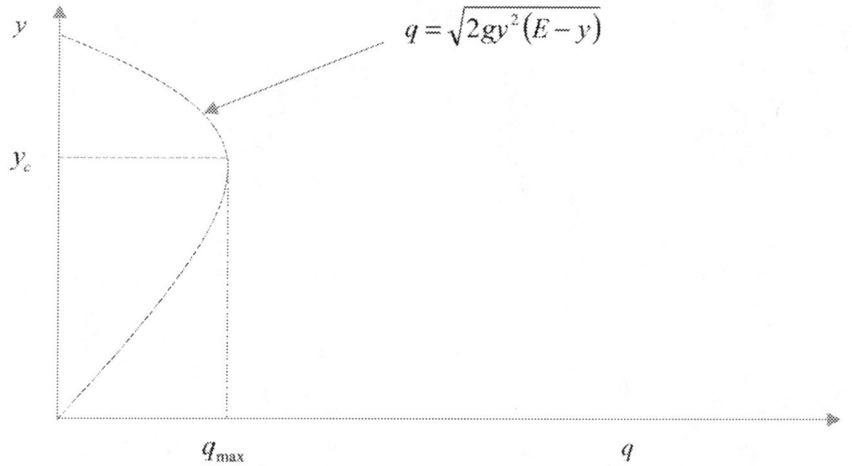

Figure 5.7 Discharge versus depth for constant specific energy.

5.3.3 Normal depth used for design and analysis

The depth of flow usually approaches the normal depth asymptotically as shown in Figure 2.10. Normal flow is needed in analyzing nonuniform flow because it indicates the depth toward which the flow is tending. In pipes, normal flow is calculated for a given discharge, diameter, and slope. The discharge Q_f, which the pipe would carry if it were running full, is found using the given diameter and slope. The actual discharge Q_p is used to find Q_p/Q_f and the graph gives the corresponding value of y/d. This can also be obtained from normograms shown in Figure 5.8 or from design charts in Figure 5.9 (variation of flow and velocity with depth in circular pipes). Figure 5.9 and 5.13 are partial flow diagrams which give only approximate results, particularly for high velocities. Therefore, in practical designs it is always good practice to incorporate a factor of safety to ensure that the storm sewers never flow full. In many cases, computed flows may vary from actual flows due to wave formation and several other unique factors which are usually not well defined in small channels.

Important: The normograms are very helpful in design and auditing drainage projects.

5.4 FLOW PROFILES/REGIMES FOR A SUSTAINABLE DESIGN

For design purposes, the design follows assumptions—an extensive sound evaluation process where flow through the hydraulic structures can be assumed to be normal, critical, subcritical, and supercritical. The depth versus specific energy at constant discharge is given in Figure 5.10.

Primary goal: Following the evaluation process (Chapter 2) and by fixing most appropriate Froude number, flood waves as they propagate through the structure must be predicted to be conserved within structure and therefore should not get out in open channels or be pressurized in conduits. Hence, dimensions output by Manning's equation must obey program conditions set for a reliable structure, using a software.

From Equation (5.3),

$$\frac{dE}{dy} = 1 - F_r^2$$

Suppose,

Scenario A: Critical flow; $F_r^2 \approx 1 \Rightarrow \dfrac{dE}{dy} = 0$, bed slope = friction slope; normal flow.

Scenario B: Subcritical flow; imagine $F_r^2 = 0.5 \Rightarrow \dfrac{dE}{dy} = 0.5$

Scenario C: Supercritical flow; imagine $F_r^2 = 1.5 \Rightarrow \dfrac{dE}{dy} = -0.5$

Critical depth depends not on slope; therefore, for **rectangular channels** a quick estimate can be established following Equations (5.10) and (5.11).

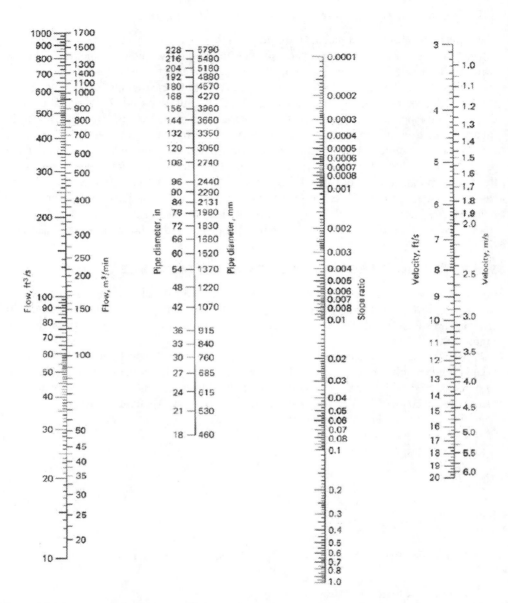

Figure 5.8 Normograms for solution of Manning's equation for circular pipes flowing full (n = 0.013).

Hydraulics design principles 165

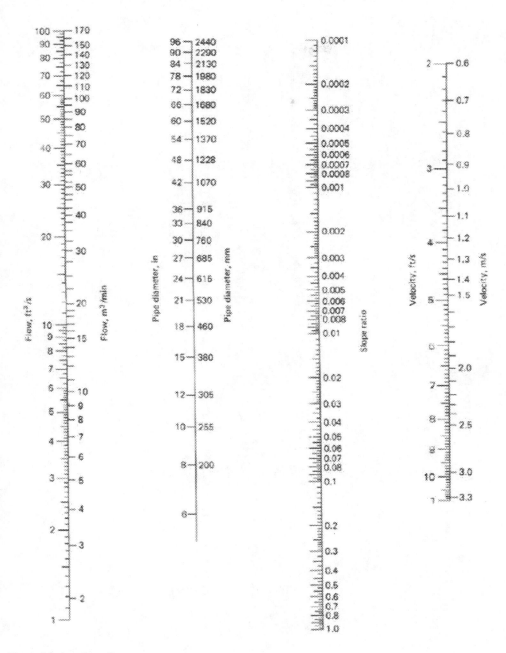

Figure 5.8 (continued)

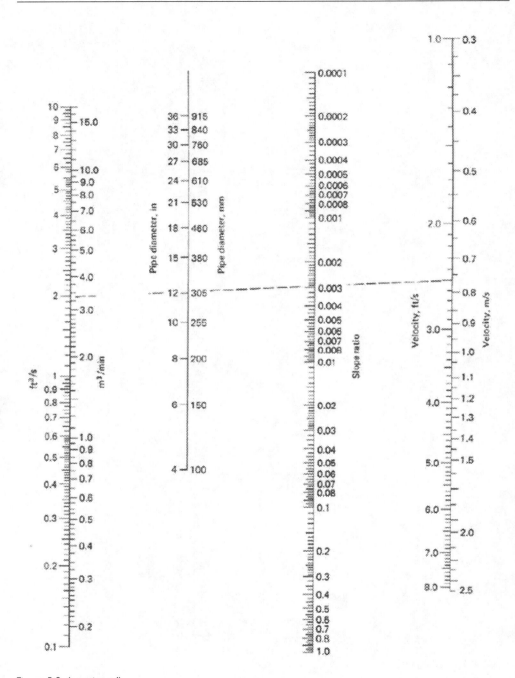

Figure 5.8 (continued)

Hydraulics design principles

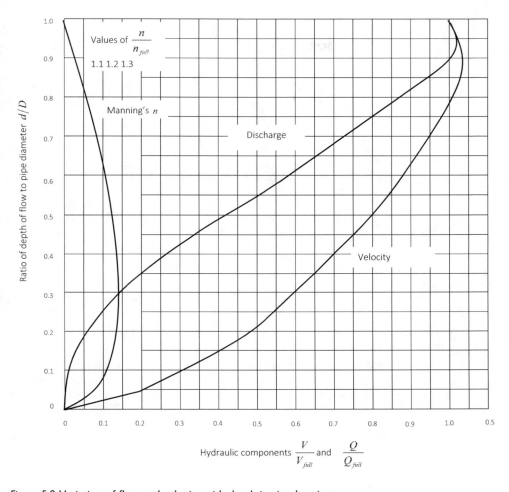

Figure 5.9 Variation of flow and velocity with depth in circular pipes.

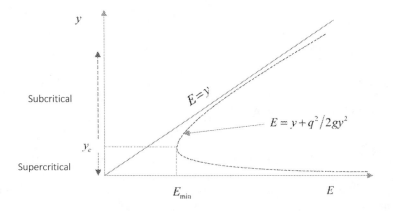

Figure 5.10 Depth versus specific energy at constant discharge.

$$y_c = \sqrt[3]{\frac{q^2}{g}} \qquad (5.10)$$

Also,

$$y_c = \frac{2}{3}E \qquad (5.11)$$

$F_r^2 = \dfrac{\alpha V^2}{g(A/B)}$ for all geometric shapes and for rectangular flows.
$V = Q/by = q/y$, where q is discharge per unit width and b is the breadth.

$$F_r^2 = \frac{\alpha V^2}{g(A/B)} = \frac{\alpha V^2}{g(by/b)} = \frac{\alpha V^2}{gy} = \frac{\text{inertial forces}}{\text{gravitational forces}}$$

Considering scenario A
Critical flow, $F_r^2 \approx 1$; for $\alpha = 1$

$$\therefore \frac{V^2}{gy} = 1, \Rightarrow V = (gy)^{0.5}$$

Using Manning's equation in the dimensions computation for a U-channel flow of breadth b and height y to carry a peak flow of Q m^3/s for typically a 25-year average recurrence interval (ARI) for primary structures, we can generate Equations (5.12) and (5.13).

$$V = \frac{1}{n} R^{2/3} S^{1/2} \text{ (Metric units)}$$

$$\Rightarrow (gy)^{0.5} = \frac{1}{n}\left(\frac{by}{2y+b}\right)^{2/3} S^{0.5} \qquad (5.12)$$

And

$$Q = VA$$

$$\Rightarrow Q = (gy)^{0.5}(by) \qquad (5.13)$$

In normal steady flow conditions, $S = S_0$ in Equation (5.12). Table 5.1 provides Manning's roughness coefficients for pipe flows.

Considering scenarios B and C
In same way, $F_r^2 > or < 1$ generates an initial condition with a primary goal of conserving the flows. Check dams in channels and backdrop manholes on conduit lines are applied for steep terrain/sloping (Figure 5.11), following the evaluation process in Chapter 2. The Froude F_r number choice depends on the evaluation process, that is, channel length, site steepness,

control size, flow quantity, and other localized constraints. The evaluation process is explicitly provided in Chapter 2.

Therefore, Equations (5.12) and (5.13) are modified as well for $F_r^2 > and <1$ govern the sizing of hydraulic structures, that is, conduits and drains. The geometric shapes parameters such as the hydraulic radius can always be input to modify the equations. The choice of the Froude number influences the routing technique (see Chapter 6) and vice versa. That consequently influences the size of hydraulic structures and the overall embedded carbon. The primary goal is to conserve the flood waves at a reasonable cost. As explained in Section 5.1, the flood waves must be conserved, and this requires an iterative process to achieve an optimal solution which is sustainable—back-and-forth algorithms must be built. Therefore, the choice of the Froude number and routing technique is key to economically conserve flood waves within the system.

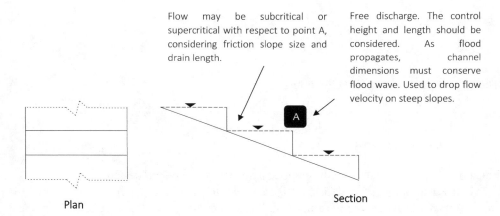

Figure 5.11 Drain with controls.

Table 5.1 Manning's roughness coefficient, n, for pipe flows

Type of pipe	Manning's n	
	Min.	Max.
Glass, brass, or copper	0.009	0.013
Smooth cement surface	0.010	0.013
Wood-stave	0.010	0.013
Vitrified sewer pipe	0.010	0.017
Cast iron	0.011	0.015
Precast concrete	0.011	0.015
Cement mortar surfaces	0.011	0.015
Common clay drainage tile	0.011	0.017
Wrought iron	0.012	0.017
Brick with cement mortar	0.012	0.017
Riveted steel	0.017	0.020
Cement rubble surface	0.017	0.030
Corrugated mortar storm drain	0.020	0.024

Figure 5.12 Pipe flowing partially full.

Important: The length of a channel affects its carrying capacity in circumstances where the available energy is fixed. **See Figure 6.8**.

Conduits design

For conduits (Figure 5.12), and considering Manning's equation, the ratio of partial flow to full flow in pipes is

$$\frac{Q_P}{Q_f} = \frac{n_f A_P R_P^{2/3}}{n_P A_f R_f^{2/3}} \tag{5.14}$$

The ratio of partially full velocity to full velocity is

$$\frac{V_P}{V_f} = \frac{n_f R_P^{2/3}}{n_P R_f^{2/3}}, \text{ in both of which } R \text{ is hydraulic radius, } A/P \tag{5.15}$$

Equations (5.14) and (5.15) are the chief equations used in conduit design.

In both Equations (5.14) and (5.15), R is the hydraulic radius equal to A/P. Equations (5.14) and (5.15) are always solved to compare full flow to partially flowing sewers. In developing hydraulic models, normograms are always relied on wherever there is interconnectivity. Figures 5.9 and 5.13 are extensively used to design and audit projects. Figure 5.8 represents normograms for solution of Manning's equation for circular pipes flowing full (n = 0.013). Figure 5.9 represents the variation of flow and velocity with depth in circular pipes. Both figures are used for design and auditing of projects. The primary assumption is that flow tends to normal flow conditions.

In the case Q_p/Q_f is known, y and V_p/V_f can be obtained. Figure 5.13 provides the variation of R and A with depth for a circular pipe.

5.5 THE MOMENTUM PRINCIPLE

From Newton's second law of motion, **the rate of change of momentum** is equal to the **force acting on an object**, that is

$$\sum F = \Delta MV \tag{5.16}$$

Some stormwater runoff flow problems fall short of energy equations, and momentum equations are deemed necessary. The fast change of momentum due to flash floods, for

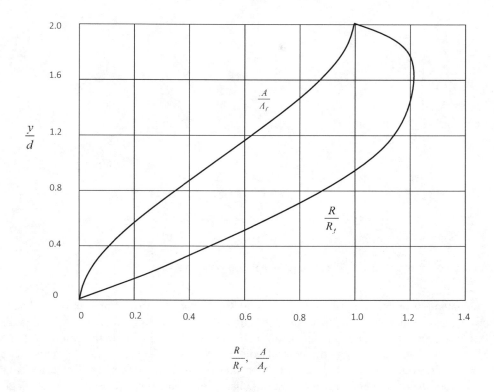

Figure 5.13 Variation of R and A with depth in circular pipes.

example, is responsible for the impact force on properties. The world is increasing experiencing flash floods as result of climate change.

Key terms

1. Stationary control volume.
2. Incompressible fluid.
3. Rate of change of momentum equals the total force acting on the fluid element.
4. Pressure acts at the centroid of cross-section, A.

Starting point: Define a stationary control volume as shown in Figure 5.14. The first step is to define a stationary control volume. The total forces acting on the stationary control volume **considering** *x*-**direction** components are as follows:

a. Hydrostatic forces. These are forces due to fluid pressure.
b. Friction forces.
c. Gravity forces.
d. Reactive forces due to pressure on the bed and walls.

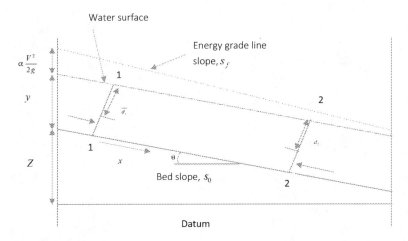

Figure 5.14 Forces on fluid in a control volume for open channel flow.

Momentum principle is applied by defining a stationary control volume in open channel problems. Component of the total force on the fluid in the control volume in a specified direction is equated to the rate of change of momentum of the fluid within the control volume. Momentum constantly enters the control volume **at 1.1** and leaves **at 2.2**. In many cases, gravity forces and boundary forces (friction and reactive forces) are neglected and thus hydrostatic forces are equated to the rate of change of momentum. Reason: they are negligible as compared to hydrostatic forces.

Hydrostatic forces are derived from force per unit area acting on the centroid of the stationary control volume at the entrance and exit. The pressure will be hydrostatically distributed in the control volume. Therefore, the force may be written as the product of the pressure at the centroid of the cross-section, below the surface, and the area of cross-section, that is
At 1.1 and 2.2,
Hydrostatic forces are

$$F = h\rho g A$$

Where,

h : Effective water depth, denoted as y, Figure 5.14
ρ Density of water
g Acceleration due to gravity
A Cross-sectional area

Therefore,

$$F_1 = \rho g A_1 \overline{d_1} \quad \text{and} \quad F_2 = -\rho g A_2 \overline{d_1}$$

Resultant hydrostatic force,

$$F_1 + F_2 = \rho g \left(\overline{d_1} - \overline{d_2} \right)$$

Rate change of momentum flow,

$$\Delta M = \rho Q V$$

Velocity is considered uniform but in reality, it is not. In reality, velocity varies across the cross-section and **momentum flux** would be written as,

$$\int \rho V^2 dA$$

This is practically written as,

$$\beta \rho Q V$$

V is the mean velocity $= \dfrac{Q}{A}$ or $V \int \dfrac{1}{A} dA$

The Boussinesq velocity distribution coefficient, $\beta = \int v^2 \dfrac{dA}{(V^2 A)}$

The coefficient has a value of about 1.12 for ordinary channel flows.

Equating forces

Equating all x-component forces to the rate of change of momentum, we obtain the following:

$$\rho g \overline{d_1} A_1 - \rho g \overline{d_2} A_2 + \text{boundary and gravity forces} = \beta_2 \rho Q V_2 - \beta_1 \rho Q V_1$$

Therefore, the sum of hydrostatic forces and momentum flux for a specific cross-section of flow has a composite characteristic of flow at that section called 'pressure-flux-momentum'.

Collecting terms relating to inflow and outflow sections and dividing each term by ρg we obtain the following:

$$\overline{d_1} A_1 + \dfrac{\beta_1 Q V_1}{g} + \dfrac{\text{boundary and gravity forces}}{\rho g} = \overline{d_2} A_2 + \dfrac{\beta_2 Q V_2}{g}.$$

In several applications boundary and gravity forces are negligible, hence the result is called specific force:

$$F = \overline{d} A + \dfrac{\beta Q^2}{g A} \tag{5.17}$$

That implies that at Sections 1 and 2, specific force $F_1 = F_2$. Specific force assumes boundary and gravity forces are negligible in most applications. Thus, pressure and momentum flux are equated. The specific force has an analogous property to specific energy E in that for a given flow there is a critical depth which makes F a minimum. For larger values of F, two values of depth give the same specific force. In Figure 5.15, for larger F values, two depth values will

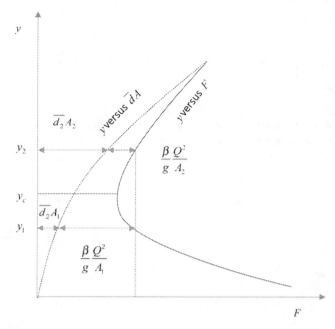

Figure 5.15 Relationship of depth of flow with specific force.

give equal specific force. Depth at the lower end is always supercritical flow and at the upper end flow is subcritical. **For rectangular cross-section,**

$$F = \frac{1}{2} y^2 b + \frac{\beta Q^2}{gby} \tag{5.18}$$

Employing the Froude number,

$$F = y^2 b \left(\frac{1}{2} + F_r^2 \right)$$

Since,

$$F_r^2 = \frac{Q^2}{\left(gb^2 y^3\right)} \tag{5.19}$$

$$\Rightarrow \frac{F}{b^3} = \left(\frac{y}{b}\right)^2 \left(\frac{1}{2} + F_r^2\right) \tag{5.20}$$

5.5.1 Application of momentum equations

The momentum equation is used to determine the resultant force exerted on the boundaries of a flow passage by a stream of flowing fluid as the flow changes its direction or the magnitude of velocity or both. Momentum equations are applied in the following scenarios:

a. Analyzing junctions, manholes, inlets, and outlets/outfalls

The analysis of junction applies the momentum function, and it is a highly theoretical approach which can be obtained from books specialized on hydraulics. It is of course important to appreciate that the momentum function is versatile and can be applied to analyze nodes within in a hydraulic model.

b. Channels receiving flow along their length—lateral inflows

Lateral inflows are shown in Figure 5.16. The flow increases linearly along the channel and it is being collected in a node downstream of the channel. Assuming the bed slope is zero, the discharge is also zero at a point most remote from the collecting channel outfall, that is, upstream of the channel. There are no gravity force in the x-direction since the bed slope is zero. Assume that the channel is of uniform width and friction and boundary forces are neglected. Since the friction and boundary forces are also neglected and the wall is prismatic, there are no x-direction forces generally.

Therefore, considering the upstream and downstream of a channel,

$$F_1 = F_2 \tag{5.21}$$

From Equations (5.20) and (5.21), because friction is neglected, and the channel is prismatic,[5] thus,

$$\left(\frac{y_1}{b}\right)^2 \left(\frac{1}{2} + F_{r_1}^2\right) = \left(\frac{y_2}{b}\right)^2 \left(\frac{1}{2} + F_{r_2}^2\right) \tag{5.22}$$

The upstream end of the channel has no flow, thus $F_{r_1} = 0$, and hence

$$y_1 = y_2 \sqrt{\left(1 + 2F_{r_2}^2\right)} \tag{5.23}$$

y_2 and F_{r_2} are controlled by the conditions downstream of the outlet. These will be known from the design of the downstream channel. Often, the outlet is in the form of a step so that y_2 is the critical depth, and provided the step is not drowned, then downstream conditions have any influence. In this case,

$$F_{r_2}^2 = F_c^2 = 1.0$$

Also,

$$y_1 = y_c \sqrt{3}$$

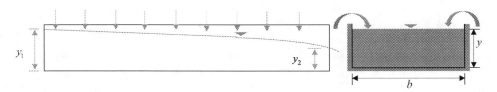

Figure 5.16 Channel with spatially increasing flow.

Equation (5.5) provides the variation of flow depth with distance for nonspatially increasing flow (no lateral inflows). Equation (5.24) is a complete equation for spatially increasing flow. In order to account for friction or nonzero bed slopes, the Equation (5.24) is integrated arithmetically.

$$\frac{dy}{dx}\left(gA^2 - \frac{Q^2}{D}\right) = gA^2\left(S_0 - S_f\right) - 2Q\frac{dQ}{dx} \qquad (5.24)$$

In which Q, A, and D all vary with x. The full derivation of this equation can be obtained from books specialized on hydraulics.

c. Channels discharging flow along their lengths
The equation of spatially decreasing flow is given by Equation (5.25):

$$\frac{dy}{dx}\left(gA^2 - \frac{Q^2}{D}\right) = gA^2\left(S_0 - S_f\right) - Q\frac{dQ}{dx} \qquad (5.25)$$

Spatially decreasing flow occurs in channels with side-weirs for overflowing stormwater both in sewerage systems and at treatment works, specifically for partially combined or combined systems. In such cases, however, there are additional difficulties. The sideways flow itself depends on the depth, y so in principle another equation relating dQ/dx to y is needed. Also, the sideways flow retains some x-direction momentum.

d. Hydraulic jumps
A hydraulic jump is a phenomenon in the science of hydraulics which is frequently observed in open channel flow such as rivers and spillways.[1] When liquid at high velocity discharges into a zone of lower velocity, a rather abrupt rise occurs in the liquid surface. For example, hydraulic jumps are typical features of river rapids where the water swirls and foams around rocks and logs. In open channel flows, rocks, logs, sediment, and garbage may cause hydraulic jumps.

Generally, when stormwater runoff at high velocity discharges into a zone of lower velocity, an abrupt rise occurs in the water surface. That is called a hydraulic jump. The rapidly flowing water is abruptly slowed and increases in height, converting some of the flow's initial kinetic energy into an increase in potential energy, with some energy lost through turbulence to heat. In an open channel flow, this manifests as the fast flow rapidly slowing and piling up on top of itself similar to how a shockwave forms. The interruption of flow patterns greatly reduces the kinetic energy of the water. The momentum equation can be used to determine the depth of flow across a transition in which the energy loss is unknown. In real-life applications, for open channel flow, a hydraulic jump occurs in the following situations:

- Downstream of the sluice gate
- At the foot of the spillway
- When the gradient suddenly changes from a steep to a flat slope
- When a rapid flow meets a streaming flow having a larger depth

Hydraulics design principles 177

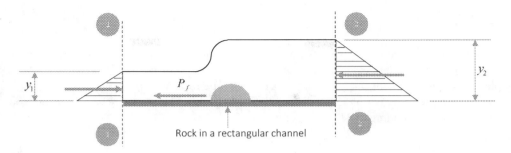

Figure 5.17 Demonstrating a hydraulic jump.

The assumptions in the analysis of a hydraulic jump are as follows:

- The slope of the channel is very small, so the corresponding weight component in the direction of flow is negligible.
- No friction on the sides and channel bed.
- The flow is uniform.
- The hydraulic jump occurs suddenly.
- The pressure distribution is hydrostatic before and after the jump.

Figure 5.17 shows an object such as a rock in the course of flow in a rectangular open channel. Water creates a hydraulic jump when it meets the rock as it moves from point 2 to point 1. The depth of flow at point 2 is y_2 and at point 1 it is y_1. P_f is the force that the object induces on the flow and P_f/b is the force per unit width of the channel.

Neglecting boundary and gravity forces because in most applications they are negligible, Equation (5.16) can be written as:

$$\gamma \frac{y_1^2}{2} \times b - P_f - \gamma \frac{y_2^2}{2} \times b = Q\rho \Delta V$$

$$\gamma \frac{y_1^2}{2} \times b - P_f - \gamma \frac{y_2^2}{2} \times b = qb\rho \left(\frac{q}{y_2} - \frac{q}{y_1} \right)$$

$$\frac{P_f}{b} = \gamma \left[\left(\frac{q^2}{y_2} + \frac{y_2^2}{2} \right) - \left(\frac{q^2}{y_1} + \frac{y_1^2}{2} \right) \right] \qquad (5.26)$$

In case, $P_f/b = 0$, then the right-hand side of the equation equals to zero and

$$\frac{q^2}{y_1} + \frac{y_1^2}{2} = \frac{q^2}{y_2} + \frac{y_2^2}{2} \qquad (5.27)$$

The function $\left(q^2/y + y^2/2 \right)$ is called the *momentum function*

$$M = \frac{q^2}{y} + \frac{y^2}{2} \tag{5.28}$$

Therefore, once you know the specific discharge and depth upstream of a hydraulic jump, say at point 2, you can calculate the corresponding depth downstream at point 1. If M is plotted versus y for constant q, we obtain a curve of the form shown in Figure 5.18. As with the specific energy Equation (5.9), differentiating M with respect to y, we can determine the value of y corresponding to M_{min}. Therefore, at M_{min},

$$y = (q^2)^{1/3} \tag{5.29}$$

This indicates that the condition of minimum value of the momentum function corresponds to critical depth. The momentum function can often be used to advantage to determine the depth of flow across a transition in which the energy loss is unknown. The application of this relation to hydraulic jumps is a typical example of such problems.

e. Flash floods, property damage due to impact of flash floods

The impact from flash floods can be evaluated using the momentum equation. When flash floods hit properties, sometimes they create hydraulic jumps, or produce excessive impact that sweep away properties, collapse buildings, and retaining walls. The momentum equation can be applied to evaluate the impact of a given discharge on a stationary property, that is

$$F = Q\rho\Delta V \tag{5.30}$$

Sometimes, flash floods create hydraulic jumps when they strike objects especially in the horizontal plane. In that case, a hydraulic jump can be seen when flash flood water radiates outward as a result of flash floods striking the horizontal surface of a property. The flash floods may initially flow in a smooth sheet with consistent current patterns. In this region, the speed of the water exceeds the local wave speed. Friction against the object surface slows the flow until an abrupt change occurs. At this point, the depth increases as flash floods accumulate in the transition region and flow becomes excessively turbulent.

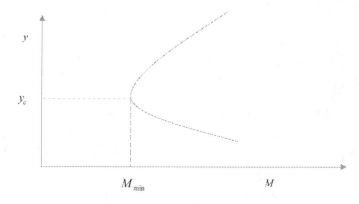

Figure 5.18 Momentum function versus depth at constant discharge.

f. Auditing projects

Both energy and momentum equations can be used to audit projects to investigate the performance of conduit or a drain through a systematic review of the drainage system. The hydraulic efficiency of the storm sewer can be audited taking field measurements and employing energy and momentum techniques. In some cases, technical audits accompany value for money audits and forensic audits. In drainage schemes, momentum and energy equations are useful techniques to forecast the hydraulic efficiency and capacity of a storm sewer. Depending on the size and complexity of the drainage project, momentum equations are versatile and can be used to estimate the hydrostatic pressures induced by a standing flood in a collection system and how the flood attenuates and propagates once released. Other applications of the momentum equation include analyzing flow though bend pipes, fluid flow though stationary and moving plates or vanes, and nonuniform flow through sudden enlarged pipes.

5.6 CASE STUDIES WITH WORKED EXAMPLES

5.6.1 Case study (5.1): Typical design project, proposed warehouse park on plot 31, Namanve, Kampala, Uganda (2021)

5.6.1.1 Overview

From the topographical survey produced (Figure 5.19), appropriate drainage routes for the proposed warehouse park on plot 31, Namanve, accompanied the architectural and structrual engineering proposals. The facilities proposed were truck service bay, pump stations, staff ffacilities block, truck washing bay, offices, cafteria, fuel tanks, and underground water tank. The 0.0202 km^2 estate would be covered with concrete pavers and roofs. Other data collected involved the rainfall received in the area and utility maps. This case study provides the drainage system evaluation process, design criteria, analysis and conclusion, and recommendations.

5.6.1.2 Evaluation

The estate was found to be free from underground utilities such as water pipelines, telecom, electricity, and gas, thus making it feasible to construct storm sewers. An investigation of the presence of underground utilities in the area was conducted through consultations with the council and reading available maps. The estate was not a flat surface as shown in the topographical survey (Figure 5.19). Its terrain required reducing by somewhat voluminous earthworks to level so that approved fill material would be imported and properly compacted. Thus, new levels would be obtained. For this particular design, a weighted runoff coefficient of 0.850 was adopted for the entire estate, holistically judged from the architectural master plan.

The estate would be independent of external flows and only runoffs generated from the estate would be safely routed off the impermeable surfaces to the existing council drains situated out of the estate. There would be foul water from the truck's washing bay and oil spillage would be expected. Any excess oil would be placed in containers and transported off-site. However, the foul water from the truck's washing bay and any excess oils would need oil and grease trap installed on the storm sewer line near the source. The oil/water separator would be besides the truck service bay, hence catering for sustainability goals going forward. See Section 11.2.11. Therefore, the drainage system was partially assumed to be a

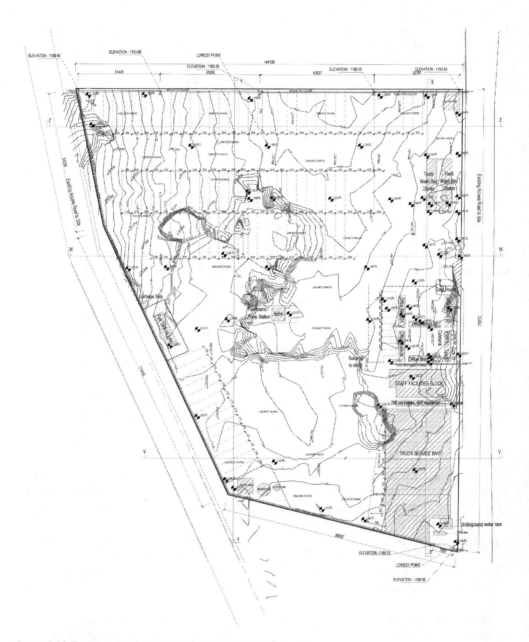

Figure 5.19 Topographical survey with some existing features.

combined system although foul water would be negligible/minimal, ably accommodated in the drainage system on both dry and wet days. It is a statutory requirement to ensure that the water discharged into the public storm sewer is free of any oil.

Intensity-duration-frequency (IDF) curves provided in Figure 5.20 were used for the rational method to determine the peak flows from each subbasin. From the drainage design

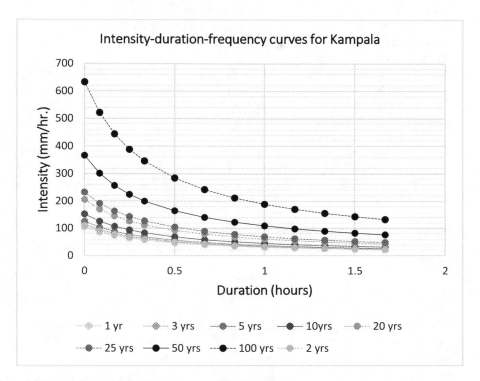

Figure 5.20 Intensity-duration-frequency curve for Kampala city.

manuals, that is, Uganda's Ministry of Works and transport drainage design manual (2010), an average recurrence interval of 25 years was considered. From the geotechnical investigation report, the water table was reported at 5-m depth throughout the estate. This made storm sewer routes viable throughout the estate because they would not pose any environmental hazard in that regard. That means the groundwater table would not be affected because excavations would stop at a maximum depth of about 1.25 m. The inlets would be located at manholes and throughout grated U-channels. Storm sewers were found feasible when located at the estate boundary because all runoff would end there as it gets out of the estate to the discharge points. The outfalls were located at the existing immediate council drains. There would be a precast concrete culvert sewer line and U-drain grated channels as shown in Figure 5.21 (Option A) or U-drains only in Figure 5.22 (Option B). Therefore, two options were evaluated.

5.6.1.3 General assumptions, conditions, and guiding design criteria

a. Normal flow conditions
    - Friction slope ≈ bed slope; Froude number = 1.
    - Critical depth ≈ normal depth.
    - Assumed critical slopes throughout.

b. Grated manholes to provide inlets for runoff and will control runoff at the entrance.
c. Grated U-concrete channels—grating to control entrance for runoff.
d. Neglected approach velocities to inlets.
e. Normal trench loading cases.
f. Inlet control conditions. Neglected head losses due to pipe and exit and entrance types.
g. Excavate up to 1.20-m depth.
h. No intermediate controls except manhole junctions connecting storm sewers and drains.
i. Neglected energy losses due to entrance and exit.
j. Neglected lateral inflows—and considered peak flows from subcatchments/catchment (routing the same hydrograph through each link).
k. The inflow hydrograph to each inlet node in a subbasin has the same peak flow for the subbasin.
l. Each manhole has a gutter inlet type.
m. Steady flow routing conditions. See Section 6.5.
n. Application of SuDS: Space constraints dictated the application of SuDS, that is, type and relevance.
o. SuDS considered for aesthetics (primarily for landscaping purposes) and harvested water used during dry days for use in the truck wash bay. This was neglected for overland flow sewer/drain sizing.
p. Oil/water separator provided for the oil from automobiles/near the truck service bay.
q. 1-D analysis considered.
r. Approximate design methods adopted.
s. Manning's n = 0.013.
t. Rational method, IDFs, and Kirpich's formula for time of concentration adopted.

5.6.1.4 Option A

See Figure 5.21.

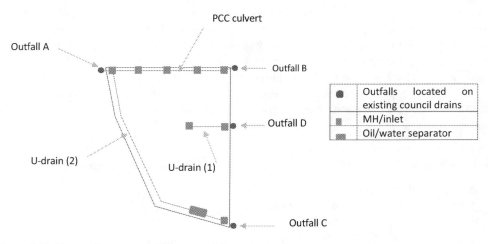

Figure 5.21 Option A—proposed location of drainage sewers.

5.6.1.5 Option B

See Figure 5.22.

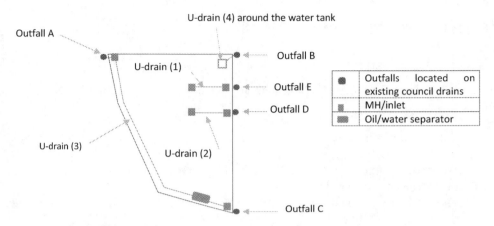

Figure 5.22 Option B—proposed location of drainage sewers.

5.6.1.6 Design

See Figure 5.23.

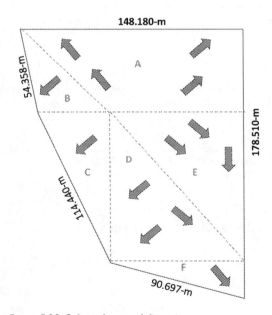

Total area = **0.0202-km²**
Using the Rational Method; discharges vary with catchment area.
$Q = 0.278CiA$

	Runoff flow direction
Sub-basin A	0.0067-km²
Sub-basin B	0.0017-km²
Sub-basin C	0.0034-km²
Sub-basin D	0.0034-km²
Sub-basin E	0.0034-km²
Sub-basin F	0.0017-km²

Figure 5.23 Subcatchment delineations.

184 Integrated Drainage Systems Planning and Design for Municipal Engineers

Figure 5.24 Isometric view of drainage sewers.

5.6.1.6.1 Option A—Part A: Storm sewers

Storm sewers located on the **148.180-m long estate wing** are visualized in Figure 5.24 (isometric view).

- Catchment/contributory area = 0.0067 km²
- n = 0.013
- Weighted runoff coefficient = 0.85 obtained from proposed site layout/physical plan.
- Subcatchment average slope = 1.857%
- Time of concentration for subcatchment A (L = 90 m, S = 1.875%), T_c = 2.882 minutes (by Kirpich's formula)
- From IDF curves (Figure 5.20), design intensity = 230 mm/h
- Peak discharge by rational method = **0.364 m³/s (21.848 m³/min)**
- The manhole (MH) rim elevation will be determined/varied on site as appropriate since it influences the available headwater at manhole.
- The peak flow is assumed to be accommodated throughout the culvert lines A to E below although in reality peak flow obtained from the subcatchment may not flow through all pipes—but considered worst scenario.
- Profile of the drainage sewer, Figure 5.25.

Culvert sizing
From normograph in Figure 5.8.

From	To	Q_p (m³/min)	Slope	Culvert diameter (m)	Velocity flowing full, (m/s)	Length of culvert line (m)	Capacity of sewer Q_p (m³/min)	% flow (Q_p/Q_f)	Ground elevations (m)		Invert elevations (m)	
C	B	21.848	0.021	0.460	2.600	38.30	25.000	87.39	1183.75	1182.50	1182.55	1180.90
B	A	21.848	0.041	0.380	3.250	31.41	23.000	94.99	1182.50	1180.50	1180.9	1179.30
C	D	21.848	0.005	0.615	1.550	43.53	28.150	77.61	1183.75	1183.50	1182.55	1182.30
D	E	21.848	0.005	0.615	1.550	32.77	28.150	77.61	1183.50	1183.00	1182.30	1182.10

From Figure 5.9, variation of flow and velocity with depth in circular pipes,

Hydraulics design principles 185

Line CD		Line DE	
$\dfrac{Q_p}{Q_f} = 0.7761$	Water surface elevation downstream = 1182.595 m Water surface elevation upstream = 1182.845 m	$\dfrac{Q_p}{Q_f} = 0.7761$	Water surface elevation downstream = 1182.395 m Water surface elevation upstream = 1182.595 m
$\Rightarrow \dfrac{y}{D} = 0.48$		$\Rightarrow \dfrac{y}{D} = 0.48$	
$D = 0.615\,\text{m}$		$D = 0.615\,\text{m}$	
$\Rightarrow y = 0.295\,\text{m}$		$\Rightarrow y = 0.295\,\text{m}$	

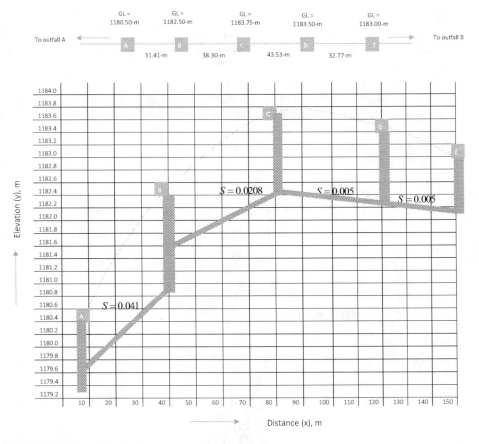

Figure 5.25 Profile of the drainage sewer.

Conclusion/recommendation

- To err on a safe side, use concrete culverts of **615-mm diameter** throughout the sewer line (A–E).
- Manhole B and A are backdrop manholes to minimize voluminous earthworks/excavations.
- The pipe size may reduce if construction slopes are found to increase on site, for some reasons.

Option A—Part B: U-Drain (1)
Grated U-drainage channel located **between office block and staff facilities block** is visualized in Figure 5.26 (isometric view)

- Slope = 1.25%.
- Contributory area **(subcatchment E)** = 0.0034 km^2
- Time of concentration from Kirpich's formula provided a design intensity off the IDF curve = 2.7776 minutes; (L = 70 m S = 1.25%).
- Design intensity = 235 mm/h., ARI = 25 years (Uganda's design manual, 2010)
- Weighted coefficient = 0.85 obtained from proposed site layout/physical plan..
- Peak discharge from catchment by rational method = 0.189 m^3/s.
- Breadth = 0.3-m (assumed breadth).
- Specific discharge for U-channel = 0.630 m^2/s.
- Thus, critical depth.

$$y_c = \sqrt[3]{\frac{q^2}{g}} = \sqrt[3]{\frac{(0.630)^2}{9.81}} = 0.343 \text{m}$$

Conclusion
A feasible drain should have an internal breadth of **300 mm and depth of 343 mm**.

Option A—Part C: U-drain (2)—steel grated
U-drain located at the longest wing of the estate. .

- Slope = 1.56%.
- Contributory area = **B + C + F + D** = 0.0102 km^2

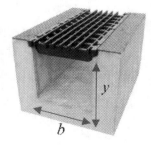

Figure 5.26 Isometric view of U-drain.

Hydraulics design principles 187

- Time of concentration from Kirpich's formula provided a design intensity off the IDF curve = 3.354 minutes; (L = 100 m and S = 1.56%).
- Design Intensity = 220 mm/h., ARI =25 years.
- Weighted coefficient = 0.85 obtained from proposed site layout/physical plan.
- Peak discharge from catchment by rational method = 0.530 m³/s.
- Breadth = 0.3 m (assumed breadth).
- Specific discharge for U-channel = 1.767 m²/s.
- Thus, critical depth.

$$y_c = \sqrt[3]{\frac{q^2}{g}} = \sqrt[3]{\frac{(1.767)^2}{9.81}} = 0.683 \text{m}$$

Conclusion
A feasible drain should have an internal breadth of **300 mm and depth of 683 mm**.

Option B—Design
From estate/catchment area, Figure 5.27.

Option B—Part A: U-drain (1), Figure 5.22.

- Slope = 1.25%.
- Contributory area **(subcatchment A)** = 0.0054 km²
- Time of concentration from Kirpich's formula provided a design intensity off the IDF curve = 1.804 minutes; (L= 40 m and S = 1.25%).
- Design intensity = 235 mm/h, ARI =25 years.
- Weighted coefficient = 0.85.

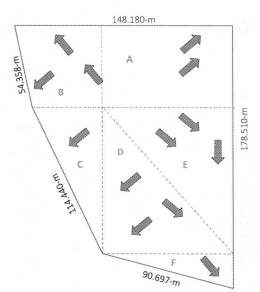

Total area = **0.0202-km²**

Using the Rational Method; discharges vary with variation in catchment area...

	Runoff flow direction
A	0.0054-km²
B	0.0044-km²
C	0.0034-km²
D	0.0034-km²
E	0.0034-km²
F	0.0017-km²

Figure 5.27 Option B—subcatchments delineations.

- Peak discharge from catchment by rational method = 0.300 m³/s.
- Breadth = 0.3 m (assumed breadth).
- Specific discharge for U-channel = 1.200 m²/s.
- Thus, critical depth.

$$y_c = \sqrt[3]{\frac{q^2}{g}} = \sqrt[3]{\frac{(1.00)^2}{9.81}} = 0.467 m$$

Conclusion
Adopt drain of internal breadth of 300 mm and depth of 467 mm.

Grated U-drainage channel located **next to the truck wash bay** is visualized in Figure 5.28 (isometric view).

Option B—Part B: U-drain (2), Figure 5.22
As designed in Option A—breadth 300 mm and depth 343 mm.

Option B—Part C: U-drain (3), Figure 5.22
U-drain located on the longest wing is visualized in Figure 5.29.

- Slope = 1.56%.
- Contributory area = B + C + F + D = 0.0129 km²
- n = 0.013.
- Time of concentration from Kirpich's formula provided a design intensity off the IDF curve = 3.354 minutes; (L= 100 m and S = 1.56%).
- Design intensity = 220 mm/h, ARI = 25 years (Figure 5.20).
- Weighted coefficient = 0.85 obtained from proposed site layout/physical plan.
- Peak discharge from catchment by rational method = 0.671 m³/s.
- Breadth = 0.3 m (assumed breadth).
- Specific discharge for U-channel = 2.235 m²/s.
- Thus, critical depth.

$$y_c = \sqrt[3]{\frac{q^2}{g}} = \sqrt[3]{\frac{(1.707)^2}{9.81}} = 0.799 m$$

Figure 5.28 Isometric view of U-drain, precast unit.

Conclusion

Adopt drain of internal breadth of 300 mm and depth of 799 mm.

Option B—Part C: U-drain (4)

- Slope = 1.25%.
- Contributory area = 0.0004 km²
- n = 0.013.
- Time of concentration from Kirpich's formula provided a design intensity off the IDF curve = 1.381 minutes; (L = 28.28 m and S = 1.25%).
- Design intensity = 238 mm/h, ARI = 25 years.
- Weighted coefficient = 0.85 obtained from proposed site layout/physical plan.
- Peak discharge from catchment by rational method = 0.022 m³/s.
- Breadth = 0.2 m (assumed breadth).
- Specific discharge for U-channel = 0.112 m²/s.
- Thus, critical depth.

$$y_c = \sqrt[3]{\frac{q^2}{g}} = \sqrt[3]{\frac{(0.112)^2}{9.81}} = 0.109m$$

It is important to make a small drain around the tank to drain excess runoff to outfall B.
Breadth = 200 mm
Depth = 109 mm

Option B—Part D: Oil/water separator

- Oil/water sepearator, see Section 11.2.11, separator designed accordingly.

Check design with Manning's equation, U-drain (3), Figure 5.22.

Figure 5.29 U-drain located on the longest wing.

From Manning's equation $V = \dfrac{1}{n} R^{2/3} S^{1/2}$

Critical velocity $V_c = \sqrt{gy_c}$

For the U-drain located at the longest wing is visualized, Figure 5.22,

Critical velocity
$V_c = \sqrt{gy_c} = \sqrt{9.81 \times 0.799} = 2.80 \text{ m/s}$

Thus, with overland slope,

$2.80 = \dfrac{1}{0.013}\left(\dfrac{0.799b}{2 \times 0.799 + b}\right)^{2/3} (0.0156)^{0.5}$

$b = 0.396$

Using peak discharge, $Q = 0.671$, and assuming $b = 0.3$ m, Manning's equation yields,

$0.671 = \dfrac{0.3y}{0.013}\left(\dfrac{0.3y}{2y + 0.3}\right)^{2/3} (0.0156)^{0.5}$

$y = 0.913 \text{ m}$

Conclusion
Find critcal slopes for all drains, for example, the critical slope for U-drain (3), Figure 5.22

$2.80 = \dfrac{1}{0.013}\left(\dfrac{0.799 \times 0.3}{2 \times 0.799 + 0.3}\right)^{2/3} (S_c)^{0.5}$

$S_c = 0.021$

Thus, the bed/construction slopes for all drains were achievable and obtained and specified accordingly. Critical slopes are necessary to scale down the sewer sizes to reduce embodied carbon. Therefore, since the overland slopes can support the achievement of critcal slopes, the bed/construction slopes were recommended accordingly. The overarching goal is to scale down carbon in the drainage network by negotiating construction slopes.

5.6.1.7 Analysis

The design assumed the peak flows would be conveyed through all hydraulic structures (storm sewer lines and drains) for a given subcatchment. The logic behind this is that the assumption considered the worst scenario for a return period of 25-years' storm and that all structures would be able to conserve the worst flood wave in the estate.

For option A—In reality, the inlets, for example, in subcatchment A, would collect varying amounts of runoff at any given time in the rainstorm. This was neglected and the maximum discharge for any given node hydrograph was equated to the subcatchment peak discharge. Consequently, the available headwater at a node, being influenced by the rim elevation ranging between 0.8 and 1.2 m derives from the peak discharge for subcatchment A, that is, volumetric flow rate. Therefore, 615-mm culverts were found to be the best solution for subcathment A storm sewer line. In subcatchments B, C, F, and D, runoff will move in a staggared manner and will all end on the longest estate wing. The peak flow generated from the subcathments was considered and it was found appropiriate to propose a U-drain that will accommodate the total peak discharge. In subcatchment E, an intermediate drain between office block and staff facilities block would drain away all runoff from the mainstream working area. In reality, there will always be differences in time of travel of runoff from one subcatchment or node or a link to another and therefore, the combined flows may fluctuate, at any given time. In some instances, this may lead to arguments that would call for a reduction in sewer sizes. However, by rule of thumb never reduce the size of pipes in the flow direction (see Section 2.7.2). The design also considered worst scenario case for which the sewers being laid at the given slopes will be able to accommodate the flood waves.

For option B—All drains were designed with self-cleansing velocities, engineering slopes as critical slopes. Option B is proposed in the case if no structures are built in subbasin A. It was found relevant, to channel all runoff (as proposed in option A—subbasin (A) to U-drain (1) and U-drain (3)). During dry days, the water tank would be used the most, drawing water for use in the truck wash bay. Although, the finished slopes would enable all runoff to drain toward U-drain (1) and (3), respectively, there would be some runoff that would drain toward the water tank as well—there exists a slight depression. To avoid stagnant waters around the tank, a small U-drain (4) was proposed round the tank.

5.6.1.8 Conclusion and recommendations

The drainage system shall be all covered whith steel gratings on the U-drains. This will make it quite safe,economical, reliable, and robust. To avoid voluminous excavations, drop manholes were proposed as appropriate for option A. The system is designed for critical flow scenario and critical depth for a return period of 25-years' storm was considered throughout and this is quite reliable as similar projects done before in the locale have stood a test of time. The site is not so steep to warranty subcritical and supercritical flows analysis. All velocities are self-cleansing and pose no risk of deposition or accumulation of particles along the drain sewer lengths. During construction, the slopes may change. However the proposed sewers are adequate enough to cater for the variations. And where slopes are quite steep, backdrop manholes should be constructed.

To minimize runoff, it was found relevant to properly landscape the estate and introduce some runoff volume reduction techniques such as rainwater harvesting jars, inflitration systems, islands with flower pots, and trees as advised by the Architect. These would help to improve the aethetics of the area, provide a cool temperature range, improve air quality, provide shades for workers, and to absorb greenhouse gases (GHGs) within the estate. It is important for the Drains and sewers not to be overworked so that they never flow full.

5.6.2 Case study (5.2): Typical audit and design project for collapsed retaining wall following flash floods, Kitende, Kampala, Uganda (August 14, 2021)

5.6.2.1 Overview

This case study investigated the probable cause of the collapsed retaining wall by auditing the drainage system of the estate. The estate area measured 0.001336 km^2. Site layout is shown in Figure 5.30.

5.6.2.2 Problem evaluation

On August 14, 2021, the area experienced a heavy storm and flash floods found way to the estate (Figure 5.30), overwhelming and overworking the drainage system. Part of the retaining wall at the lower end of the estate received direct impact of the flash floods and could not withstand the forces induced by the flash floods hence collapsing, see Figure 5.31.

The drainage scheme would have been designed to be more direct without bending the drainage route at $\approx 90°$ at the lower end of the collapsed structure. In that case, the flash floods would be routed off the estate as quickly as possible without allowing time for accumulation

192 Integrated Drainage Systems Planning and Design for Municipal Engineers

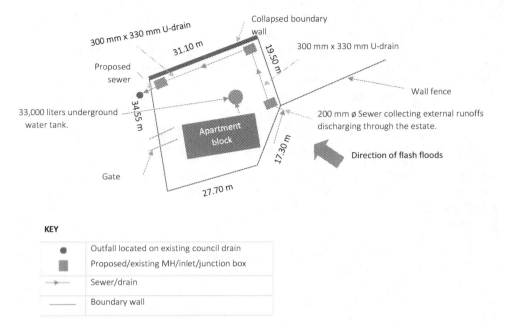

Figure 5.30 Site layout (drainage and retaining wall).

Figure 5.31 Built apartment and collapsed wall.

of huge runoff volumes ponding in the estate. However, due to haphazard building practices, this was not possible—the immediate neighbor objected to the proposal to allow discharging the runoff to his estate because it was unsafe. Otherwise, the excellent solution would be to have the drainage scheme route that routes the flash floods on a steep terrain to be direct without a ≈ 90° bend in line with velocity reduction techniques.

Since that was the case, the retaining wall design must have been robust to withstand forces induced by flash floods designed accounting for hydrostatic/static and dynamic loads. Enough care was not exercised while constructing the retaining wall. Appropriate imported soil with

Hydraulics design principles 193

Figure 5.32 Collapsed wall.

acceptable shear strength parameters was not used as backfill. The soil shear strength can be improved by compaction to achieve the desired stability and strength. Therefore, adequate compaction of the retained soil mass was essential. The probable cause of the collapse of the retaining wall was found to be due to imported fill material that did not conform to standards, inadequate design, poor compaction or no compaction of backfill, and poor workmanship. See Figures 5.31 and 5.32.

Fully saturated soils induce extra thrust on the retaining wall because the effective stresses within the soil change. However, the developer wished to fully pave and roof the estate adopting an underground tank only as a green infrastructure technique capable of minimizing runoffs. This was due to the need to secure enough parking for the tenants. That means runoff would not infiltrate into the retained soil mass.

5.6.2.3 Justification of the probable cause

a. The developer installed a 33,000-L underground water tank (SuDS) harvesting the rainwater from the built apartments whose overflow pipe discharges out of the estate. This implied the current built apartments do not contribute to the runoff discharging to the estate overland flow drainage system.
b. Auditing the capacity of the primary sewer pipe that discharged external flows to the estate, the pipe diameter was found to be 200 mm and laid at a slope $S_0 = 0.06$. Therefore, from the normograph (Figure 5.8) this provided a discharge of 4.95 m³/min (0.0825 m³/s).
c. The existing drain measured 300- by 330 mm which was adequate for the estate flows.
d. The immediate fence which had the pipe entrance believed to receive flash floods showed no flash floods effect.
e. Therefore, points (a) to (d) ruled out flash floods impact force being the probable cause. The impact force was negligible.

5.6.2.4 Drainage design

The existing drainage system was remodeled to accommodate external flows.

General design assumptions

a. Normal flow conditions; steady flow.
  - Friction slope ≈ bed slope.
  - Critical depth ≈ normal depth.
  - Assumed critical slopes throughout.
b. Grated manholes and U-drains to provide inlets for runoff and will control runoff at the entrance.
c. Neglected approach velocities to inlets.
d. No intermediate controls except manhole/junctions.
e. Neglected energy losses due to entrance and exit.
f. Neglected lateral inflows—and considered peak flows from catchment.
g. Estate designed for worst scenario, tank runoff volume reduction not considered.

5.6.2.5 Remodeling the U-drain

- Existing U-drain (width = 0.30 m, height = 0.33 m).
- Slope = 4.25%.
- Contributory area = 0.001336 km² (designing for worst case; all estate area considered).

- Time of concentration from Kirpich's formula provided a design intensity off the IDF curve, $T_C = 1.126$ minutes; (L = 40 m and S = 4.25%).
- Design intensity, i = 240 mm/h, ARI = 25 year (Figure 5.20).
- Weighted coefficient = 0.85 obtained from the proposed site layout/physical plan.
- Peak discharge from catchment by rational method = 0.076 m³/s.
- Breadth = 0.3 m (assumed breadth).
- Specific discharge for U-channel = 0.253 m²/s.
- n = 0.013 (concrete).
- Thus; critical depth

$$y_c = \sqrt[3]{\frac{q^2}{g}} = \sqrt[3]{\frac{(0.253)^2}{9.81}} = 0.187 m$$

Without external flows, a drain of breadth 300 mm and depth 187 mm is adequate for estate flows (worst scenario), but a more feasible drain should cater for external flows. The existing drain measured 300- by 330 mm which was adequate for the estate flows. However, it may not be sufficient to accommodate the flood waves due to external flows. Estimating future external flows was quite difficult; it needed collecting an area topographical survey, area physical plan, aerial photography, and topographical maps to map out the extent of the area level of urbanization/imperviousness and the estate's specific drainage path. This would come at a cost! However, checking the capacity of the primary sewer pipe (discharging external flows to the estate); entrance pipe = 200 mm dia, laid at a slope $S_0 = 0.06$; from the normogram (Figure 5.8) provided a discharge of 4.95 m³/min (0.0825 m³/s). Thus,

- New peak discharge = (0.076 + 0.0825) = 0.1585 m³/s, considering external flows as lateral inflows.
- Breadth = 0.3 m (assumed breadth).
- Specific discharge for U-channel = 0.528 m²/s.
- n = 0.013 (concrete).
- Thus, critical depth.

$$y_c = \sqrt[3]{\frac{q^2}{g}} = \sqrt[3]{\frac{(0.528)^2}{9.81}} = 0.305 m$$

Conclusion/recommendation

- With addition of external flows, a U-drain measuring 300- by 305 mm is adequate to accommodate flows. However, future external flows might increase due to growing urbanization.
- $V_c = \sqrt{gy_c} = \sqrt{9.81 \times 0.305} = 1.73 \, m/s$
- **Note:** From Manning's equation

$V = \frac{1}{n} R^{2/3} S^{1/2}$ (Metric system)

Overland slope = 4.25%
Thus, the bed/construction slope of 1.10% was achievable. Where construction slopes can be negotiated, it is a good practice to scale down the size of the sewer/drain.

- Velocity of flow in the drain is 1.73 m/s which is self-cleansing.
- The U-drain was grated throughout for safety reasons.
- Concrete culvert sewer piece (Figure 5.33) proposed at the entrance (to receive external flows) and at the exit of the drainage system properly haunched in concrete with gratings to trap garbage/sediment at the junctions. Velocity reduction techniques would be appropriate at the exit point—to provide concrete steps/check dams at the exit. This was dictated by the site conditions.
- The manhole/junction box rim elevations would be determined/varied on site as appropriate since it influences the available headwater at manhole/junction.

5.6.2.6 Analysis

The drainage design assumed the peak flows would be conveyed through all hydraulic structures (sewer and drain lines) for the estate. The logic behind this was that the assumption considered the worst scenario for a return period of 25-years' storm and that all structures would be able to conserve the worst flood wave in the estate. In reality, the inlets would collect varying amounts of runoff at any given time in the rainstorm. This was neglected and the maximum discharge from the hydrograph was equated to the catchment peak discharge. Consequently, the available headwater at a node junction, being influenced by the rim elevation ranging between 0.5 and 0.8 m derives from the peak discharge for the catchment.

In reality, there will always be differences in time of travel of runoff from one node or a link, that is, drain/sewer/junction to another and therefore, the combined flows may fluctuate at any given time. In some instances, this may lead to arguments that would call for a reduction in sewer/drain sizes. However, the design considered the worst scenario case for which the drain and sewers being laid at the constructed slopes will be able to accommodate the flood waves. See Figure 5.34 illustrating the drainage system and proposed retaining wall details, 3-D model not to scale.

Entrance of external flows

Exit of drainage system

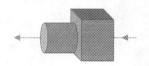

Figure 5.33 Entrance and exit pipes.

Hydraulics design principles 197

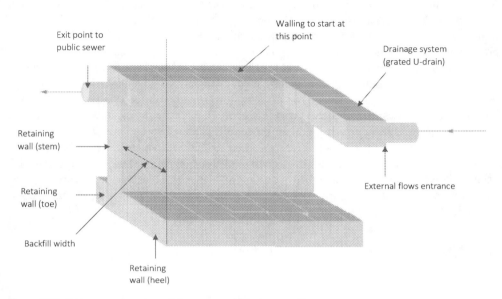

Figure 5.34 3-D representation of drainage and retaining wall.

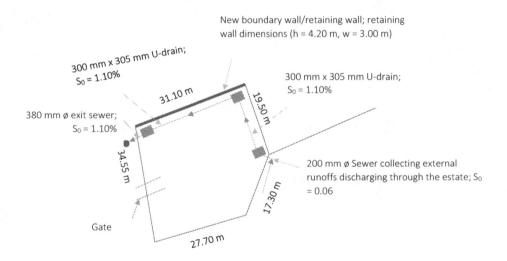

Figure 5.35 Modified estate's drainage system.

5.6.2.7 Conclusion and recommendations

The drainage system would be all covered with steel gratings on the U-drains. This would make it quite safe, economical, reliable, and robust. The drainage system was designed for critical flow scenario and critical depth for a return period of 25 years was considered throughout and this is quite reliable as similar projects done before in the locale had stood a test of time. The site (within the estate) was leveled and did not warranty subcritical and supercritical flows. Construction slopes would be negotiated to nearly equate to critical slopes. All velocities were self-cleansing and posed no risk of deposition or accumulation of particles along the drain sewer lengths. However, during construction, the slopes may slightly vary. However the proposed sewers/drains were adequate enough to cater for the variations.

Despite the need to secure enough parking, it was found relevant to properly landscape the estate and introduce some runoff volume reduction techniques such as additional rainwater harvesting jars, infiltration systems, islands with flower pots, and trees. These would help to improve the aethetics of the estate, provide a cool temperature range, improve air quality, provide shades for tenants/workers, minimize runoff, and to absorb GHGs within the estate. It is important for the sewers not to be overworked so that they never flow full.

Negotiating a drainage easement was recommended so that the estate does not receive huge external flows. The lesson to take forward as a draiange engineer is that haphazard building construction can be dangerous especially now that the world is facing climate change accompanied by floods.

5.6.2.8 Final drainage scheme route and retaining wall

See Figure 5.35.

5.6.3 Worked example 5.1: Application of Equation (5.2)

A rectangular channel of width b = 60.96 m and 7315.2-m long with a flat bottom carries a flow of 169.90 m³/s. The channel terminates in a free discharge. Plot the water profile, n = 0.035.

Solution:
The depth at the free discharge will be critical.

$$y_c = \left(\frac{q^2}{g}\right)^{1/3} = \left(\frac{(169.90/60.96)^2}{9.808}\right)^{1/3} = 0.925m$$

In general, at any point along the channel,

$$R = \frac{A}{P} = \frac{60.96y}{60.96 + 2y}$$

$$S_0 = 0$$

$$S_f = \left(\frac{Vn}{R^{2/3}}\right)^2 = \left[\frac{(0.035)(169.90/60.96)}{y(60.96y/(60.96+2y))^{2/3}}\right]^2$$

Nominating values of y, we can calculate successively the values shown in the table below. Increments are chosen to be relatively small near the discharge, where the depth is changing rapidly. The depth at the upstream end is most readily determined graphically as shown in the plot of depth versus distance in Figure 5.36.

$$\frac{\Delta(y+V^2/2g)}{\Delta x} = S_0 - \left(\frac{Vn}{R^{2/3}}\right)$$

For uniform channels S_0 is constant. Thus, all terms in the equation are simple functions of y. We can therefore, select appropriate values of y and calculate the intervals Δx between successive depths. This procedure is commonly called the step method.

y	V	E	ΔE	S_f	S_{favg}	Δx	x
0.925	3.013052	1.387716		0.012918			0
1.025	2.719096	1.401834	0.014118	0.009213	0.011066	1.275871	1.275871
1.125	2.477399	1.437819	0.035985	0.006784	0.007999	4.498812	5.774683
1.225	2.275162	1.488831	0.051012	0.005129	0.005957	8.563979	14.33866
1.325	2.103452	1.55051	0.061679	0.003965	0.004547	13.56435	27.90301
1.425	1.955841	1.61997	0.06946	0.003124	0.003545	19.59513	47.49815
1.525	1.827589	1.695239	0.075269	0.002503	0.002813	26.7533	74.25145
1.625	1.715122	1.774931	0.079692	0.002033	0.002268	35.1373	109.3887
1.725	1.615695	1.858051	0.083121	0.001673	0.001853	44.84676	154.2355
1.825	1.527164	1.94387	0.085818	0.001393	0.001533	55.98232	210.2178
1.925	1.44783	2.031841	0.087971	0.001171	0.001282	68.64546	278.8633
2.025	1.376333	2.121549	0.089708	0.000993	0.001082	82.93831	361.8016
2.125	1.311564	2.212676	0.091127	0.000849	0.000921	98.96359	460.7652
2.225	1.252617	2.304972	0.092296	0.000731	0.00079	116.8245	577.5897
2.325	1.198741	2.398241	0.093269	0.000634	0.000683	136.6244	714.2141
2.425	1.149309	2.492325	0.094084	0.000553	0.000594	158.4673	872.6814
2.525	1.103791	2.587098	0.094773	0.000486	0.000519	182.4571	1055.139
2.625	1.061742	2.682457	0.095359	0.000428	0.000457	208.6979	1263.836
2.725	1.022779	2.778317	0.09586	0.00038	0.000404	237.294	1501.13
2.825	0.986575	2.874609	0.096292	0.000338	0.000359	268.3494	1769.48
2.925	0.952846	2.971275	0.096666	0.000302	0.00032	301.9684	2071.448
3.025	0.921347	3.068266	0.096991	0.000271	0.000287	338.2551	2409.703
3.125	0.891864	3.165541	0.097275	0.000244	0.000258	377.3135	2787.017
3.225	0.864209	3.263066	0.097525	0.000221	0.000233	419.2472	3206.264
3.325	0.838218	3.360811	0.097745	0.0002	0.000211	464.16	3670.424
3.425	0.813744	3.45875	0.097939	0.000182	0.000191	512.1552	4182.579
3.525	0.790659	3.556862	0.098112	0.000166	0.000174	563.3358	4745.915
3.625	0.768848	3.655129	0.098266	0.000152	0.000159	617.8048	5363.72
3.725	0.748208	3.753533	0.098404	0.000139	0.000146	675.6646	6039.384
3.825	0.728647	3.85206	0.098528	0.000128	0.000134	737.0173	6776.402
3.925	0.710082	3.950699	0.098639	0.000118	0.000123	801.9647	7578.366

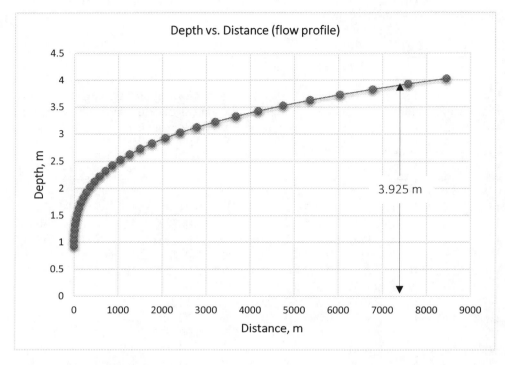

Figure 5.36 Water profile in flat-bottomed channel.

y	V	E	ΔE	S_f	S_{favg}	Δx	x
4.025	0.692441	4.049438	0.098739	0.000109	0.000113	870.6083	8448.975

Note

- In the case of the flat-bottomed channel, it is obvious that the depth of the water will increase without limit as the length of the channel increases.
- In channels for $S_0 \neq 0$, the depth will increase until normal depth is attained and will remain at that level until some other factor causes it to change.

5.7 CONCLUSION

The choice to use energy equations or momentum equations depends on the project at hand. In many cases, the energy equation permits solution of a large number of open channel flow problems. However, there are some situations which require use of the momentum equation. Both equations are derived from the first principles of energy and momentum conservation in a system. Depending on the level of simplifications to suit the application, these equations are used to solve a number of open channel flow problems.

The total energy or head in a fluid is the sum of kinetic and potential energies. Potential energies are pressure energy and elevation energy. Momentum equations are largely employed to analyze hydraulic jumps in open channels, nonuniform flow through sudden enlarged pipes, analyzing bends, the impact of floods, and channels that receive runoff along their length and those that discharge flow along their length. The rate of change of momentum is equal to the force applied to the fluid.

Rectangular channels are the most versatile and highly conform to sustainability requirements because their hydraulic radius is the most appropriate, cost-effective, and technically responsive to applications. Rectangular channels can easily fit within available drainage easements, which means they can help lower the construction cost. To minimize earthworks, the hydraulic radius is a key component as it influences the hydraulic efficiency of the drainage structure and at the same time influences cost.

Notes

1. To account for friction head losses.
2. Manning's roughness coefficient.
3. For friction slopes calculation.
4. To account for head losses due to inlet/outlet conditions.
5. Prismatic means that the channel profile is arbitrary, but it does not change the axial (x) direction which is typical in the case of artificial channel systems.

Reference

1. Hager, W. H. (1992). *Energy dissipators and hydraulic jump*. Kluwer Academic Publishers.

Further reading

Water supply and sewerage, 6th Edition, Terence J. McGhee. McGraw-Hill international editions, civil engineering series, ISBN 0-07-100823-3.

Chapter 6

Flow routing techniques

NOTATION

ρ	:	Density of water
A	:	Cross-sectional area
V	:	Volume
g	:	Acceleration due to gravity
Q	:	Discharge
q	:	Lateral inflow
x	:	Distance
t	:	Time
S_0	:	Bed slope
S_f	:	The energy/friction slope
S_e	:	Eddy loss slope
W_f	:	The wind shear factor
B	:	Average top width of water surface
β	:	Velocity distribution coefficient
v_x	:	x-component of the velocity of the lateral flow
F_g	:	Gravity force acting on the control volume
F_f	:	Friction force
F_e	:	Contraction/expansion force
F_w	:	Wind shear force
F_p	:	Unbalanced pressure force
Z	:	Stage
H_w	:	Available headwater
v	:	Flow velocity
c	:	Wave celerity
y_{Hw}	:	Headwater depth
y_{Tw}	:	Tailwater depth
K_e	:	Head loss coefficient
L	:	Length of pipe/channel

6.1 INTRODUCTION

Flow routing is **a procedure to determine the time and magnitude of flow** (i.e., the flow hydrograph) at a point on a watercourse from known or assumed hydrographs at one or more points upstream. The procedure is specifically known as flood routing if the flow is a flood.[1] Routing is a technique used to predict the changes in shape of a hydrograph as water moves through a river, channel, or a reservoir. Flow routing has various applications in solving stormwater problems, which is very helpful to the drainage engineers and planners. There are two sets of methods for flood routing calculations in open channels. These methods are classified as **hydraulic** and **hydrologic** schemes depending on whether the methods are based on empirical or physical process equations of motion. The hydraulic methods are based on solving Saint Venant equations. These equations can be solved numerically by the **finite difference** or **finite element** methods. Hydrologic methods are based on the storage continuity equation and another equation which usually expresses the storage volume as a linear or nonlinear function of inflow and outflow discharges. Hydrologic models were previously used in flood routing, but because they do not explicitly involve any spatial variability and strictly express the flow routing variable as a function of only time, hydraulic models are preferred. The Muskingum routing method is one of the commonest hydrologic routing methods. The scope of routing techniques presented in this chapter is limited to **hydraulic methods**. The governing equations of the hydraulic routing methods are the Saint Venant equations.

Flow routing in open channels is a technique for determining the propagation of flow from one point in the channel to another. The term channel is used in a wide range and includes rivers, streams, bayous, partially flowing pipes and tunnels, brooks, creeks, canals, sewers, gutters, borders, and furrows. [2] The **flow variables** whose propagation characteristics are of interest are **discharge, velocity, depth, cross-section, volume**, and **duration**. The propagation characteristics of interest are peak, time to peak, duration of the hydrograph, and attenuation. Although routing can be from upstream to downstream or from downstream to upstream, the emphasis has been largely on routing from upstream to downstream. Flow routing in open channels entails wave dispersion, wave attenuation or amplification, and wave retardation or acceleration. These wave characteristics constitute the hydraulics of flow routing or propagation and are greatly affected by the geometric characteristics of channels, the characteristics of sources and/or sinks, as well as initial and boundary conditions. Unsteady 1-D open channel flow is usually conducted in prismatic channels. In general, channels are heterogeneous and nonuniform with regard to geometric, morphologic, and hydraulic characteristics. In other words, these characteristics vary longitudinally as well as transversely in space. Furthermore, they also vary in time, especially over macro scales. However, prismatic channels are good candidates for flow routing especially while developing stormwater systems.

Urbanization modifies the region's hydrology. Increased imperviousness generates more runoff. Urbanization leads to changes in areas' peak discharges and short time to peak. Therefore, routing a flood wave through a hydraulic structure helps to predict how the discharge will be conserved throughout the drainage system, taking into consideration different scenarios like junctions, manholes, and lateral inflows. Flow routing **connects excess water from precipitation and runoff to the stream to other surface water** as part of the hydrological cycle. Simulating flow helps elucidate the transportation of nutrients through a stream system, predicts flood events, informs decision makers, and regulate water quality and quantity issues.

6.2 HYDRAULIC ROUTING TECHNIQUES

The following are considered hydraulic routing techniques:

a. Steady flow
b. Dynamic wave
c. Kinematic wave
d. Diffusion wave

The routing techniques predict how the drainage system would perform under several scenarios. By defining boundary conditions, stream or storm sewer hydrographs can be generated. Based on the economics, safety, and reliability, the system can be resized to arrive at the optimal solution.

6.3 APPLICATIONS OF HYDRAULIC FLOW ROUTING TECHNIQUES

Modeling flow is important for planning, designing, regulating, and managing our water resources (Nwaogazie, 1986). [3] Flow routing techniques can generate hydrographs for node junctions and node manholes. Also, hydrographs for sewers and drains can be generated. Flow routing connects excess water from precipitation and runoff to the stream to other surface water as part of the hydrologic cycle. Simulating flow helps elucidate the transportation of nutrients through a stream system, predicts flood events, informs decision makers, and regulates water quality and quantity issues. Constant volume and changing volume techniques use the continuity equation, while the third technique, the kinematic wave approximation, uses the continuity and momentum equations. Inputs and outputs differ for each flow routing technique, with dynamic wave being the most complex model. Each option has a set of applications it is best suited for. Numerical errors such as distortion and instability are potential errors that can affect the accuracy of model outputs. Routing methods should be chosen based on input data availability and scope of the problem being addressed.

6.3.1 Evaluating the impact of flash floods

Flash floods occur with little warning time due to the following reasons:

a. Excessive rainfall of high intensity.
b. Sudden release of high discharges of water from streams and rivers, dams, and by breaches of structural failures of dams, tanks, and levees, gushing out large volumes of water in a short time.

Due to climate change, the impact of flash floods globally has been a serious concern in a number of court cases and insurance claims. Flash floods are highly impactful due to self-weight of the floods and fast change of momentum within flood waves. The impact of flash floods on properties, environment, and lives is calculated to process helpful information in several matters. See Case study 12.1 about the impact of flash floods to property. The dynamic wave model can be used to analyze the flash floods caused by **dam breaks**. For example, the downstream flood wave caused by the 1976 failure of the Teton dam in Idaho, USA, was rebuilt by the dynamic wave model. On June 5, 1976, Teton Dam in southeastern Idaho failed releasing more than 28,316.85 m^3/s. In the end, 11 people died, about 25,000 people were left homeless, and there was millions of dollars of property damage. The reclamation's dam safety program evolved out of this disaster. Effects of floods include the following:

- Loss of lives.
- Large concentration of debris, grit, sediment, and toxic pollutants, which can be health hazards.
- Erosion due to high velocities.
- Property damages when floods enter buildings.
- Loss of livestock and crops on farmlands when flash floods hit these places.

6.3.2 Predicting the arrival time of floods

Numerical solutions of the Saint Venant equations are used to predict the flood arrival time and its magnitude (i.e., the flood hydrograph) at various locations along a river once the flood hydrograph at an upstream location is provided. As a case study, an accurate prediction of the travel of the flood wave during the unprecedented floods in the Yamuna in 1978 supported the timely evacuation of parts of Delhi. The typical distances considered in that case are in hundreds of kilometers, times of the order of 2 to 3 days with routing time steps of one hour, and the typical flood discharges may be of the order of thousands of cumecs (m^3/s).

6.3.3 Urban drainage design

Flood routing is used in the planning and design of urban stormwater drainage systems. Routing estimates how much time is needed to route the food wave off the impervious surface and safely discharge it into the receiving waters. Urbanization modifies the hydrologic characteristics of a region. Therefore, the hydrographs after urbanization show a higher peak and a lower time-to-peak, for the same rainfall pattern due to the quicker and higher runoff because of the increased impervious area. As impervious area increases, therefore, drainage of stormwater becomes challenging. The kinematic wave model is used to route the flood hydrographs through the urban drainage system to examine the adequacy of the system.

6.3.4 Analyzing flood control system's safety

In engineering designs for flood control (e.g., flood control reservoirs), a design flood hydrograph is routed to ensure the safety of the design. The design flood, such as a 50-year flood, is chosen based on the perceived acceptable risks. The flood hydrograph used for routing is constructed based on the design duration, which depends on the hydrology of the region.

6.4 DERIVATION OF HYDRAULIC FLOW ROUTING TECHNIQUES

The hydraulic methods are based on solving the Saint Venant equations based on the continuity and momentum equations. Continuity equation can be explained as conservation of mass/volume. The extent of simplification of the Saint Venant equations derives the particular method used.

This section presents a theoretical framework for forecasting the propagation of a flood wave though an open channel. Consider a flow wave traversing from **point 1** to **point 2** through a channel as shown in the Figure 6.1. As the wave propagates through the channel, the discharge or flow varies with time as represented in the hydrographs in Figure 6.2.

Although in rare cases, a monoclinical rising wave (a uniformly progressing translatory wave of stable form) could travel along a small section of a stream, in real situations, the flow through channels and pipes is usually unsteady and the discharge and depth of flow vary as the wave propagates through the structure. In practical applications, determining the variations of depth and discharge with time at a specific point as the flow propagates through the channel is essential. The computation of the flow wave along a channel with time and space is called 'flow routing'. Given the hydrograph at the upstream end of the channel, it can

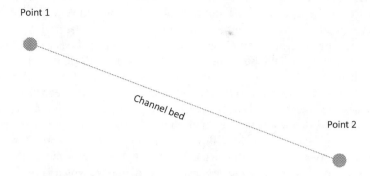

Figure 6.1 Flow moves from point 1 to point 2 in the channel.

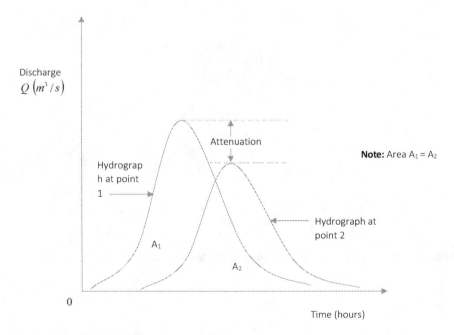

Figure 6.2 Hydrographs for the flow at point 1 and 2.

Note: Area $A_1 = A_2$

be routed downstream and this is excellently conducted by solving the Saint Venant equations as the applicable technique.

Several hydraulic models are based on solving the Saint Venant equations and its simplifications. This involves studying the changes in depth and discharge with time along the length of a channel. Considering an unsteady flow in a channel, which is usually the case in storm sewers, the velocity of flow and depth gradually change with gradual changes in time. The governing equations for flow movement are derived from continuity and momentum equations, that is, conservation of mass and momentum, respectively. A 1-D unsteady open channel flow is usually considered. Note that flow routing would be strictly 3-D in nature, mainly because of spatial heterogeneities and nonuniformities in the horizontal and vertical planes. Therefore, the governing equations are also 3-D. However, because of the lack of data on the spatial variability of roughness, sources and sinks, and initial and boundary conditions and the difficulties encountered in solving them, a one-dimensional form is often employed.[2]

The Saint Venant equations are also referred to as shallow water equations because the horizontal scales are significantly larger than the flow depth. To derive the continuity and momentum equations, define a control volume shown in Figures 6.3, 6.4, and 6.5.

Imagine unsteady 1-D open channel flow in a prismatic open channel flowing with an incompressible stormwater runoff as depicted in Figure 6.5. Prismatic means that the channel profile is arbitrary, but it does not change the axial (x) direction, which is typical in the case of artificial channel systems. Let $y(x,t)$ denote the water depth, which can vary along both

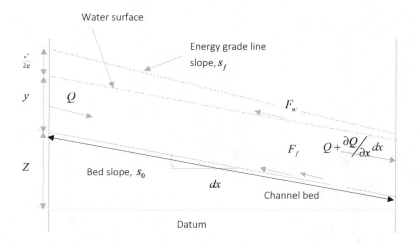

Figure 6.3 Control volume defining continuity and momentum equations (profile view).

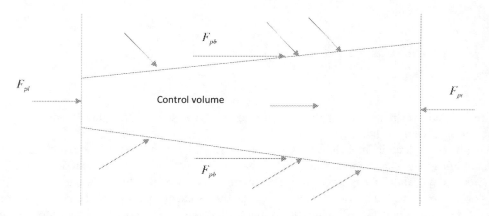

Figure 6.4 Control volume defining continuity and momentum equations (plan view).

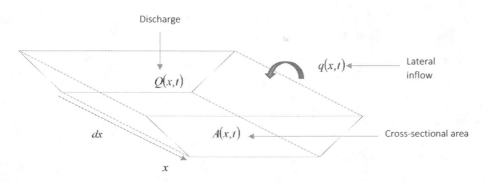

Figure 6.5 3-D representation of the control volume.

channel length (x) and time (t). It is clear that the flow rate is $Q(x,t)$. $Q(x,t) = A(y(x,t)v(x,t))$ The wetted cross-section varies. The same flow rate can occur for multiple y, v values, for example, low water depth with high velocity or high water level with low flow velocity. The wetted perimeter P will also be important as it represents the area along which friction forces act on the water. Flow routing can be used to obtain flow parameters at a *point* in the channel, that is

a. The depth, y
b. The velocity, v
c. The time, t
d. The distance, x

To find y and v, we assume/start with Δt and Δx.

6.4.1 Deriving the continuity equation

Consider an elemental control volume of length dx in a channel (Figure 6.3).

a. The **total inflow** to the control volume is the sum of the flow Q entering the control volume at the upstream end of the channel and the *lateral inflow* q entering the control volume as a distributed flow along the channel.
b. The dimensions of q are those of flow per unit length, so the rate of lateral flow is qdx. The mass inflow rate to the control volume is $\rho(Q+qdx)$.
c. The **mass outflow** rate from the control volume $= \rho(Q + \partial Q/\partial x\, dx)$, where $\partial Q/\partial x$ is the rate of change of channel flow with distance.
d. The volume of the channel element is $A dx$, where A is the average cross-sectional area. The rate of change of mass stored within the control volume is, thus, $\partial(\rho A dx)/\partial t$.
e. The continuity of mass is written as:

Outflow + change in storage = inflow

which is written for the control volume as:

Outflow + change in storage = inflow

Inflow / Outflow rate = density × discharge

- Mass inflow rate = $\rho(Q+qdx)$
- Change in storage = $\dfrac{\partial V}{\partial t} = \dfrac{\partial(\rho A dx)}{\partial t}$
- Mass outflow rate = $\rho\left(Q+\dfrac{\partial Q}{\partial x}dx\right)$

For water, the density ρ may be assumed constant. Dividing throughout by ρdx, the continuity equation can be written as:

$$\frac{\partial Q}{\partial x}+\frac{\partial A}{\partial t}-q=0. \tag{6.1}$$

6.4.2 Deriving the momentum equation

a. Saint Venant proposed the mathematical model of a water flow in rivers, based on the laws of conservation of momentum and mass of fluid. Saint Venant shallow water equations are derived from the Navier-Stokes equations. The Saint Venant equations (the shallow water equations) are often used in theoretical and applied studies of the unsteady water motion in free channels. The Siant Venant shallow water momentum equation is derived from Newton's second law, which states that the rate of change of momentum equals the net force. The derivation is quite long, a broad outline is provided. The detailed derivation can be obtained from books specialized on hydraulics. See also Reference 6.
b. The forces considered for the control volume are as follows:
   - The gravity force, F_g.
   - The friction force, F_f.
   - The contraction/expansion force, F_e.
   - The wind shear force on the water surface F_w.
   - The unbalanced pressure force, F_p (net force due to F_{pl}, F_{pb}, F_{pr} as shown in Figure 6.4).
c. The net outflow of momentum from the control volume is computed from the mass inflow, $\rho(Q+qdx)$, and the mass outflow, $\rho(Q+\partial Q/\partial x dx)$, considering the average velocity, V, in the control volume and the x-component v_x of the velocity of the lateral flow q.
d. The volume of the elemental channel considered is Adx, its momentum is $\rho AdxV$ or ρQdx. Therefore, the rate of change of momentum storage in the control volume is $(\rho \partial Q/\partial t dx)$.
e. A momentum coefficient β is introduced to account for nonuniform distribution of velocity at a section when computing the momentum. The complete form of the momentum equation is written as:

$$\frac{\partial Q}{\partial t} + \frac{\partial\left(bQ^2/A\right)}{\partial t} + qA\left(\frac{\partial y}{\partial x} - S_0 + S_f + S_e\right) - \beta q v_x + W_f B = 0. \tag{6.2}$$

Where S_0 is the bed slope, S_f is the energy slope, S_e is the eddy loss slope, W_f is the wind shear factor, and B is the average top width of water surface in the control volume. With $\beta = 1$, neglecting lateral flow, wind shear, and eddy losses, the momentum equation (6.2) can be simplified to

$$\underset{(1)}{\left(\frac{1}{A}\right)\left(\frac{\partial Q}{\partial t}\right)} + \underset{(2)}{\left(\frac{1}{A}\right)\frac{\partial(Q^2/A)}{\partial x}} + \underset{(3)}{g\frac{\partial y}{\partial x}} \underset{(4)\ (5)}{- g(S_0 - S_f)} = 0 \tag{6.3}$$

Note:

- Local acceleration term (1)
- Convective acceleration term (2)
- Pressure force term (3)
- Gravity force term (4)
- Friction force term (5)

Equations (6.2) and (6.3) are together called the Saint Venant equations in the honor of Saint Venant who developed these equations in 1871. It is not possible to solve Equations (6.1) and (6.2) together analytically, except in some very simplified cases. Numerical solutions are possible and are used in most practical applications, especially in finite difference and finite element methods. Depending on the accuracy desired, alternative flood routing equations are generated by using the complete continuity equation (except the lateral flow term, in some cases) while eliminating some terms of the momentum equation. Based on the terms retained in the momentum equation, the flood wave is called the kinematic wave, the diffusion wave, and the dynamic wave, as shown in Equation (6.4) below:

$$\left(\frac{1}{A}\right)\left(\frac{\partial Q}{\partial t}\right) + \left(\frac{1}{A}\right)\frac{\partial(Q^2/A)}{\partial x} + g\frac{\partial y}{\partial x} - g(S_0 - S_f) = 0 \tag{6.4}$$

Kinematic wave
Diffusion wave
Dynamic wave

The kinematic wave is thus represented by $g(S_0 - S_f) = 0$, or $S_0 = S_f$.

The diffusion wave is represented by $g\dfrac{\partial y}{\partial x} - g(S_0 - S_f) = 0$ resulting in $\dfrac{\partial y}{\partial x} = (S_0 - S_f)$.

And the dynamic wave is represented by the complete momentum equation (6.4). In these quantities, Q is the discharge (m³/s), A is the area, q is the lateral flow per unit length of the channel (m³/s/m), x is the distance along the channel, y is the depth of the flow, g is the acceleration due to gravity, S_0 is the slope of the channel, S_f is the friction slope.

In the momentum equation

  a. The local acceleration term describes the change in momentum due to change in velocity over time.
  b. The convective acceleration term describes the change in momentum due to change in velocity along the channel.
  c. The pressure force term denotes a force proportional to the change in water depth along the channel.
  d. The gravity force term denotes a force proportional to the bed slope.
  e. The friction force term denotes a force proportional to the friction slope.

In most practical applications, either the wave resulting from the simplest form of the momentum equation, that is, the kinematic wave, or from the complete momentum equation, that is, the dynamic wave, is used. For the kinematic wave, the acceleration and pressure terms in the momentum equation are neglected, and hence the name kinematic, referring to the study of motion exclusive of the influence of mass and force. The remaining terms in the momentum equation represent the steady uniform flow.

In other words, we consider the flow to be steady for momentum conservation but take into consideration the effects of unsteadiness through the continuity equation. Analytical solution is possible for the simple case of the kinematic wave, where the lateral flow is neglected and the wave celerity is constant. However, the backwater effects (propagation upstream of the effects of change in depth or flow rate at a point) are not reproduced through a kinematic wave. Such backwater effects are accounted for in flood routing only through the local acceleration, convective acceleration, and the pressure terms, all of which are neglected in the kinematic wave.[4] Therefore, for more accurate flood routing, numerical solutions of the complete dynamic wave equation are used.

6.4.2.1 The grid

The scheme developed is in rectangular Cartesian coordinate system for the grid point (i, j) with two time levels at n and $n+1$.[5] Finite difference numerical methods use the grid to find the final step for $(t+1)$ time discretization. See Figure 6.6.

$$\frac{\partial Q}{\partial x} = \frac{Q_i^t - Q_{i-1}^{t+1}}{\Delta x}$$

$$\frac{\partial Q}{\partial t} = \frac{Q_i^{t+1} - Q_i^t}{\Delta t}$$

6.5 STEADY FLOW

Steady flow routing assumes that within each computational time step, flow is uniform and steady. It translates the inflow hydrographs at the upstream end of the conduit to the downstream end, with no delay or change in shape. In steady flow routing, the Manning's equation simply relates the flow rate Q to flow area A (or depth y)—almost no routing is done in steady routing flow.

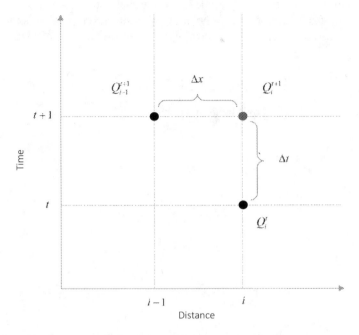

Figure 6.6 Mesh points for a numerical solution of the wave equation.

Unlike other approaches, the steady flow routing approach does not account for channel storage, backwater effects, entrance/exit losses, flow reversal, or pressurized flow. In developing hydraulic models, it can only be applied to dendritic conveyance networks, where each node has only a single outflow link (unless the node is a divider in which case two outflow links are required). This form of routing is insensitive to the time step employed and is really only appropriate for preliminary analysis using long-term continuous simulations.

6.6 KINEMATIC WAVE ROUTING TECHNIQUE

A kinematics problem begins by describing the geometry of the system and declaring the initial conditions of any known values of position, velocity, and/or acceleration of points within the system. Then, using arguments from geometry, the position, velocity, and acceleration of any unknown parts of the system can be determined. The kinematic wave approximation requires definition of the hydraulic characteristics of all system elements such as overland flow, collector channels, and main channel. Values for roughness, hydraulic radius, area, and slope must be collected and these must be fed into the software program being used. The calculation procedure involves a solution of the **finite-difference approximation** to the kinematic wave equation. The solution is not necessarily correct, since the kinematic wave approach does not permit modeling of backwater effects. The model may be considerably simplified by neglecting minor collectors.

From Equation (6.4), for a kinematic wave, the momentum equation reduces to

$$S_0 = S_f \tag{6.5}$$

Equation (6.5) means that the energy grade line is parallel to the channel bottom and that the flow is relatively steady and uniform. The above simplification of the momentum equation can be shown to be equivalent to the following relationship between discharge, Q, and area of flow, A,

$$A = \alpha Q^\beta \tag{6.6}$$

Equation (6.6) can be satisfied by Manning's equation,

$$Q = \frac{\sqrt{S_f}}{n} R^{2/3} A \quad \text{(Metric units)} \tag{6.7}$$

Equation (6.7) can be rearranged as

$$A = \left(\frac{nP^{2/3}}{\sqrt{S_f}}\right)^{3/5} Q^{3/5} \tag{6.8}$$

Thus, comparing Equation (6.8) with Equation (6.6),

$$\alpha = \left(\frac{nP^{2/3}}{\sqrt{S_f}}\right)^{3/5} \quad \text{and} \quad \beta = 3/5 = 0.6 \tag{6.9}$$

Together with the continuity Equation (6.1), these equations represent the kinematic wave flow routing approach. Suppose an observer is to move with the kinematic wave at the same speed as the kinematic wave celerity, in that case, they would observe the flow rate increase at the rate equivalent to lateral inflow rate, q, as shown in Equation (6.10).

$$\frac{dQ}{dx} = \frac{\partial Q}{\partial x} + \frac{\partial Q}{\partial t}\frac{dt}{dQ} = \frac{\partial Q}{\partial x} + \frac{1}{c_k}\frac{\partial Q}{\partial t} = q \tag{6.10}$$

For conditions where there is no lateral inflow, that is, $q = 0$, $dQ/dx = 0$, the kinematic waves do not attenuate. They translate downstream without dissipation. Since, at any cross-section, Q is functionally related to A as $A = \alpha Q^\beta$, the continuity equation can be rewritten as

$$\frac{\partial Q}{\partial x} + \frac{dA}{dQ}\frac{\partial Q}{\partial t} = q \tag{6.11}$$

And

$$\frac{\partial A}{\partial t} = \frac{dA}{dQ}\frac{\partial Q}{\partial t} = \alpha\beta Q^{\beta-1}\left(\frac{\partial Q}{\partial t}\right) \tag{6.12}$$

Or

$$\frac{\partial Q}{\partial x} + \alpha\beta Q^{\beta-1}\left(\frac{\partial Q}{\partial t}\right) = q \qquad (6.13)$$

Considering a linear solution, the equation is linearized by substituting an average of known solutions for the coefficient of the nonlinear term. Thus, the following solution is obtained:

$$\frac{Q_{i+1}^{j+1} - Q_i^{j+1}}{\Delta x} + \alpha\beta\left(\frac{Q_{i+1}^j + Q_i^{j+1}}{2}\right)^{\beta-1}\left(\frac{Q_{i+1}^{j+1} - Q_{i+1}^j}{\Delta t}\right) = \left(\frac{q_{i+1}^{j+1} + q}{2}\right) \qquad (6.14)$$

$$Q_{i+1}^{j+1} = \frac{\left(\dfrac{\Delta t}{\Delta x}Q_i^{j+1} + \alpha\beta\left(\dfrac{Q_{i+1}^j + Q_i^{j+1}}{2}\right)^{\beta-1} Q_{i+1}^j + \Delta t\left(\dfrac{q_{i+1}^{j+1} + q_{i+1}^j}{2}\right)\right)}{\left(\dfrac{\Delta t}{\Delta x} + \alpha\beta\left(\dfrac{Q_{i+1}^j + Q_i^{j+1}}{2}\right)^{\beta-1}\right)} \qquad (6.15)$$

In Equation (6.15), the subscript refers to the space coordinate, and the superscript refers to the time coordinate. The solution advances on a timeline from upstream to downstream. For illustrative purposes, the kinematic wave worked example is provided below.

6.6.1 Worked example—Kinematic wave routing technique

A large rectangular flood control channel of width $b = 60.96$ m is 7315.2-m long, has a bed slope of $S_0 = 0.01$, and a Manning's roughness factor $n = 0.035$. The inflow hydrograph to the channel is tabulated below. Implement a linear finite-difference solution of the kinematic wave equations to route the inflow hydrograph to the end of the channel. The initial conditions correspond to uniform flow along the channel at a rate of 56.63 m³/s. Use Δt of 3 minutes (180 seconds) and Δx of 914.4 m.

Inflow time (min)	Inflow rate (m³/s)
0	56.63
12	56.63
24	84.95
36	113.27
48	141.58
60	169.90
72	141.58
84	113.27
96	84.95
108	56.63
120	56.63

Table 6.1 is the application of Equation (6.15) to route the hydrograph given above. When we solve Equation (6.7) for the flow depth, it can be shown that the flow depth, y, corresponding to $Q = 141.58$ m³/s is 0.890 m, that is

216 Integrated Drainage Systems Planning and Design for Municipal Engineers

Table 6.1 Integrating downstream

		Distance (m), x								
j		1	2	3	4	5	6	7	8	9
		0	914.4	1828.8	2743.2	3657.6	4572	5486.4	6400.8	7315.2
i	Time (min)									
1	0.00	56.63	56.63	56.63	56.63	56.63	56.63	56.63	56.63	56.63
2	3.00	56.63	56.63	56.63	56.63	56.63	56.63	56.63	56.63	56.63
3	6.00	56.63	56.63	56.63	56.63	56.63	56.63	56.63	56.63	56.63
4	9.00	56.63	56.63	56.63	56.63	56.63	56.63	56.63	56.63	56.63
5	12.00	56.63	56.63	56.63	56.63	56.63	56.63	56.63	56.63	56.63
6	15.00	63.71	59.32	57.64	57.01	56.77	56.68	56.65	56.64	56.65
7	18.00	70.79	63.77	59.98	58.13	57.28	56.91	56.75	56.68	56.71
8	21.00	77.87	69.35	63.61	60.22	58.39	57.47	57.02	56.81	56.86
9	24.00	84.95	75.65	68.36	63.37	60.29	58.54	57.59	57.10	57.16
10	27.00	92.03	82.40	74.00	67.55	63.09	60.27	58.61	57.67	57.71
11	30.00	99.11	89.42	80.32	72.65	66.84	62.80	60.21	58.63	58.62
12	33.00	106.19	96.58	87.10	78.54	71.50	66.20	62.52	60.11	59.99
13	36.00	113.27	103.82	94.21	85.04	76.99	70.48	65.62	62.23	61.95
14	39.00	120.35	111.10	101.50	92.00	83.19	75.61	69.57	65.09	64.59
15	42.00	127.43	118.40	108.91	99.27	89.94	81.50	74.37	68.75	67.99
16	45.00	134.51	125.69	116.38	106.74	97.13	88.04	79.96	73.24	72.20
	48.00	141.59	132.97	123.86	114.33	104.60	95.09	86.26	78.54	77.23
	51.00	148.67	140.24	131.34	121.98	112.28	102.53	93.15	84.59	83.04
	54.00	155.75	147.50	138.81	129.64	120.06	110.24	100.51	91.31	89.56
	57.00	162.83	154.75	146.26	137.30	127.89	118.12	108.22	98.57	96.69
	60.00	169.91	161.98	153.68	144.94	135.73	126.09	116.17	106.25	104.28
	63.00	162.83	162.39	157.81	150.99	142.83	133.77	124.11	114.17	112.11
	66.00	155.75	159.23	158.49	154.54	148.31	140.51	131.60	122.01	119.82
	69.00	148.67	154.25	156.48	155.45	151.67	145.72	138.12	129.34	127.03
	72.00	141.59	148.32	152.64	154.13	152.83	149.05	143.20	135.72	133.32
	75.00	134.51	141.92	147.63	151.07	152.00	150.44	146.58	140.75	138.35
	78.00	127.43	135.27	141.90	146.79	149.56	150.02	148.20	144.22	141.92
	81.00	120.35	128.51	135.75	141.67	145.88	148.08	148.14	146.05	141.92

84.00	113.27	121.68	129.36	136.02	141.31	144.93	146.64	146.33	143.97
87.00	106.19	114.83	122.84	130.02	136.13	140.85	143.95	145.22	144.55
90.00	99.11	107.98	116.25	123.83	130.53	136.11	140.32	142.95	143.81
93.00	92.03	101.13	109.64	117.52	124.67	130.90	135.99	139.73	141.92
96.00	84.95	94.30	103.02	111.16	118.65	125.37	131.15	135.79	139.09
99.00	77.87	87.49	96.43	104.78	112.54	119.65	125.96	131.31	135.52
102.00	70.79	80.70	89.86	98.42	106.40	113.80	120.52	126.43	131.38
105.00	63.71	73.94	83.34	92.08	100.27	107.90	114.94	121.29	126.81
108.00	56.63	67.22	76.86	85.80	94.16	101.99	109.28	115.96	121.95
111.00	56.63	63.17	71.46	79.98	88.26	96.16	103.61	110.55	116.88
114.00	56.63	60.68	67.26	74.90	82.79	90.57	98.05	105.13	111.72
117.00	56.63	59.15	64.14	70.66	77.89	85.34	92.70	99.81	106.54
120.00	56.63	58.20	61.87	67.24	73.65	80.58	87.67	94.67	101.44

$$141.58 = \frac{\sqrt{0.01}}{0.035}\left(\frac{60.69y}{60.69+2y}\right)^{2/3}(60.69y)$$

Thus, by trial and error $y = 0.890$m. Clearly $y << b$ and the channel is very wide. Therefore, Equation (6.8) yields

$$\alpha = \left(\frac{nP^{2/3}}{\sqrt{S_f}}\right)^{3/5} = \left(\frac{0.035 \times 60.69^{2/3}}{\sqrt{0.01}}\right)^{3/5} = 2.752 \quad \text{and} \quad \beta = 3/5 = 0.6$$

Using these values of α and β in Equation (6.15) yields the solution shown in Table 6.1 and in Figure 6.7.

For example, for $j = 5$, and $i = 1$, Equation (6.15) is

$$Q_2^6 = \frac{\left(\frac{\Delta t}{\Delta x}Q_1^6 + \alpha\beta\left(\frac{Q_2^5+Q_1^6}{2}\right)^{\beta-1}Q_2^5 + \Delta t\left(\frac{q_2^6+q_2^5}{2}\right)\right)}{\left(\frac{\Delta t}{\Delta x}+\alpha\beta\left(\frac{Q_2^5+Q_1^6}{2}\right)^{\beta-1}\right)}$$

Where

$\alpha = 2.752$, $\beta = 0.6$, $\Delta t = 180$ seconds, and $\Delta x = 914.4$ m. Also, $Q_1^6 = 63.71$ m³/s, $Q_2^5 = 56.63$ m³/s, $q_2^6 = 0$, $q_2^5 = 0$. These values produce $Q_2^6 = 59.32$ m³/s. The integration proceeds by increasing i until the end of the reach and then resetting i to 1 and increasing j.

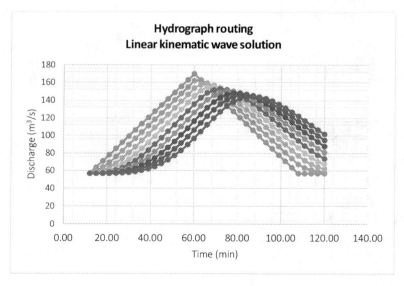

Figure 6.7 Hydrograph routing for the linear kinematic wave routing.

6.7 DYNAMIC WAVE ROUTING TECHNIQUE

The dynamic wave routing solves the complete 1-D Saint Venant flow equations and therefore produces the most theoretically accurate results. These equations consist of the continuity and momentum equations for conduits and a volume continuity equation at nodes in a hydraulic system.

Unsteady flow conditions are properly catered for by applying the dynamic wave routing method. Unsteadiness results from rapid and abrupt or gradual changes in depth of flow as the flow propagates through the channel. The Froude number of the approach velocity is of significant importance in routing to cater for unsteadiness in hydraulic modeling. In many cases, looped drainage networks in a hydraulic model are designed with minimal controls so that the approach flow velocity Froude number for the entire channel is mantained a constant.

Using conduit illustration, imagine partially full flow; critical flow conditions at outlet; in outlet control conditions; neglecting lateral inflows; neglecting wind shear stress and eddy losses, considering concentration time, node A feeds B through link C; maximum available energy (headwater depth) is fixed at manholes/nodes. The node could be fed by several links (both drains and pipes)—Figure 6.8. **Note:** The length of a link C affects its carrying capacity in circumstances where the available energy is fixed at both nodes A and B. Look at nodes A and B as manholes connecting a concrete storm sewer pipe.

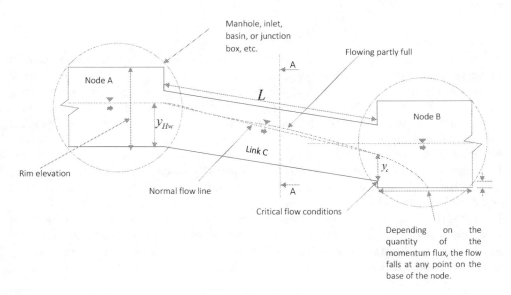

Figure 6.8 Typical conduit/node sectional illustration.

Section A–A

The governing hydrodynamic flow equations in 1-D, with a momentum coefficient equals to 1 and neglecting lateral inflows and wind shear, take form

$$\frac{\partial A}{\partial t} + \frac{\partial Q}{\partial x} = 0 \qquad (6.16)$$

Through manipulating variables, Equation (6.4) can be expressed as shown in Equation (6.17) in which the friction slope is introduced from Manning's equation. Equation (6.17) in its complete form is complex to solve except in very simplified cases where some components are dropped. In many cases, software programs are used setting initial boundary conditions.

$$\frac{\partial Q}{\partial t} + \frac{\partial \left(Q^2/A\right)}{\partial x} + gA\frac{\partial Z}{\partial x} + \frac{gn}{AR^{4/3}}Q|Q| = 0 \qquad (6.17)$$

Temporal spatial   Gravity   Friction

Where

Q	:	Discharge
t	:	Time
A	:	Cross-sectional area
n	:	Friction coefficient
g	:	Acceleration due to gravity
x	:	Distance
R	:	Hydraulic radius
Z	:	Stage

Equation (6.16) expresses mass conservation, and Equation (6.17) expresses momentum conservation. By defining boundary conditions, say at $t = 0$, $x = 0$, $Q = 1.2$ m³/s at node A and perhaps critical depth at exit of link C, the dynamic wave routing method is applied using software such as MATLAB or Autodesk Storm and Sanitary Analysis and hence generate hydrographs for nodes and links in the hydraulic model system. The incoming hydrograph is always routed downstream—it is better to route downstream. Because the links are always interconnected in storm sewer networks, node A would receive varying discharges at any time in the storm. The duration of the storm is always fixed in the model, say 2 hours (120 minutes). That means the discharge entering node A will keep changing. The inflow hydrograph at node A can be generated which can be successfully routed through link C. The two governing Equations (6.16) and (6.17) are usually solved by a second-order finite-difference methods. Different numerical schemes are used to solve Saint Venant equations. If both the **temporal acceleration** and **spatial acceleration** terms are removed, Equation (6.17) represents steady and uniform flow and it reduces to a simple, Manning's-based equation for flow, that is, Equation (6.18):

$$Q = \frac{1}{\sqrt{n}} AR^{2/3} \left(\frac{\partial Z}{\partial x}\right)^{1/2} \qquad (6.18)$$

Equations (6.16) and (6.17) are nonlinear analytical solutions, which are both partial differential equations; a finite-difference method is used to solve these two governing equations. Third- or higher-order accurate numerical methods are used to solve these equations. Several explicit and implicit finite difference schemes are used to solve 1-D Saint Venant equations for solving flow problems in open channels. These include the following: MacCormack discretization scheme, Lax-Wendroff discretization scheme, Preissmann implicit model, and the method of characteristics (MOC). This chapter will explain the method of characteristics, which is one of the most reliable methods used in the modern day.

6.7.1 How to solve the full 1-D Saint Venant equations using the method of characteristics (finite-difference scheme)

The method of characteristics aims to reduce a partial differential equation to a family of ordinary differential equations along which the solution can be integrated from some initial data given a suitable hypersurface. It is a technique typically applicable to first-order equations, although it is valid for hyperbolic partial differential equations.

Step 1. Rearranging the continuity and momentum equations in a matrix form

Let us consider the continuity and momentum equations as shown in Equations (6.19) and (6.20), respectively, again introducing the density ρ.

$$\frac{\partial \rho A}{\partial t} + \frac{\partial \rho A v}{\partial x} = 0 \tag{6.19}$$

Momentum equation

$$\frac{\partial \rho A v}{\partial t} + \frac{\partial \rho A v^2}{\partial x} = -A\left(\frac{\partial P}{\partial x} + \rho g\left(S_0 - S_f\right)\right) \tag{6.20}$$

If density is considered constant, the continuity becomes

$$\frac{\partial A}{\partial t} + \frac{\partial Q}{\partial x} = 0 \tag{6.21}$$

While the momentum equation becomes

$$\frac{\partial Q}{\partial t} + \frac{\partial}{\partial x}\left(Q^2/A\right) = -A\left(\frac{1}{\rho}\frac{\partial P}{\partial x} + g\frac{\partial Z}{\partial x} + \frac{\lambda}{2d}v|v|\right) \tag{6.22}$$

Therefore, moving from pressure P to water height y, we compare the average pressure along the water depth.

$$\overline{P} = \frac{F}{A} = \frac{1}{A}\int_{y'=0}^{y} P(y')b(y')dy' = \frac{1}{A}\int_{y'=0}^{y} \rho g(y-y')b(y')dy' := \frac{\rho g}{A}I_1 \tag{6.23}$$

For a rectangular channel with constant breadth b, we have

$$\overline{P} = \frac{\rho g}{A}\int_{y'=0}^{y} \rho g(y-y')b(y')dy' := \frac{\rho g b}{A}\left[yy' - \frac{y^2}{2}\right]_0^y = \frac{\rho g}{2by} = \frac{\rho g y}{2} \tag{6.24}$$

The final form of the continuity and momentum equations (Saint Venant equations) is

$$\frac{\partial}{\partial t}\begin{pmatrix} A \\ Q \end{pmatrix} + \frac{\partial}{\partial x}\begin{pmatrix} Q \\ \frac{Q^2}{A} + gI_1 \end{pmatrix} = \begin{pmatrix} 0 \\ gA(S_0 - S_f) \end{pmatrix} \tag{6.25}$$

For a rectangular channel with constant width b, we have $I_1 = \dfrac{by^2}{2} = \dfrac{Ay}{2}$

Notice that this is a conservative set of equations in the form of

$$U_t + F_x = S \tag{6.25}$$

That is to say, temporal derivative + flux = source terms.

The equation can be rewritten as $U_t + A(U)U_x = S$, rewritten as a quasi-linear set of partial differential equations, we get a coefficient **matrix** $A(U)$.

With matrix A as shown below:

$$A = \frac{\partial F_i}{\partial U_j} = \begin{pmatrix} 0 & 1 \\ -\dfrac{Q^2}{A^2} + g\dfrac{dI_1}{dA} & \dfrac{2Q}{A} \end{pmatrix} = \begin{pmatrix} 0 & 1 \\ v^2 + gy & 2v \end{pmatrix} \tag{6.26}$$

For the rectangular channel,

$$\frac{dI_1}{dA} = \frac{y}{2} + \frac{A}{2b} = y \tag{6.27}$$

The eigenvalues of matrix A are

$$\lambda_{1,2} = v + \sqrt{2gy} \tag{6.28}$$

Thus, the wave celerity for a rectangular channel is given by

$$c = \sqrt{gy} \tag{6.29}$$

Step 2. Differential form and obtaining characteristic equations

We now replace the flow rate Q and wetted cross-section area A by velocity v and water depth y.

Thus, energy equation becomes

$$\frac{\partial y}{\partial t} + v\frac{\partial y}{\partial x} + y\frac{\partial v}{\partial x} = 0 \tag{6.30}$$

The momentum equation becomes

$$\frac{\partial v}{\partial t} + v\frac{\partial v}{\partial x} + g\frac{\partial y}{\partial x} = g(S_0 - S_f) \tag{6.31}$$

Also, replace the water depth by the wave celerity:

$$y = \frac{c^2}{g} \tag{6.32}$$

Thus,

$$y_t = \frac{2cc_t}{g} \quad \text{and} \quad y_x = \frac{2cc_x}{g} \tag{6.33}$$

Replacing Equations (6.30) and (6.31) with the wave celerity, Equation (6.30) becomes

$$2c_t + 2vc_x + cv_x = 0 \tag{6.34}$$

Equation (6.31) becomes

$$v_t + vv_x + 2cc_x = g(S_0 - S_f) \tag{6.35}$$

Adding and subtracting the two equations give the following:
Addition:

$$(v + 2c)_t + (v + c)(v + 2c)_x = g(S_0 - S_f) \tag{6.36}$$

Subtraction:

$$(v - 2c)_t + (v - c)(v - 2c)_x = g(S_0 - S_f) \tag{6.37}$$

Equations (6.36) and (6.37) are called characteristics equations/lines. From Equations (6.36) and (6.37), we note that the slope of the characteristic lines is

$$\frac{dx}{dt} = v \pm c \tag{6.38}$$

And the Riemann invariants are

$$v + 2c \tag{6.39}$$

Step 3. Numerical technique for internal points

In the case of open channel flows, there are several differences between the slightly compressible cases (pressurized pipeline systems):

The time step should be calculated in such a way that characteristic lines should not get out of the grid.

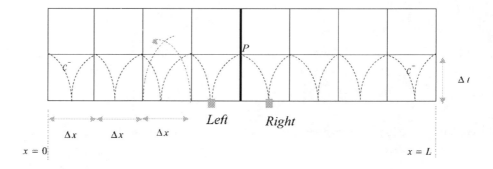

Figure 6.9 Characteristic grid for open channel flow.

- The wave celerity is not constant.
- The flow velocity can increase beyond the wave celerity.

That means the slope of the characteristic lines vary at each grid point, see Figure 6.9. Hence, we have to choose appropriate time steps Δt to ensure that neither (none) of the characteristic lines travels beyond the adjacent grid point per time step.

We calculate the time needed to travel a distance Δx with $v \pm c$ velocity for each grid point.

Thus,

$$\Delta t_i^+ = \frac{\Delta x}{|v_i + c_i|} \tag{6.40}$$

And

$$\Delta t_i^- = \frac{\Delta x}{|v_i - c_i|} \tag{6.41}$$

The final time step for all grid points is that of

$$\Delta t = \min_i \left(\Delta t_i^+, \Delta t_i^- \right) \tag{6.42}$$

This is identical to the Courant-Friedrichs-Lewy (CFL) stability criterion, that is

$$\Delta t = CFL \min_i \left(\frac{\Delta x}{|v_i| + c_i} \right) \tag{6.43}$$

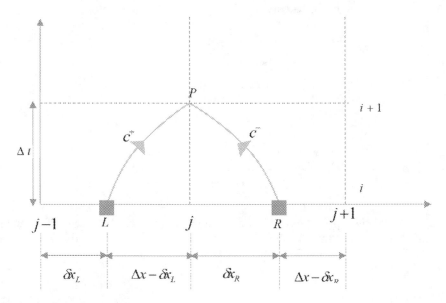

Figure 6.10 Updating internal points.

Where CFL is a **safety factor** for which the most common choice is **0.9**. See Figure 6.10. The time step is chosen in a way that neither of these characteristic lines goes beyond a grid point. Therefore, we limit the time step by the above formula.

From Figure 6.10, when updating the internal points, we need to find the 'L' (left) and the 'R' (right) points, from which the characteristic lines 'hit' grid point j and the next time step t_{j+1}. Then, we interpolate velocity and celerity linearly, that is

Velocity

$$v_L(\delta x_L) = v_{j-1}\left(1 - \frac{\delta x_L}{\Delta x}\right) + v_j \frac{\delta x_L}{\Delta x} \qquad (6.44)$$

Celerity

$$c_L(\delta x_L) = c_{j-1}\left(1 - \frac{\delta x_L}{\Delta x}\right) + c_j \frac{\delta x_L}{\Delta x} \qquad (6.45)$$

We also prescribe that

$$\frac{\Delta t}{\Delta x - \delta x_L} = \frac{1}{v_L(\delta x) + c_L(\delta x)} \qquad (6.46)$$

That is when the characteristic line started from point 'L' with the interpolated c_L and v_L values meets the grid point x_j at the next time step.

Solving these equations for δx_L gives

$$\delta x_L = \Delta x \frac{1 - \frac{\Delta x}{\Delta t}\left(c_{j-1} + v_{j-1}\right)}{1 + \frac{\Delta x}{\Delta t}\left(c_j - c_{j-1} + v_j - v_{j-1}\right)} \tag{6.47}$$

Similarly,

$$\delta x_R = \Delta x \frac{1 - \frac{\Delta x}{\Delta t}\left(c_j + v_j\right)}{-1 + \frac{\Delta x}{\Delta t}\left(c_j - c_{j+1} + v_j - v_{j+1}\right)} \tag{6.48}$$

Once the left and right points have been computed, we progress by means of the standard method of characteristics (MOC), that is, compute the **Riemann invariants**.

$$\alpha_L = (y + 2c)_L \tag{6.49}$$

$$\beta_R = (y - 2c)_R \tag{6.50}$$

Then we use **forward Euler scheme** to update them, that is

$$\alpha = \alpha_L + \Delta t g\left(S_0 - S_f\right)_L \tag{6.51}$$

And

$$\beta = \beta_R + \Delta t g\left(S_0 - S_f\right)_R \tag{6.52}$$

Lastly, we compute the new primitive variables and the new time level:

$$v_j^{i+1} = \frac{(\alpha + \beta)}{2} \quad \text{and} \quad c_j^{i+1} = \frac{(\alpha - \beta)}{4} \tag{6.53}$$

Note that the above technique assumes that the flow is subcritical, that is, point 'L' is located between grid point $j-1$ and j, while point 'R' is located between j and $j+1$. In the case of supercritical flow, this is not the case. However, the computational technique can still be used provided that the left and right points are located correctly.

Step 4. Setting boundary conditions

At the left boundary we have β_R value known, hence our first equation is always

$$\beta_R = v - 2c \tag{6.54}$$

Where v and c stand for the flow velocity and wave celerity at the first node point, at the unknown (new) time level, respectively.

As long as the flow is **subcritical**, that is, $-c < v < c$, this characteristic quantity is available, and the second equation stems from the boundary condition as follows:

(a) **Prescribed velocity:** In this case v is known and we have

$$c = \frac{1}{2}(v - \beta_R) \tag{6.55}$$

(b) **Prescribed water depth:** We have y prescribed, hence,

$$c = \sqrt{gy} \quad \text{and} \quad v = \beta_R + 2c \tag{6.56}$$

(c) Prescribed volumetric flow rate (see Figure 6.11).

Let $Q = byv$ be the prescribed volumetric flow rate. We can solve numerically the equation set:

$$\beta_R = v - 2c \tag{6.57}$$

$$Q = byv \tag{6.58}$$

$$c = \sqrt{gy} \tag{6.59}$$

Let us consider the case when the inflow is from a tank (detention pond, rainwater tank, or manhole) in which the total pressure is presented, that is

$$y_t = y + \frac{v^2}{2g} \tag{6.60}$$

This is the energy equation.

y_t is taken as the water tank depth or node depth or detention pond depth. In Figure 6.8, we assume that the node A can be approximated to a tank, receiving inflow hydrographs. Discharges are volumes in motion instantaneously changing. This is termed volumetric flow rate. In the design, nodes are located first and populate them with survey data, that is, invert levels. We thus feed the available headwater (as volumetric flow rate) as a prescribed boundary condition to route downstream. Raising the invert as shown in Figure 6.8 at the outlet creates critical flow conditions hence prescribing the initial boundary conditions at the exit. Often, the outlet at the node B (Figure 6.8) is in the form of a step so that y_B is the critical depth, and provided the step is not drowned, downstream conditions will not have any influence. Figure 6.11 shows the boundary conditions in the characteristic equations solving strategy with a regular grid. Therefore, we solve the energy equation together with the characteristic equation to give the equation below:

The solution is

$$v = \frac{1}{3}\left(\beta_R + \sqrt{2}\sqrt{6gy_t - \beta_R^2}\right) > 0 \text{ That is if } y_t > \frac{\beta_R^2}{4g} \tag{6.61}$$

The inlet velocity reaches the wave celerity if $v = c$, which occurs at $y_t = \dfrac{\beta_R^2}{2g}$

At this point,

$$c = -\beta_R$$

In the case of outflow, we need to replace $y_t = y + \dfrac{v^2}{2g}$ by $y_t = y$, which gives

$$v = \beta_R + 2\sqrt{gy_t} < 0 \text{ That is if } y_t < \dfrac{\beta_R^2}{4g} \tag{6.62}$$

The outflow reaches critical velocity if

$$y_t < \dfrac{\beta_R^2}{9g}$$

Beyond this point, we remove the y_t boundary condition and prescribe critical depth at the **outlet**.

$$y = \dfrac{\beta_R^2}{9g} \tag{6.63}$$

To sum up the above, we present the boundary conditions in Tables 6.2 and Table 6.3. In all cases,

$$2c = v - \beta_R \tag{6.64}$$

Since P is known, Equation (6.38) must be searched with the coordinates x_L and x_R, and Equations (6.36) and (6.37) are numerically solved in order to obtain v_p and y_p into all grid points. The contouring points have no negative feature (c+ and c−) so that the channel

Table 6.2 Open channel flow left boundary condition

Condition	$y_t < \dfrac{\beta_R^2}{2g}$	$\dfrac{\beta_R^2}{2g} < y_t < \dfrac{\beta_R^2}{4g}$	$\dfrac{\beta_R^2}{4g} < y_t < \dfrac{\beta_R^2}{2g}$	$y_t > \dfrac{\beta_R^2}{2g}$
Flow $v =$	Transcritical flow $-\sqrt{gy}$	Subcritical outflow $\beta_R + 2\sqrt{gy_t}$	Subcritical inflow $\dfrac{1}{3}\left(\beta_R + \sqrt{2}\sqrt{6gy_t - \beta_R^2}\right)$	Supercritical inflow Prescribed
$y =$	$\dfrac{\beta_R^2}{9g}$	$\dfrac{c^2}{g}$	$\dfrac{c^2}{g}$	Prescribed

Note: β is calculated from the characteristic equation. Similar derivation leads to the right-hand side boundary conditions, listed in Table 6.3.

Table 6.3 Open channel flow right boundary condition

Condition	$y_t < \dfrac{\alpha_R^2}{9g}$	$\dfrac{\alpha_L^2}{9g} < y_t < \dfrac{\alpha_L^2}{4g}$	$\dfrac{\alpha_L^2}{4g} < y_t < \dfrac{3\alpha_L^2}{2g}$	$y_t > \dfrac{3\alpha_L^2}{2g}$
Flow	Transcritical flow	Subcritical outflow	Subcritical inflow	Supercritical inflow
$v =$	$\sqrt{gy}$	$\alpha_L - 2\sqrt{gy_t}$	$\dfrac{1}{3}\left(\alpha_L - \sqrt{2}\sqrt{6gy_t - \alpha_L^2}\right)$	Prescribed
$y =$	$\dfrac{\alpha_L^2}{9g}$	$\dfrac{c^2}{g}$	$\dfrac{c^2}{g}$	Prescribed

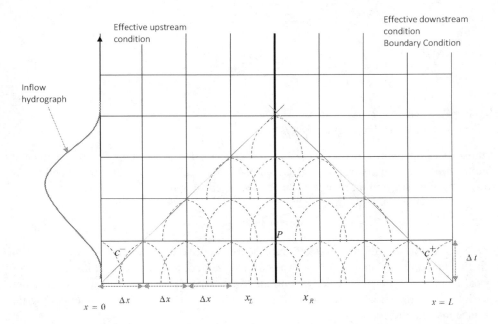

Figure 6.11 Boundary conditions in the characteristic equations solving strategy with a regular grid.

correctly ends. This calculation is applied only to 'interior points' but not for points where the contour has only one characteristic curve. At $x = 0$ there is only the negative curve (c−) and at $x = L$ there is only a positive curve (c+). Figure 6.11 gives the boundary conditions in the characteristic equations solving strategy with a regular grid. The solution depends on the initial condition at $t = 0$. In these contouring points, the characteristic equations for v and y, that is, Equations (6.36) and (6.37) must be supplemented by another equation from the boundary condition (Figure 6.11). Generally, at $x = 0$, the inflow hydrograph is used as an input boundary condition, and at the extreme downstream $x = L$, a relationship between flow and elevation (curve-key) is also used as a boundary condition.[7]

6.7.2 Analysis of dynamic wave routing technique

6.7.2.1 Understanding energy and momentum equations as we route flows in the design

In Figure 6.8, the depth of water above the culvert inlet bottom is known as the headwater depth. This depth represents the amount of energy available to convey or move water through the culvert. Headwater depths are a function of the entrance shape, along with the depth and velocity immediately inside the culvert. Headwater depths are determined by summing the energy losses associated with entrance shape, exit expansion and friction of the culvert. This is described in the basic energy balance equation:

$$y_{Hw} + \frac{v_{Hw}^2}{2g} + \Delta Z = y_{Tw} + \frac{v_{Tw}^2}{2g} + y_{\text{Friciton loss}} + y_{\text{Entrance loss}} + y_{\text{Exit loss}} \qquad (6.65)$$

Where:

- y : Depth (m)
- v : Velocity (m/s)
- ΔZ : Change in elevation (m)
- Tw : Tailwater
- Hw : Headwater
- g : Acceleration due to gravity (m/s²)

In most cases the approach velocity (v) is low and the approach velocity head is neglected. Similarly, the exit velocity can be neglected in the energy equation if the upstream and downstream channels are similar, reducing the headwater calculation to

$$y_{Hw} + \Delta Z = y_{Tw} + y_{\text{Friciton loss}} + y_{\text{Entrance loss}} + y_{\text{Exit loss}}$$

Where y_{Hw} is the sum of all losses and represents the difference in water surface elevation at the inlet (headwater) and outlet (tailwater). In some design analysis, bend losses, junction losses, or grate losses are neglected. Entrance loss depends on the geometry of the inlet. This loss is expressed as the velocity head immediately inside the culvert reduced by the entrance loss coefficient, K_e.

$$H_L = K_e \frac{v^2}{2g}$$

Where:

- H_L : Head loss (m)
- K_e : Entrance loss coefficient
- v : Velocity in the culvert barrel (m/s)
- g : Acceleration due to gravity (m/s²)

The headwater depth using the following equation is

$$y_{Hw} = (1 + K_e) \frac{v^2}{2g} + y$$

The total head loss is calculated as the sum of the entrance loss, exit loss, and friction loss. When the stormwater runoff propagates through the conduit, reaching node B, it mixes with slowly moving runoff within the node which creates a hydraulic jump which may force runoff back into the conduit (link C). From node A, depending on the quantity of the momentum flux, the flow falls at any point on the base of the node B (see Figure 6.8). As water accumulates in the node, it creates a swirl when speeding runoff meets the slowly moving water in the node. This creates a hydraulic jump which may cause backwater effects in link C. One of the purposes of the nodes is to drop the velocities. Therefore, the approach velocities for links may be approximated to zero. The invert levels are designed in such a way that the flow freely discharges at the exit of link C for simplicity. This creates an initial boundary condition, critical depth.

When developing storm sewer network models, the primary design assumptions (boundary conditions) is to assume the channel/conduit suddenly closes at the extreme end (worst scenario design), thereby making the velocity there equal to zero, creating backwaters and subcritical flows in the link. Or assume it terminates into free discharge creating critical flow conditions. This is very crucial in looped systems because of the interconnectivity of channels from different directions. That is why we assume that at a certain time (t), the node could get nearly full as a result of flows coming from other link(s).

The purpose of flow routing is to locate the flow profile, while examining how the flow instantaneously propagates with time and space along the channel. The purpose is to find the optimal hydraulic radius throughout the looped drainage network that conserves the flood waves (see Section 5.1), thus reducing embedded carbon as much as possible and practicable. Therefore, simplification of energy equations is vital while establishing initial boundary conditions. In looped network systems, backwaters are inevitable, causing extensive subcritical flows; the channels and drains may flood or surcharge depending on the size of friction/energy and bed slopes. Iterations are therefore carried out so that the optimal solution is obtained. Therefore, for the flood waves to be conserved, the time at which the flows reach the nodes from different interconnecting links keeps varying with flow and at some point the nodes may get nearly full. This may create backwater effects, flooding, and surcharging. The designer has to ensure that the flow waves are conserved and the links never flood or surcharge hence minimizing carbon as much as possible.

In Figure 2.9, node A acts as a control for all links. Supercritical flows are associated with controls at the upstream end of a channel, while subcritical flows are associated with controls at the downstream end. Looped storm sewer network assumes that the nodes (inlets, ponds, manholes, and junctions) may get nearly full, fully closed, and suddenly release the accumulated flow at some point the velocity $v = 0$. In that particular scenario, the downstream velocity for links 1 and 2 is assumed to be zero **(worst scenario design assumption)**, where the flow instantaneously becomes stationary at the downstream of the link. In the same way, the approach velocity at node A is neglected for link 3. Therefore, at junction node A (Figure 2.9), the links 1 and 2 are simulated and analyzed considering worst scenario that is when the downstream flow gets subcritical with a dominance of backwater effects as a result of node A flowing nearly full. When the node is full, it floods.

The Froude number proposed in Section 2.7 must allow flood waves to be properly conserved in the hydraulic system throughout the storm duration. The quantity of flow must be conserved all throughout as water propagates though the channel. The point is to design

232 Integrated Drainage Systems Planning and Design for Municipal Engineers

an optimal hydraulic system that will consume materials with the minimum amount of embedded carbon.

The looped storm sewer networks have junctions, manholes, and sustainable drainage systems (SuDS) such as detention ponds and inlets serving as control structures. All depth variations as flow propagates in the channel, including hydraulic jumps and backwaters, are fully catered for when the full Saint Venant equation is solved. In that case, we can obtain the optimal depth required to conserve the flood waves without flooding or surcharging. As an ideal, the designer can thus be confident that the runoff can be properly accommodated and conserved in the hydraulic system. When working out the hydraulic model, the designer can always adjust the Froude numbers until they obtain the most optimal solution weighing between the hydraulic efficiency of the system, the cost of constructing the hydraulic radius, and the associated carbon emissions. This is of course influenced by the type, capacity, and technology of the integrated SuDS upstream of the channels.

Section 9.6 gives an insightful example of carbon estimation where a storm sewer pipe is reduced from 910- to 610 mm, and a bioretention is designed to accommodate the excess flows. Section 3.8.2.3 provides the design criteria for a bioretention garden.

6.7.3 MATLAB simulation

Numerical techniques are solved in MATLAB. The techniques in Section 6.7.1, Figure 6.10, assume that the flow is primarily subcritical, that is, point 'L' is located between grid point $j-1$ and j, while point 'R' is located between j and $j+1$. In the case of supercritical flow, this is not the case. However, the computational technique can still be used provided that the left and right points are located correctly. Provided a (MATLAB-style) interpolator is given in the form of,

$$I(x_d, y_d, x_g),$$

which interpolates the data points (x_d, y_d) to grid x_g.

Let x_g denote the grid points. After a time step Δt, the location of the positive and negative characteristic lines starting from grid point $x_{g,i}$ will be

$$x_{g,i}^+ = x_{g,i} + \Delta t (v_i + c_i)$$

And

$$x_{g,i}^- = x_{g,i} + \Delta t (v_i - c_i)$$

Once the new locations are computed, we search for those points, which after the translation will be located at the original $x_{g,i}$ point. To find these points we interpolate 'backwards',

$$x_L = I(x_{g,i}^+, x_{g,i}, x_{g,i})$$

And

$$x_R = I(x_{g,i}^-, x_{g,i}, x_{g,i})$$

This approach is universal in the sense that it easily copes with cases when the flow is supercritical.

6.8 COMPARISON OF HYDRAULIC ROUTING TECHNIQUES

Hydraulic flow routing techniques are based on the continuity and momentum equations. The best routing approach is selected depending on the nature of project. The comparison of kinematic and dynamic wave routing for pipe storm sewer systems is provided below. Choosing between kinematic routing method and dynamic method for routing flow is purely out of understanding the project at hand because it influences the cost of hydraulic structures. The selection of the most appropriate routing technique depends on the complexity of the drainage network. The dynamic wave routing method solves the full Saint Venant's equation of shallow water equations and caters for backwater effects, channel storage, backwater, entrance/exit losses, flow reversal, and pressurized flow. The kinematic wave assumes the flow is uniform and that the friction slope is approximately equal to the channel slope. The dynamic wave describes the full 1-D Saint Venant equation and is valid for all channel flow scenarios. The dynamic method allows for pressurized flow such that maximum flow is greater than the full normal flow value. It can also account for channel storage, backwater, entrance/exit losses, flow reversal, and pressurized flow. This makes the dynamic method the best approach for drainage master planning projects.

The kinematic wave routing method when used to design looped drainage networks may lead to wider channels when used to route flood waves in looped drainage networks implying more embedded carbon. The reason is that the friction slope is equated to the energy slope, and therefore, higher values of depth (y) are obtained to be able to conserve the flood waves. The name kinematic refers to the study of motion exclusive of the influence of mass and force. The temporal acceleration and spatial acceleration are removed in the kinematic wave routing.

The example given in Section 6.6.1 for solving the kinematic wave is for pretty large-scale flood control channels or rivers. However, in the design of looped storm sewer networks, the lengths and widths of the channels are in the tens and hundreds of meters. With the advent of computers, this can be well simulated and correct approximations of optimal hydraulic radii obtained.

6.8.1 Kinematic wave routing

a. It solves the continuity equation and a simplified form of the momentum equation for each conduit.
b. The maximum flow is equal to the full normal flow value (no pressure flow).
c. It is the typical approach for master drains (big channels) especially flood control structures.
d. It applies the continuity equation: $A_1 V_1 = A_2 V_2$ or $Q_1 = Q_2$.
e. It results in attenuated hydrographs as flow is routed through a conduit.
f. It assumes that slope of the water surface equals the slope of the conduit.
g. Flow in excess of the full normal flow value is either lost to the system, can pond atop the node junction, or can be diverted.
h. It allows flows to vary spatially and temporally within a conduit/pipe.
i. **However,** it does not account for the following:
   - Backwater effects
   - Entrance/exit losses
   - Flow reversal
   - Pressurized flow

6.8.2 Dynamic wave routing

a. It solves the complete 1-D Saint Venant equations, that is, continuity and momentum equation for each conduit and volume continuity equation for each node.
b. It allows for pressurized flow such that maximum flow is greater than the full normal flow value.
c. It is the typical approach for urban design projects—looped networks.
d. It is theoretically less conservative because of less simplifications.
e. Flooding occurs when the water depth at a node junction exceeds the maximum available depth.
f. It can also account for the following:
- Channel storage
- Backwater
- Entrance/exit losses
- Flow reversal
- Pressurized flow

6.9 STORMWATER RUNOFF HYDRAULIC MODELING AND SIMULATION

Due to the complexity of dealing with 3-D models, developing 1-D hydraulic models in software platforms simplifies the analysis and it is the commonest approach used to develop and simulate hydraulic models. 1-D hydraulic models are usually quickly dealt with. The model predicts the hydraulic grade lines (HGL) and energy grade lines (EGL).

Software such as Autodesk Storm and Sanitary Analysis is a **subbasin-node-link–based model** that performs hydrology, hydraulics, and water quality analysis of stormwater and wastewater drainage systems, including sewage treatment plants and water quality control devices. Subbasins contribute rainfall runoff and water quality pollutants, which then enter nodes. Figure 6.12 provides model for a network of storm sewer links and nodes. A node can represent the junction of two or more links (e.g., a manhole), a storm drain catch basin inlet, the location of a flow or pollutant input into the system, or a storage element (such as a detention pond, retention pond, treatment structure, or lake). From nodes, flow is then routed (or conveyed) along links. A link represents a hydraulic element (e.g., a pipe, open channel

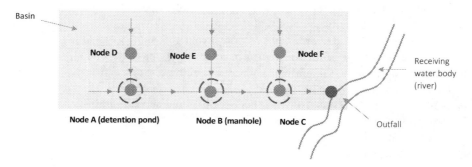

Figure 6.12 Typical hydraulic model sketch (1-D analysis of storm sewers).

stream, swale, pump, standpipe, culvert, or weir) that transports water and water quality pollutants). The hydraulic model is built following goals set in Section 2.5.1.

Storage routing techniques based on the continuity equation and the assumption that flow occurs at normal depth are used in many urban drainage models. These procedures neglect all kinematic effects, including both the variable slopes, which occurs during the passage of a flood wave and backwater effects of downstream controls. These models can be simulated with kinematic wave technique or dynamic wave technique.

A computer-based stormwater model makes it possible to estimate stormwater runoff characteristics from various rainfall events considering the land use and soil type of an area. Large paved surfaces translate into more stormwater runoff from rainfall events, compared to a naturally forested area or an area integrated with SuDS that absorb and hold a large percentage of the rainfall. Pipes, points representing junctions, inlets, manholes, outlets, outfall, and drains are input into the model as point and line features with the attributes (i.e., inverts, sizes, and geometry) populated using survey data and record drawings obtained from previous projects and field surveys conducted within the catchment. Specifically, the model development process consists of the following general components: existing storm water drainage network, subbasin drainage delineation, and scenario definition.

The roughness factors for the hydraulic structures are specified and input into the model. Nodes which represent junctions and manholes are identified and links representing sewers and drains annotated on the hydraulic models. Invert levels can be fed into the system, clearly taking into consideration the required excavations to level. The rim elevations at the node influences the available headwater for the adjoining sewers downstream. 1-D analysis pictorial representation is shown in the Figure 6.12. The model is populated by survey data. 1-D hydraulic models can be developed in several software populated using survey data.

Scenario definition. The configuration of the drainage network is the first stage of the model development. Changes to hydrologic characteristics (land use, soils, and rainfall) are applied to the modeled subbasins, and hydraulic (physical flow constraints) changes are applied to the conveyance components for each subbasin. Define primary and secondary structures and model them accordingly. With the same rainfall event applied to each subbasin, the resulting stormwater performance is obtained for the various subbasins/basin.

Existing stormwater drainage network. Existing drainage system can be remodeled by inputting new parameters. New alternatives can be proposed to estimate how the system performs and how much stormwater in the basins or subbasins is contributing to the proposed drainage system.

Model validation and verification. Model verification and validation are the two primary processes for quantifying and building credibility in numerical models. Verification is the process of determining that a model implementation accurately represents the developer's conceptual description of the model and its solution. Verification and validation is an important part of the simulation process. As you complete each project, you will become more adept at the process. There are several key phases to any simulation project to ensure success:

a. **Plan** the model through a functional specification.
b. **Build** the model.
c. **Verify and validate** the model to ensure we can trust it to solve the drainage problem.
d. **Run** validated model through various alternatives.
e. **Analyze** results to determine the best path forward.

In order to ensure that the model is correct, it is critical that it be verified and validated against actual performance. Without this step, there is no guarantee that the obtained results will be valuable. Verification and validation of the model is a skill that takes time to learn and often develops over each and every project.

Verification. Ensuring that the model behaves in a way it was intended is most likely the easier of the two tasks. If the model is constructed in segments (subbasins/basins), then each segment should be verified separately as it is completed. It is a good idea to verify the model as each of the segments are put together. The final verification must be performed with the completed model. Adding some animation constructs can aid in the verification process. Key elements can be displayed on the screen to observe important interactions occurring in the model. During the verification stage, checks on the function of the model under extreme conditions are conducted. One condition would be to create a single entity and follow it through the model logic. It is also good to consider using deterministic times that will allow you to more easily predict outcomes for simple simulation runs. Hydraulic models are usually simulated following anticipated area storm duration. Often modelers test the developing model at various stages under the same set of conditions. Occasionally, make extended runs to ensure that the randomness in the model does not create circumstances that were not previously considered, trying to create a variety of different situations to verify the model.

Validation. Ensuring that the model behaves the same as the real system can be challenging. If the system currently exists, then some kind of comparison can be made to ensure that the model represents the real world. If the system does not exist, but similar ones do, then the simulation results can be compared to the similar system, and at least a partial validation can be performed. If there is no real system to compare with the simulation, then true validation cannot be performed. If this is the case, the designer has to be sure to dedicate time to the verification process with interaction with the proposed system experts.

6.9.1 Available software used to develop stormwater models

The two governing routing Equations (6.1) and (6.4) may be solved by a variety of finite-difference techniques, both explicit and implicit, provided the boundary conditions are properly described. The software platforms available can be harnessed to quicken the analysis that would take many days/hours to arrive at the optimal solution that is environmentally sustainable and well-engineered to accommodate flows.

The purposes of routing flows in the hydraulic model is to correctly forecast the flows at any point in the hydraulic system at any time. When the model is simulated for the storm duration of a given return period and frequency, the flows keep fluctuating in the entire catchment basin or hydraulic model. It is, therefore, important to evaluate the behavior of the model that is designed to represent the real-life situation as explained in Section 6.9. Designers usually model the existing and the postdevelopment conditions subjecting the models to different conditions, in imperviousness, storms of different frequencies, and recurrence intervals to observe how the model changes. The overarching goal of the designer is to optimize the resource constraints and make the design environmentally sustainable and reliable.

Software such as Autodesk Storm and Sanitary Analysis presents several opportunities that enable the designer to model water quality volumes, pollutant loads, along with SuDS such as detention systems within the storm sewer networks, thus making the design lively and sustainable. The following software platforms are available for drainage engineers:

a. Autodesk Storm and Sanitary Analysis
b. InfoDrainage from Innovyze
c. MicroDrainage from Innovyze
d. ILLUDAS. ILLUDAS is the U.S. version of the RRL method developed in Great Britain. The model requires that the drainage basin be broken down into subareas, tributary to collection system inlets.
e. HEC-1 is one of a family of hydrologic/hydraulic computer models developed by the U.S. Army Corps of Engineers. In its most recent version, it has the ability to model the components of urban drainage systems.
f. EPA stormwater management model (EPA SWMM)
g. AutoCAD CIVIL 3D
h. MATLAB

6.10 CONCLUSION

The decision to apply a specific hydraulic routing method depends on the type of project because each hydraulic routing technique influences the size and hydraulic efficiency of the structures consequently impacting the cost of the drainage scheme. High construction cost means high carbon emissions. As outlined in Section 6.8, the advantages of flow routing techniques guide the decision to use which technique to route a flood wave.

References

1. Chow, V. T, Maidment, D. R, & Mays, L.W. (1988). *Applied Hydrology*. Singapore: McGraw Hill International Editions.
2. Singh, V. P. (2004). *Flow routing in open channels: Some recent advances*. Department of Civil and Environmental Engineering, Louisiana State University, Baton Rouge, LA, 70803–6405.
3. Nwaogazie, I. L. (1986). Comparative analysis of some explicit-implicit streamflow models. *Advances in Water Resources*, 10, 69–77. doi: 030917088702006909
4. Wujumdar, P. P. *Flood wave propagation. The Saint Venant equations*. Department of Civil Engineering, Indian Institute of Science, Bangalore.
5. Sitterson, J., Knightes, C., & Avant, B. (2018). *Flow routing techniques for environmental modeling*. Office of Research and Development National Exposure Research Laboratory. Athens, GA, 30605 United States Environmental Protection Agency.
6. Krylova, A. I., Antipova, E. A., & Perevozkin, D. V. (2017). The derivation of the Saint–Venant equations. Bull. Nov. *Comp. Center, Num.Model. in Atmosph., etc.*, 16, 21–35..
7. Barros, R. M., et al. (2014). *IOP Conference Series: Earth and Environmental Science*, 22, 042019.

Further reading

1. Autodesk® storm and sanitary analysis software, User's Guide, 2014.

Chapter 7
Useful topics in soil mechanics

NOTATION

τ : Shear stress
σ : Normal stress
σ' : Effective normal stress
σ_v : Total normal stress on a horizontal plane
σ_h : Horizontal stress
c : Cohesion intercept
c' : Cohesion intercept under drained conditions
c_u : Cohesion intercept under undrained conditions
ϕ : Internal angle of shearing resistance
ϕ' : Internal angle of shearing resistance under drained conditions
ϕ_u : Internal angle of shearing resistance under undrained conditions
k : Permeability

7.1 INTRODUCTION

Knowledge of soil mechanics is an essential component for a drainage engineer. Basic understanding of how drainage conditions affect the foundations is essential to build sustainable drainage structures and associated structures like retaining walls, pavements, and buildings. While planning and designing sustainable drainage systems, underground water movement can be detrimental if not well planned for. The water table level is always a big factor while planning for all civil engineering structures because it poses several constraints imploring planners to be vigilant, exercising enough care for environmental protection. A satisfactory solution to subsurface drainage problems requires a knowledge of geology and an insight into soil mechanics.

On the other, integrated drainage systems planning incorporates greening activities (e.g., traditional landscaping), gray infrastructure, and sustainable drainage systems (SuDS). The greening techniques require careful selection of soils meeting agricultural specifications to support growth of native plants. Gray infrastructure and SuDS require careful evaluation of soils. In developing storm sewer solutions, soil shear strength parameters are an essential consideration. Soil is used as backfill in trenches and retaining walls, bedding material, or pipe surroundings. In that case, basic knowledge of soil mechanics becomes pertinent to the drainage engineer or planner. For instance, the soil that creates a retaining wall's foundation, or base, must be examined to ensure it meets the shear strength parameters required to support the wall. Also, saturated backfill soil, especially clay, is detrimental to the retaining wall and must be avoided because it can cause excessive hydrostatic pressures exerted on the wall causing it to bulge, overturn, or slide. Retaining walls are essential components in planning for effective integrated stormwater management solutions because SuDS constantly require dealing will soils and site-specific constraints. These include the topography/slope for several SuDS, for example, infiltration systems and bioretention gardens.

The importation of soils for use in the construction of SuDS like infiltration systems, bioretention and planter boxes require tests such as infiltration test, permeability, pH test, organic matter content, and other closely related agriculture soil parameters. In developing SuDS, a thorough site investigation may be needed to evaluate soil composition, the soil strata, and the possibility of encountering contaminated land.

7.2 SUBSURFACE DRAINAGE

7.2.1 Introduction

Subsurface drainage is commonly concerned with pavement construction on roads and highways. In roads construction, particularly, the aim of subsurface drainage is to remove detrimental quantities of groundwater to ensure a stable road bed and side slope conditions are achieved. A satisfactory solution to subsurface drainage problems requires the designer to have the knowledge of geology and a good insight into soil mechanics.

Subsurface drainage is primarily concerned with removing water that percolates through or is contained in the underlying subgrade. Groundwater, as distinguished from capillary water, is free water that occurs in saturation below the ground surface. Many variables and uncertainties exist regarding the real subsurface conditions. In many cases, the need for the installation of subsurface drainage systems can only be established on site during the construction stage.

Subsurface drainage is used to remove underground water, most of the time by using perforated or slit pipes. In contrast to overland drainage and building drainage, subsurface drainage is used to remove (deal with) underground water most of the time. This water is due to the high water table, or exceptional inclement weather conditions. For example, on roads and other pavement constructions, this water can occur due to gravity flow or capillary rise.

Subsurface drainage is concerned with removing water that percolates through or is contained in the underlying subgrade. This water is typically the result of a high water table or exceptionally wet weather. In roads and building structures, this subsurface water can accumulate under the pavement structure by two chief means:

a. **Gravity flow**. The top soil absorbs the surrounding water and then it flows by gravity to subsurface layers.
b. **Capillary rise**. Capillary rise is the rise in a liquid above the level of zero pressure due to a net upward force produced by the attraction of the water molecules to a solid surface (e.g., soil). Capillary rise can be substantial, moving up to 6 m or more. Often, capillary rise is a problem in areas of high groundwater tables. The capillary rise principally depends on type of soil and porosity of materials. In general, the smaller the soil grain size, the greater the potential for capillary rise. The capillary rise mechanism is

far more important to be considered while designing a building or a road. Capillary rise can happen without hydrostatic pressure. Therefore, it is essential to adopt preventive measures against capillary rise to maintain the load carrying capacity and durability of civil engineering structures. It is well known that historic or old buildings are most affected by capillary water. However, nowadays, with the development of modern construction technologies, there are wide verities of methods that slow down and prevent capillary water in building and road foundation.

7.2.2 Common subsurface drainage terminologies

It is important to note the following terminologies repetitively used in subsurface drainage:

a. **Permeability** is a measure of how easily a fluid can flow through a porous medium such as a geotextile fabric or soil. It has nothing to do with the fluid itself as it measures the ability of a porous medium to allow fluids to pass through it.

b. **Hydraulic conductivity** is a property of vascular plants, soils, and rocks that describes the ease with which a fluid (usually water) can move through pore spaces or fractures. It depends on the intrinsic permeability of the material, the degree of saturation, and on the density and viscosity of the fluid.

c. **The hydraulic gradient** is the driving force that causes groundwater to move in the direction of maximum decreasing total head. It is generally expressed in consistent units, such as meters per meter.

d. **Hydraulic head or piezometric head** is a specific measurement of liquid pressure above a vertical datum. It is usually measured as a liquid surface elevation, expressed in units of length, at the entrance (or bottom) of a piezometer.

e. **Seepage** is flow of water through soils. Seepage takes place when there is difference in water levels on the two sides of the structure such as a dam or a sheet pile, when water enters the ground surface at the upstream side of a retaining structure like a dam and comes out at the downstream side. Seepage is the lateral movement of subsurface water (Wopereis et al., 1994).[1]

f. **Infiltration** is when water enters the ground surface but does not come out, thus increasing the moisture content of the soil.

g. **Percolation** is the vertical movement of water beyond the root zone to the water table.

h. **Groundwater** is water found underground in the cracks and spaces in soil and rock. It is stored in and moves slowly through geological formations of soil, sand, and rocks called aquifers. The area where water fills the aquifer is called the saturated zone (or saturation zone). The top of this is called the water table. The water table may be located only one-third meter below the surface or it can be many meters down. The rate at which groundwater can be extracted depends on the effective hydraulic head and on the permeability, depth, slope, thickness, and extent of the water-bearing formation. Groundwater will always flow wherever a hydraulic gradient exists. Field exploration is the most satisfactory method of determining the discharge capacity. Generally, exploration should be undertaken during the rainy season when subsurface problems are most likely to be evident.

7.2.3 Primary materials used in construction of subsurface drainage systems

To deal with the excess subsurface water effectively and efficiently, **three main materials** are used as outlined below.

7.2.3.1 Nonwoven geotextile fabrics

The mostly used geotextile fabrics in drainage works is the nonwoven type because nonwoven geotextiles combine **the strength of woven fabrics** with **high permeability**, making them a better choice in SuDS, which require both **separation** and **filtration**. They are commonly applied in infiltration systems and bioretention gardens. Nonwoven fabrics comprise needle-punched polypropylene that allows separation and filtration to occur simultaneously. In SuDS construction, processed gravel applied to the upper layer would often sink into softer subsoils. With the advent of nonwoven geotextiles, the layers can now be properly separated avoiding mixing and contamination. Previously, before geotextiles, aggregates would simply disappear into the softer subsurface material.

The geotextiles are very effective and the geotextile industry has rapidly grown over the recent years. The effectiveness of the fabrics in separating soil layers and allowing excess water to permeate through the fabrics has seen the geotextile grow rapidly. In some cases, they are referred to as 'moisture regulating fabrics'. Nonwoven products are often used to wrap French drains or in conjunction with other subsurface drainage solutions. Synthetic fiber filter fabrics should also be considered where there is a definite engineering and cost advantage, but preference should be given to the use of local natural materials.

Nonwoven geotextile is used for soil separation and permeability. The side trenches soil would mix with the aggregates; so, the geotextile separates the two; the pipe is also not submerged into the high water table; because of high permeability, water passes (permeates) through geotextile to the aggregates and then falls by gravity through the perforated pipe.

7.2.3.2 Perforated pipes

Perforated pipes are used to remove excess subsurface water. They are constructed while holes are pointing down for effective drainage. Cost, ease of handling, and ease of laying should be the main considerations when specifying subsurface pipes, which may consist of perforated, slotted, or open jointed concrete, clay, pitch fiber, or plastic pipes.

7.2.3.3 Filter materials

Permeable filter materials for subsurface drains consist of sand and/or crushed stone plus gravel of suitable grading. Grading specifications may have to be varied to suit availability of material.

7.2.4 Subsurface drain types

To deal with the excess subsurface water effectively and efficiently, **two main drain types** are used as outlined below.

7.2.4.1 French drains

A French drain is a trench filled with gravel or rock, or both, containing a perforated pipe at or near the bottom that redirects surface water and groundwater away from an area. The pipe has perforated holes or slots in the pipe. The pipe fills with water when the ground becomes saturated. Perforated pipe can be made of a variety of materials, which include steel, iron, uPVC, cement, and clay, but the most popular type is high-density polyethylene (HDPE).

The perforated pipe is laid when holes are pointing downward to allow for effective drainage of excess water. However, French drains, which consist of a trench filled with unspecified crushed rock, have been found to be unreliable and are not recommended.

The standard pipe subsurface drain consists of a half-perforated pipe pointing downward at the bottom of a narrow trench, which is backfilled around the pipe with filter material. The filter material is always wrapped in a filter fabric especially when it does not meet the grading requirements.

The pipes may be perforated, slotted, and porous or open jointed. This type should be used as a single subsurface drain along the toe of a cut to intercept seepage, along the toe of an embankment on the side from which groundwater originates and across the roadway at the downhill end of a cut (Figure 7.1).

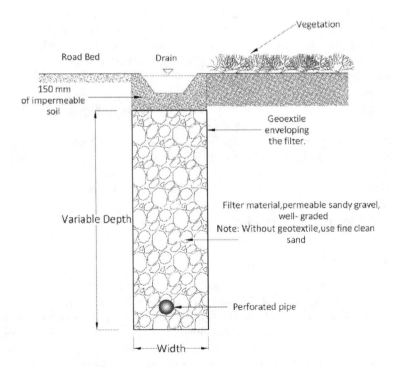

Figure 7.1 Typical French drain.

7.2.4.2 Stabilization trenches

These are usually wider trenches with sloping sides and have a blanket of filter material up to about 900-mm thick on the bottom and side slopes of the trench. Filter fabric may be required to line the trench. At least one subsurface pipe of 200-mm diameter should normally be laid at the bottom of the trench. Stabilization trenches may be required under side cut fills to stabilize waterlogged fill foundation areas that are well defined. They are usually linked to herring-bone configurations and to filter blankets.

7.2.5 Road construction subsurface drain categories

In road construction, different filter materials are used in subsoil, subpavement, and foundational drains. Subsurface drains in roads and highways have three categories.

7.2.5.1 Subsoil drains

Subsoil drains are intended for the drainage of groundwater or seepage from the subgrade and/or the subbase in cuttings and fill areas.

7.2.5.2 Subpavement drains

These are intended for the drainage of the base and subbase pavement layers in flexible pavements. They may also function to drain seepage or groundwater from the subgrade. Subpavement drains are designed to drain water from base and subbase pavement layers in flexible pavements, and to drain seepage or groundwater from the subgrade. The designer ensures that subsoil or subpavement drains are provided on both sides of the formation in the following locations[2]:

- Highly moisture susceptible subgrades, that is, commonly displaying high plasticity or low soaked California Bearing Ratios'(CBRs).
- All urban/street developments—regardless of abutting land zoning.
- Locations of known hillside seepage, high water table, isolated springs, or salt-affected areas.
- Cut formations where the depth to finished subgrade level is equal to or greater than 400 mm below the natural surface level.
- Irrigated, flood-prone, or other poorly drained areas.

7.2.5.3 Foundation drains

Foundation drains are intended for the drainage of **seepage, springs**, and **wet areas** within and adjacent to the foundations of the road formation. In buildings, foundation drains are pipes that are installed under the foundation or basement floor to collect water and move it off-site to prevent the basement from filling with water.[3] Foundation drains that are directly connected to the sanitary sewer work well and require no maintenance, but they also add a lot of clear water to the sewer, which greatly increases the chances of a basement backup for you and your neighbors. They are warranted for use where excessive groundwater is encountered within the foundation of an embankment or the base of cutting, or used to intercept water from entering these areas. The need to provide foundation drains may be apparent from the results of the geotechnical survey along the proposed road formation alignment, and in this case, the location shall be shown on the drawings.

However, more commonly, the need to provide foundation drains is determined during construction, and hence in this situation requirements and locations cannot be ascertained at the design stage. For noncorrugated pipes, an absolute minimum grade of 0.5% is acceptable. The minimum trench width of foundation drains is about 300 mm, with a variable trench depth to suit the application and ground conditions on site. Outlets are spaced at maximum intervals of 150 m. Where practicable, cleanouts are provided at the commencement of each run of foundation drain and at intervals not exceeding 80 m. Where not practicable to provide intermediate cleanouts, outlets are spaced at maximum intervals of 100 m.

7.2.6 Drainage mats in road construction

Drainage mats help water to flow away from the foundation. Without it, water can stay trapped against the concrete or pool at the bottom.[2] Drainage mats are used to drain water away from structures that are in contact with soil, as well as in surface drainage systems, green roofs, foundation wall protection systems, landfill sites, and tunnel construction.

Types of drainage mats

- **Type A drainage mats** are intended to ensure continuity of a sheet flow of water under fills to collect seepage from a wet seepage areas, or for protection of vegetation or habitat downstream of the road reserve where a fill would otherwise cut the flow of water. Type A drainage mats are constructed after the site has been cleared and grubbed and before commencement of embankment construction.
- **Type B drainage mats** are constructed to intercept water that would otherwise enter pavements by capillary action or by other means or fills and to intercept and control seepage water and springs in the floors of cuttings. Type B drainage mats are constructed after completion of the subgrade construction and before construction of the pavement. The need to design for the provision of drainage mats should be apparent from the result of the geotechnical survey along the proposed road formation alignment.

7.2.7 Filter material differences

Filter materials are gravel-filled trenches that collect and move water. Several types of filter material exist.

a. Type A filter material for use in subsoil, foundation, and subpavement drains and for Type B drainage mats.
b. Type B filter material for use in subsoil, foundation, and subpavement drains.
c. Type C filter material comprising crushed rock for use in Type A drainage mats.
d. Type D filter material comprising uncrushed river gravel for use in Type A drainage mats.

The type of filter material specified to backfill the subsurface drainage trenches (subsoil, foundation, and subpavement drains) depends on the permeability of the pavement layers and/or subgrade and the expected flow rate. Generally, Type A filter material is used for the drainage of highly permeable subgrade or pavement layers such as crushed rock or coarse sands, while Type B filter material is used for the drainage of subgrade and pavement layers of lower permeability such as clays, silts, or dense graded gravels.

7.3 SUBSURFACE DRAINAGE DESIGN CRITERIA

The objective in the design of the subsurface drainage system is to control moisture content fluctuations in the pavement and/or subgrade to within the limits assumed in the pavement design. In areas with a history of salinity problems, subsurface drainage may be prescribed to keep the groundwater table lower in the strata to avoid progressive deterioration of the health of topsoil and upper layers due to salinity levels increased by rising and/or fluctuating groundwater tables. In terms of integrated drainage design, there are two primary purposes of considering subsurface drainage in any project development:

a. Purposes of cropping in landscaping.
b. Purposes of developing a reliable road, parking, or compound foundation.

7.3.1 Subsurface drainage special design guidelines or considerations

a. The minimum inside pipe diameter for a pipe subsurface drain is normally not be less than 100 mm.
b. A 75-mm diameter pipe may be considered under special circumstances.
c. Surface drainage should not be permitted to discharge into a subsurface drain.
d. The discharge from a subsurface drain into a stormwater drain of culvert is permissible, provided the possibility of water backing up into the subsurface drain is avoided.
e. In general, the subsurface drain grade should not be flatter than 0.5%. If this slope cannot be achieved, an absolute minimum grade of 0.25% may be accepted.
f. The choice of depth and spacing of the subsurface drain is flexible and depends on the permeability of the soil, the level of the water table, and the amount of drawdown needed to ensure stability.
g. Wherever practical, a subsurface drain pipe should be set in the impervious layer below the saturated layer.

7.3.2 Maintenance design considerations for subsurface drainage

a. The start and outlet of pipe for the subsurface drains should be indicated on the surface by means of markers. These are necessary for the maintenance personnel.
b. Design of subsurface drains is associated with soil type and seepage flow rate. The interrelationship between the permeable filler material, the filter fabric, and the subsurface drain pipe selected is important in the design of an effective subsurface drain. Cost could have a major bearing on the type of subsurface drain used. All possibilities should therefore be investigated.

7.4 DRAINAGE AND GROUNDWATER RECHARGE

Groundwater recharge, sometimes referred to as **deep percolation or deep drainage**, refers to a hydrologic process in which water moves downward from surface water to groundwater. It is either human-induced or a natural part of the hydrologic (water) cycle. Human-induced groundwater recharge occurs through spreading basins, injection wells, irrigation practices, and waste disposal. Groundwater recharge is considered the primary approach in which water enters an aquifer. This process commonly occurs in the vadose zone below plant roots and it is expressed as a flux to the water table surface. Integrated drainage systems planning aims at achieving sustainability. In that case, several types of SuDS, for example, bioretention gardens, permeable pavers, planter boxes, and infiltration systems highly recharge groundwater. That is one of the sustainable benefits that SuDS provide to recharge the aquifers.

Dealing with groundwater sometimes cannot be avoided. While SuDS promote groundwater recharge, design of reliable building structures may need removal of excess groundwater through constructing adequate drainage systems. In that case, subsurface drains as outlined in Section 7.2 must be constructed. The rate at which groundwater can be removed depends on the effective hydraulic head and on the permeability, topography, depth, thickness, and extent of the water-bearing formation. Therefore, field exploration is the most satisfactory method of determining the discharge capacity. Generally, exploration must be undertaken during the rainy season when subsurface problems are evident. The concepts of hydraulic head, hydraulic gradient, and hydraulic conductivity are essential in designing systems to control groundwater movement.

7.5 MICRO-TUNNELING, PIPE JACKING, AND UNDERGROUND DRAINAGE SYSTEMS

Micro-tunneling and pipe jacking are similar techniques of trenchless technology used for pipeline installation. These techniques can be used for installations of pipes ranging from diameters of 150- to 3000 mm. The difference between micro-tunneling and pipe jacking is that micro-tunneling is well-suited to situations where a pipeline has to conform to rigid line and level criteria whereas a pipe jack is a system designed to directly install the pipe behind an open-face tunnel boring machine (TBM). Tunneling and underground drainage systems are warranted under the following circumstances:

a. When the scheme route is feasible when more direct than bending.
b. To avoid traffic interruptions.
c. To create a cost-effective route under buildings.
d. When open-cut construction seems very expensive.
e. When daily life should not be disturbed while operating.
f. When the risk of disturbing structures.
g. Where there is a risk of ground collapse during the installation of pipes.
h. Where excavations are undesirable, such as in developed locations or areas with a high risk of ground collapse.

Once tunneling is evaluated as the most feasible option, then adequate site investigations must be conducted. This necessitates collecting geological maps for the area and conducting geotechnical tests for use in tunneling or micro-tunneling. The cost of tunneling must be evaluated together with several other constraints. For example, in areas where the water table is high, the constructed subdrains connecting to pumps at the shaft may become inadequate for drainage and compressed air working may be the most appropriate. In some cases, ground freezing, chemical injection, or wellpoint dewatering may be resorted to as alternatives. However, in such circumstances tunneling may be become unfeasible.

In case micro-tunneling is required, a thorough site investigation should be conducted and several boreholes sited. Pipe jacking is usually conducted in micro-tunneling (it is a trenchless technology product) exercise. Pipe jacking or micro-tunneling is a technique for installing underground pipelines, ducts, and culverts without use of open trenches. Powerful hydraulic jacks are used to push specially designed pipes through the ground behind a shield at the same time as excavation is taking place within the shield.[2]

- a. **Micro-tunneling.** The term micro-tunneling applies to remotely controlled, steerable trenchless machines using either slurry, auger, or vacuum spoil removal to install pipelines (www.pipeliner.com.au)[4]. Micro-tunneling machines are controlled from the surface, with location and operation of the machine being continuously monitored, usually by means of a laser guidance system. Most micro-tunneling drives are straight between shafts, although, increasingly in recent years, various companies have developed guidance systems that enable curved drives to be completed, particularly on longer length, larger-diameter bores. Micro-tunneling is well-suited to situations where a pipeline has to conform to rigid line and level criteria.
- b. **Pipe jacking.** A pipe jack is defined as a system of directly installing pipe behind an open-face TBM. The pipe is then hydraulically jacked from a launch shaft, so that it forms a continuous string in the ground. The pipe, which is specially designed to withstand the jacking forces likely to be encountered during installation, forms the final pipeline once the excavation operation is completed. On shorter drives, the pipe jack can also be installed using manual excavation. This technique is used in applications where excavations are undesirable, such as in developed locations or areas with a high risk of ground collapse. Thrust boring may be cheaper than traditional (conventional) tunneling for relatively short lengths.

Pipe jacking systems are more often than not supplied with jacking frames, which are designed to provide the level of jacking pressure likely to be required by the TBM being used. A TBM is a machine used to excavate tunnels with a circular cross-section through a variety of soil and rock strata. They may also be used for micro-tunneling. They can be designed to bore through anything from hard rock to sand. Tunnel diameters can range from 1- (done with micro-TBMs) to 17.6 m. The requirements for the jacking frame on any project are determined by the ground conditions, length of drive, and the type of TBM being used. An essential feature of pipe for both micro-tunneling and pipe jacking is that the entire joint is contained within the normal pipe wall thickness. Pipe jacking and micro-tunneling are both commonly used for mainline or trunk pipelines.

7.6 CONVENTIONAL TUNNELING

Conventional (traditional) tunneling refers to the construction of underground openings of any shape with a cyclic construction process composed of the following steps:

a. Excavation, by using the drill and blast methods (explosives) or very basic mechanical excavators.
b. Mucking—removal of blasted debris or soil from tunnel interior to sides outside the tunnel entrance.
c. Placement of the primary support elements such as steel ribs or lattice girder soil or rock, soil or rock bolts, sprayed or cast in situ concrete.

Conventional tunneling describes the cyclic construction of underground tunnels of any shape where there is access to the excavation face. In the past, conventional (traditional) tunneling excluded the use of TBM. However, in recent years, the use of TBMs has become more common.

7.7 SOIL BEHAVIOR IN SHEAR

Almost every civil engineering project begins with a hole in the ground. That means basic knowledge of soil shear strength is essential to all civil engineers for designing roads, bridges, water systems, and hydraulic structures. Good knowledge of soil mechanics specifically on shear strength and stiffness modulus is essential while taking decisions on foundation design, evaluation, and construction of geotechnical constructions for roads and drainage infrastructure. These are used to forecast behavior of soil under permanent deformation. Actions taken to modify or stabilize soil aim to improve soil shear strength. A knowledge of the resistance of soil to failure in shear is essential in the analysis of soil masses and in the design of foundation structures.

Detailed information about soil behavior in shear is out of the scope of this book. However, a good analysis of the shear strength parameters is presented, which are very helpful for a drainage engineer planning to design and develop robust and reliable gray drainage infrastructure. A drainage engineer needs a relatively good understanding of the purpose of investigating soil shear strength parameters (internal angle of shearing resistance, φ and cohesion c) to correctly design and develop drainage structures. The planning and design of SuDS and storm sewers encounters different topography, geology, and soils. It is therefore, vitally important to evaluate slope stability, pore water pressures, shear strength, and soil types.

Section 8.4 provides an analysis of the loads due to the earth overburden acting on a buried pipeline. Shear strength parameters such as the angle of internal friction (angle of shear resistance) is applicable in calculating the loads due to the earth overburden. It is also important in retaining wall design. In retaining wall design, it is specifically applied to find the active and passive pressure conditions induced by the Earth. It is used in Rankine's formula for calculating the horizontal component of stress which is helpful in retaining wall design and estimating the loading due to the earth overburden on buried pipes.

7.7.1 Horizontal component of stress

Active pressure is the condition in which the Earth exerts a force on a retaining system and the members tend to move toward the excavation. Passive pressure is a condition in which the retaining system exerts a force on **the soil**. In retaining walls, particularly, Rankine's theory assumes the following:

- Movement occurs.
- The wall has a smooth back.
- The retained ground surface is horizontal.
- That the soil is cohesionless.

Thus, Rankine's formula for horizontal component of stress is given by Equation (7.1):

$$\sigma_h = k\sigma_v \tag{7.1}$$

Note:

- For soil at rest, $k = k_0$
- For active pressure, $k = k_a$
- For passive pressure, $k = k_p$

$$k_0 \approx 1 - \sin\phi \tag{7.2}$$

$$k_a = \frac{(1-\sin\phi)}{(1+\sin\phi)} \tag{7.3}$$

$$k_p = \frac{1}{k_a} \tag{7.4}$$

Rankine's theory on lateral earth pressure is the most commonly used for retaining wall design, but in several instances, the **coulomb's theory** is easier to apply for complex loading conditions.

7.7.2 Mohr–Coulomb model

Mohr–Coulomb theory is a mathematical model describing the response of brittle materials, such as soil, concrete, or rubble piles, to shear stress as well as normal stress. Most of the classical engineering materials somehow follow this rule in at least a portion of their shear failure envelope.

Shearing resistance is developed mechanically in soil element due to the interparticle contact forces and friction. Angle of shearing resistance is known as the component of shear strength of the soils, which is basically frictional material and composed of individual particles. In theory, the angle of internal friction (angle of shearing resistance) measures the ability of a unit of rock or soil to withstand a shear stress. It is the angle ϕ, measured between the normal force, N, and resultant force, R, that is attained when failure just occurs in response to a shearing stress, τ.

The shear strength is described by Mohr–Coulomb failure criterion adopted as widely accepted approach among the geotechnical engineers. In practical applications, the engineer must differentiate between undrained conditions (short-term loading, where pore water pressures are present and the design is carried out for the total stresses on the basis of ϕ_u and c_u) and drained conditions (long-term loading, where pore water pressures are dissipated and the design is carried out for effective stresses on the basis of ϕ' and c'). Short-term loading is usually taken to mean 'loads carried by the soil during construction' while long-term loading refers to the loads carried throughout the design life of the construction. This is essential to plan and design sound hydraulic structures.

Due to high permeability, coarse-grained soils (e.g., sand and gravel) quickly result into drained conditions allowing consolidation to take place rapidly, in both the short-term and long-term loading. Due to the low permeability of fine-grained soils (e.g., clay and silt), loading in the short term is usually undrained while in the long-term loading conditions it is drained.

Stress states in 2-D can be represented on a plot of **shear stress** τ against **effective normal stress** σ'. In practice, the stress states are represented by a pair of points with coordinates on a Cartesian plane. However, the Mohr circle is usually a preferred method.

The Mohr circle is usually defined by the **effective principal stresses**, and it represents the stress states on all possible planes within the soil element. The principal stress components can describe the position and size of the Mohr circle. Failure occurs at any point in the soil element where a combination of shear stress and effective normal stress develops. Figure 7.2 shows Mohr–Coulomb failure criterion.

When the element of soil reaches failure, the circle will just touch the failure envelope at a single point. In common practice, the failure envelope is usually approximated by a straight line described by the formula in Equation (7.5).

$$\tau_f = c' + \sigma' \tan \phi' \tag{7.5}$$

Where:

τ_f : Shear stress
c' : Cohesion intercept under drained conditions
σ' : Effective normal stress
ϕ' : Internal angle of shearing resistance under drained conditions

Note: c' and ϕ' are mathematical constants defining linear relationship between shear strength and effective normal stress (drained conditions). Under drained conditions, excess pore water pressures are zero because consolidation is complete. In reality, if effective normal stress is zero, then shearing resistance is also zero (unless there is a cementation or some other bonding between the particles), and the value of c' would be zero.

Figure 7.2 illustrates Mohr–Coulomb criterion for drained conditions. In drained conditions, $\phi' > 0$. For undrained conditions, $\phi_u = 0$ and $c_u > 0$. Figure 7.3 shows Mohr–Coulomb failure criterion for **undrained conditions**.

The **cohesive strength of fine-grained** soils normally increases with depth. Drained shear strength parameters are obtained from very slow triaxial tests in the laboratory. The effective internal angle of shearing resistance, ϕ', is influenced by the range and distribution of the fine particles, with lower values being associated with higher plasticity. For a normally consolidated clay, the effective (or apparent) cohesion, c', is zero but for an over consolidated clay it can be up to 30 kN/m² (Table 7.1).

In drainage design, analysis, and development, the engineer constantly encounters the need to design retaining structures to retain soils, develop embankments, and specifying backfill material for use in backfilling trenches. Retaining structures require backfill materials which must meet a certain minimum standards of shear strength parameters, that is, the effective

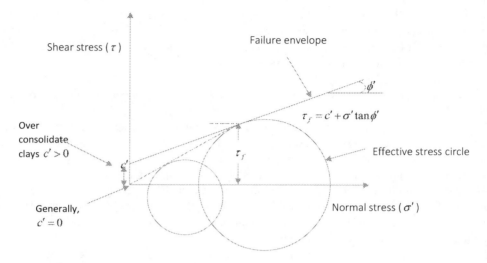

Figure 7.2 Mohr–Coulomb failure criterion.

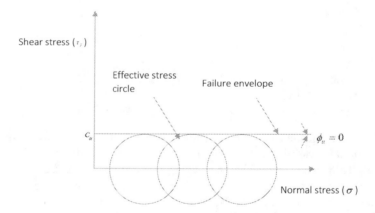

Figure 7.3 Mohr–Coulomb criterion (undrained conditions).

internal angle of shearing resistance and unit weight. Drained shear strength parameters are obtained from very slow triaxial tests in the laboratory for use in the design and construction practices of geotechnical structures. The following soil parameters are usually derived from laboratory tests and used to design geotechnical structures: bulk density, saturated density, cohesion c, internal angle of shearing resistance φ, unconfined shear strength, and permeability k. These are used in throughout Chapter 8 to specify the quality of backfill loads.

Depending on the evaluation process conducted, loading conditions and ground conditions dictate the bedding class and quality of backfill required to build reliable hydraulic structures (see Chapter 8). Retaining walls and trenches may need backfill that meets minimum shear strength parameters. The strength of the fill material is very important. The Rankine's theory

Table 7.1 Properties of drained granular and cohesive soils

Properties for drained *granular soils*

Description	SPT*, N blows	Effective internal angle of shearing resistance, ϕ'	Bulk unit weight, γ_{bulk} kN/m³	Dry unit weight, γ_{dry} kN/m³
Very loose	0–4	26–28	<16	<14
Loose	4–10	28–30	16–18	14–16
Medium dense	10–30	30–36	18–19	16–17
Dense	30–50	36–42	19–21	17–19
Very dense	>50	42–46	21	19

Properties of drained *cohesive soils*

Soil description	Typical shrinkability	Plasticity index (PI) %	Bulk unit weight γ_{bulk} kN/m³	Effective internal angle of shearing resistance, ϕ'
Clay	High	>35	16–22	18–24
Silty clay	Medium	25–35	16–20	22–26
Sandy clay	Low	10–25	16–20	26–34

* An approximate conversion from the standard penetration test to the Dutch cone penetration: $C_r \approx$ 400 N kN/m². For saturated, dense, fine or silty sands, measured N values should be reduced by N = 15 + 0.5(N−15).

on lateral earth pressure is the most commonly used for loads acting on the storm sewer pipe due to earth overburden and in retaining wall design.

The knowledge of soil strength and stiffness is essential for assessing the stability and performance of geotechnical structures. It relates ground stresses (which are in equilibrium with the applied loads) to ground strains (giving compatible deformations). The constitutive behavior of soil is highly nonlinear and dependent on the level of confining stress, but for most practical problems, it can be modeled using isotropic linear elasticity coupled with Mohr–Coulomb (stress-dependent) plasticity.

7.8 FLOOD DEFENSE STRUCTURES

7.8.1 Introduction

The most common flood types can be grouped in three categories: flash floods, river floods, and coastal floods. Flash floods refer to excessive amount of rain in a short period of time (usually within 6 hours). Flash floods can also happen as a result of the bursting of dams and canals. For example, sump pumps are constantly being used to pump groundwater away from homes and can be an excellent defense against basement seepage and flooding. They draw in the groundwater from around the house and direct it away from the structure through drainage pipes.

River floods occur when water levels in a river rise and run over the river banks as a result of heavy rains. Coastal floods occur around much larger bodies of water such as the sea or a lake and happen especially when the tide gets very high.

Flood defense structures, sometimes referred to as flood control structures or flood protection structures, include dikes, spurs, levees, and seawalls. They can further be described based on the type or location. A flood control structure is any method or a combination of methods used to control flooding. Flood control structures can be used to fight residential flooding as well as prevent larger floods. Dams, dikes, levees, and water barriers are all methods of flood protection. Flood control structures are developed to support urban programs, economic growth and trade, and food and water availability.

Flood control structures are primarily used to divert flows of water, by redirecting rivers, slowing natural changes in embankments and coastlines, or preventing inundation of vulnerable coastlines or floodplains. Dikes, spurs, leeves, and seawalls often act as the first line of defense against overflowing rivers, floods, storm surges, and rising seas. Dikes, spurs, levees, and seawalls often act as the first line of defense against overflowing rivers, floods, storm surges, and in the longer term protect communities against rising seas. More modern flood defense structures include the following:

a. Dams and spillways.
b. Diversion canals.
c. Floodplains and groundwater replenishment.
d. River defenses, for example, levees, floodwalls, bunds, reservoirs, and weirs.
e. Flood defense walls built to safeguard flash floods in urban or residential areas.
f. Costal defenses, for example, groynes, seawalls, revetments, and gabions.
g. Retention and detention ponds.
h. Moveable gates.
i. Moveable barriers.
j. Channel improvements.

7.8.2 On design of flood control structures

The design of flood control structures integrates several fields which include water resources, hydrology, geotechnical engineering, and others. The design criteria primarily depend on the type of floods to safeguard against, that is, flash floods, river floods, or ocean floods. However, the overarching goal should always be sustainability and reliability. Sustainability being governed by social, environmental, and economic conditions will always take the optimal solution. The common materials used to build flood control structures include concrete, gabions, and earthen embankments. Sometimes, during heavy rains sandbags are temporarily used to keep off flash floods from entering people's homes.

Flood control structures, especially for rivers, are designed as earthen embankments using local materials to minimize the cost of construction, while protecting the environment from harm. For sustainability purposes going forward, flood defense walls built in concrete would be the last option provided that available local materials can meet the minimum standards, that is, the desired strength of materials.

River flooding is always controlled through constructing embankments, which can vary in design and technology depending on the evaluation process conducted to decide the type of structure. River flooding control structures are grouped into three categories: river training structures, river bank protection structures, and river containing structures. The primary purpose of all types of structures is controlling floods. However, apart from controlling floods, they serve other purposes. River training, in particular, aims at controlling and stabilizing a river along a desired course with a suitable waterway, for one or more of the following purposes:

a. Flood protection.
b. Bank protection.
c. Sediment control.
d. Land reclamation.
e. Confined river channel.

A levee (also known as an embankment, dike, or bund) is the oldest method used for flood protection along rivers. It is constructed along a river bank to protect the adjacent area from getting flooded. It is one of the methods generally used for river training. The method of flood control by levees is economical as it uses locally available material and labor for their construction.

The topic of seepage is very pronounced in flood defense structures, especially earthen embankment structures. Seepage is the lateral movement of subsurface water. Flash floods, rising sea levels constantly occur and need flood defense structures. (Port defense structures) These defenses consist of earthen dikes (levees), dunes, dams, and storm surge barriers. The old standards were formulated in terms of a probability of exceedance of hydraulic load conditions (water levels, waves) that a flood defense should be able to withstand safely.

The stability of flood defense structures is usually analyzed for gravity, earthquakes, and seepage. Detailed subsurface investigations are conducted for proper foundation evaluation. If required, the particular foundation is usually treated for its strengthening or protection by developing specific design to suit the site conditions. These specific designs may include the following:

a. Reworking of the foundation soil.
b. Grouting of the armored river bed to the required extent.
c. Inverted filter or geotextile layer at the foundation of revetment/gabion.
d. Hydraulic structure.
e. Where required, provision of a suitable cutoff.

Flood control structures do not completely eliminate the risk of flooding[5]. Flooding may occur if the design water levels are exceeded. For example, all river flood protection structures are designed with a sufficient freeboard. Sufficient freeboard should be provided above the design flood level for safety against overtopping. If poorly designed, constructed, operated, or maintained, flood control structures can increase risk by providing a false sense of security and encouraging settlements or economic activity in hazard-prone areas. Therefore, climate change projections should be included in the various project lifecycle stages, including the design, construction, and maintenance of flood control structures.

7.8.2.1 Seepage control in flood control earthen embankment structures

Seepage control in embankment dams is essential when planning a flood defense system such as earthen embankments, that is, levees. If poorly designed, constructed, or maintained, earthen embankments can fail prematurely due to seepage and the overarching goal of stopping floods may not be achieved. Another failure of earthen embankments is due to erosion. There can be two kinds of erosion that can happen in an earthen embankment:

   a. Piping erosion. This is erosion that creates voids in the form of pipes by working its way back into the embankment hence weakening the stability of the dam.
   b. Erosion at the boundary between the central core and adjacent soil (of higher permeability) under a high exit gradient form the core. Protection against this erosion can be given by means of a chimney drain as shown in Figure 7.4(a) at the downstream boundary of the core.

Seepage through and beneath earth dams is controlled through three primary techniques: using a low permeability core, cutoff walls, or impermeable blankets. The central core should be of very low permeability as it is meant to reduce seepage in earthen embankments. The central core should also be of a reasonable width, not so narrow so that high hydraulic gradients are avoided.

Underseepage can be controlled by cutoff methods, such as grout curtain (Figure 7.4(b)) and impermeable blanket (Figure 7.4(c)). Control of underseepage is important when the foundation soil is more permeable than the earthen embankment.

Most embankment dam sections are nonhomogeneous with zones of different soil types. The permeability of different soils is affected by the size of the voids and the permeability of the fine soils may be so much smaller than those for coarse soils. Falling head tests are commonly used to measure permeability in fine soils, and constant head tests are used for coarse soils. These may be conducted in the laboratory on undisturbed samples removed from the ground or in situ.

Advanced designs of dams, dikes, and levees may include internal drains or barriers to trap or collect seepage water, or to dissipate the hydraulic head that drives flow through the dam. The operation of structural water barriers often involves internally varying seepage pressures, which are essential in stability analysis along with foundation design and construction methodology.

These analyses are useful for predicting the location of the phreatic surface, which is used as an input in the stability analyses, hydraulic gradients for erosion analysis, and seepage quantity and quality for environmental and water management purposes.

PLAXIS is a geotechnical software used extensively to perform finite element analyses (FEA) within the realm of geotechnical engineering, including deformation, stability, and water flow. PLAXIS finite element analysis software can analyze a variety of technical issues related to seepage control. The software can be used to analyze and design seepage and uplift control measures, such as cutoff trenches, slurry walls, grout curtains, and liners.

As mentioned, flood control structures may be designed to safeguard communities around river banks, lakes, oceans, or against flash floods. Their construction must be robust to withstand a number of forces and well-engineered to withstand seepage. Geotechnical engineers working with water resources engineers must have good understanding of integrated topics in soil mechanics and water resources such as foundation stability analysis, embankments design, seepage control, erosion control, and strength of fill materials.

Useful topics in soil mechanics 257

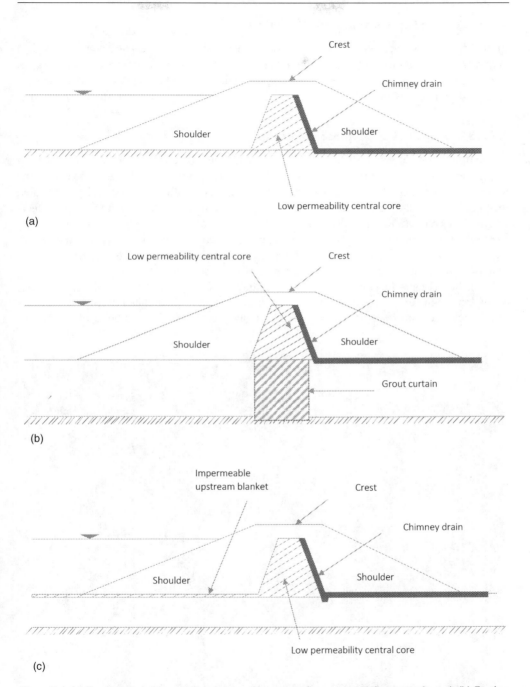

Figure 7.4 (a) Earthen embankment flood control structure (low permeability central core). (b) Earthen embankment flood control structure (grout curtain). (c) Earthen embankment flood control structure (impermeable blanket).

7.8.2.2 Control of water movement in basements

In buildings, impermeable membranes can be used to control seepage together with a geotextile filter and drain tube as shown in the Figure 7.5. Generally for this type of work, an impermeable membrane is applied in the basement and after we put the drain membrane with tube on bottom (with a geotextile filter). Then, on top we put a special profile (Figure 7.6).

7.8.2.3 Concrete flood defense structures

The simple flood defense walls use a concrete interlocking block (concrete lego blocks) with enough mass to act as a gravity flood defense wall. The foundation design depends on the topographical and site investigation information. The joints of the blocks need to be sealed to stop water leakage through them. There are many products on the market to do this effectively. Most concrete interlocking block flood defense walls are less than 2-m high ranging from two to three membranes. The membrane walls are commonly constructed from concrete lego block walls with a waterproof membrane. Figure 7.7 shows a three-membrane flood defense wall constructed out of concrete blocks with a waterproof membrane.

Flash floods have increasingly become common in recent years due to climate change. Municipal engineers and planners are increasingly tasked to plan ahead by providing flood emergency and warning systems. Among the many systems that would be in place include

Figure 7.5 Impermeable membranes used to control seepage together with a geotextile filter and drain tube. (Reproduced with permission from TEAIS ES (www.teais.es).)

Useful topics in soil mechanics 259

Figure 7.6 Artistic impression of how an impermeable membrane controls water movement in building basement. Reproduced with permission from TEAIS ES (www.teais.es).

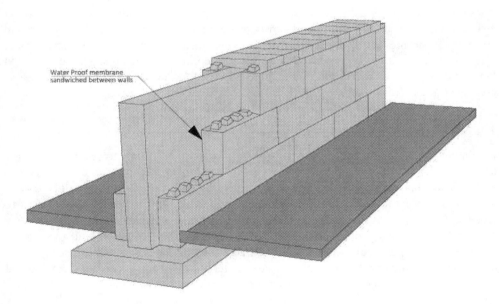

Figure 7.7 Concrete interlocking (lego) block flood defense wall.

flood defense walls well-built to mitigate seepage. Rising sea levels globally due to climate change is another issue that requires mitigation measures, adaptation strategies, and engineering controls aimed at avoiding devastating incidences when heavy rainfalls fall.

Figure 7.5 shows a typical concrete flood wall with aggregates inside the concrete barriers. The aggregates are properly compacted to give the concrete flood wall a sufficient mass to withstand the floods. It is designed as **a gravity flood defense** wall properly founded in concrete and well-engineered to control seepage, possibly with waterproof membranes or waterproof cements.

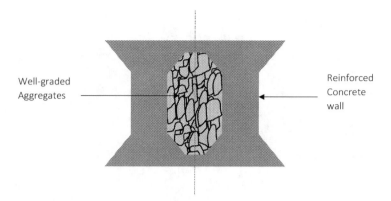

Figure 7.8 Typical coastal concrete flood defense wall.

7.8.2.4 Coastal flood defenses

These are expensive flood defense structures used at the sea coastline to protect against coastal flooding. Sea walls are a solid barrier made from concrete, masonry, or gabions and are designed to prevent high ties and storm surges reaching inland and causing flooding. They can have a variety of profiles such as sloped, stepped, or vertical, and are designed to withstand the force of waves for around 30 to 50 years. Figure 7.8 shows a typical coastal concrete flood defense wall.

7.9 RETAINING WALLS

Retaining wall structures are constructed to retain soil behind them. However, the specific needs vary depending on the project. For instance, they can be used in SuDS to make the site aesthetically appealing by raising part of the developed area. For example, small landscape stone walls can be used to surround a bioretention garden. Others may be used to control erosion from hard rains or create a terraced yard to reduce maintenance. Several SuDS may have to be constructed with retaining walls. The retaining wall must be built to standard because an improperly built wall may bulge, crack, or lean. The purpose of this section is to demonstrate the use of retaining walls on stormwater management, general landscaping and drainage projects, and how to deal with subsurface drainage behind retaining walls.

7.9.1 Goal setting for retaining wall design

Detailed design criteria are out of the scope of this book. In case, detailed designs are required, consult books on soil mechanics and geotechnical engineering. The retaining wall outline provided in this book is descriptive analysis of their applicability in stormwater management and drainage solutions. In designing retaining walls, three main goals are usually set:

   a. The wall should not overturn.
   b. The wall should not slide.
   c. The wall should not subside.

Slope stability, overturning, subsiding, and sliding are topics very helpful for a drainage systems' planner and designer. Retaining walls are usually built to hold back soil mass. However, retaining walls can also be constructed for aesthetic landscaping purposes. This is where SuDS are integrated in the landscaping architecture to make the project more aesthetically appealing.

The design of a retaining wall necessitates the engineer to have knowledge and understanding of the retaining material, the water table, the ground-bearing capacity, and several site constraints. The design also requires the site investigation to be carried out. Relevant codes of practice include the following:

a. BS 5930 for site investigations.
b. BS 8004 for foundation design.
c. BS 8002 for retaining wall design.
d. Eurocode 7 for geotechnical design.

7.9.2 Factors influencing the choice of retaining wall

The following factors are considered to determine the type and material for the retaining wall.

7.9.2.1 Soil

The foundation or base must be constructed with soil after being tested to confirm to the required standards and perhaps mechanically modified or stabilized to suit the job. The base of a retaining wall should be set below the ground level, and the taller a wall is, the further below ground level it should be set. The designer or planner would determine the soil type, bearing capacity value, stress parameters, and friction angle of the soil for use in the foundation and reinforced zone along with the retained soil zone. Foundation soil must be stable and never moist. It is essential for supporting the rest of the wall, a good base is made of compacted soil and at least a 6-in layer of compacted sand and gravel. Backfill soil needs to meet certain minimum requirements and must be well-compacted in layers (but not always) as advised by the engineer. For SuDS applications, the soil is not always compacted, and the walls could be about 0.9- to 1.2-m high. The need to compact the backfill entirely depends on the amount of lateral pressure expected to be exerted on the wall.

Wet clay soils are detrimental and must be avoided; they can get fully saturated exerting hydrostatic pressure to the wall, hence collapsing it. This is because additional water cannot find its way out through to the drains. In some instances, wet soil can expand and contract, which may damage the wall. However, sandy soils allow for good drainage once used as backfill material. It is therefore important to check for expansive soils, native (no-site) soils' chemical properties, and groundwater conditions prior to building a retaining wall.

7.9.2.2 Location

SuDS require that the designer or planner has adequate knowledge of the presence of utility lines, that is, above and below ground, which include power, portable water, and telecoms. The site topography predetermines the nature of retaining wall suitable for the job. If the wall is constructed on slope, it might need additional infill transported to site. In case there is need to cut into a hillside, planning for excess soil is essential. The designer needs to assess whether there will be additional loads on top of the wall, directly above the wall or a few meters from the wall. These loads include perimeter fence, parking lots, swimming cool, driveways, and guardrails. Temporary construction machinery and equipment also impose extra surcharge on the retaining wall during construction and must be accounted for.

7.9.2.3 Drainage

Water is the primary reason for the collapsing retaining walls. Therefore, retaining walls need good drainage, which is usually achieved by ensuring that no water builds up behind the wall. The designer needs to provide drainage solutions for potential surface water sources ensuring drainage adjacent to the wall is properly considered. Excessive hydrostatic pressure building up behind the wall can cause the wall to collapse by bulging or cracking. Provision of subsurface drainage systems is recommended where retaining walls are likely to face excessive hydrostatic pressure. In order to minimize hydrostatic pressures the groundwater could create, sometimes gravel is used as the backfill with perforated pipes draining away excess water to adjacent sewers or drains. If landscaping behind the wall is required, a 150-mm layer of native soil should also be placed over the gravel fill. Weep holes are also used to allow water to drain through the wall and help to reduce the buoyancy and uplift on the structure and make it structurally strong and stable. Weep holes are provided at the bottom side of the retaining wall allowing water to drain from the water lodge on the retaining wall. In the same way, waterproof membranes are used in basement walls; they are used to protect decorative veneer such as stucco, stone, and tile on retaining walls. Therefore, protecting the beauty of the retaining wall faces is enhanced when waterproof membranes are applied on the back of the retaining wall. Suitable waterproof membranes for local climate and soil type are available from which the most effective can be selected.

7.9.2.4 Design

The following are considered before beginning retaining wall design: wall height, slope, and footprint size. These depend on the topography and site elevation. The height of a retaining wall determines the load it can withstand and how much extra reinforcement will be necessary. Gravity causes the retained soil/material to move downward and must be counteracted by the wall to reduce lateral pressure behind the walls, which at the maximum value could overturn the wall. The type of wall to build depends on a number of factors such as required strength, type of soil/material to retain, the topography, the availability of materials, and ground conditions. The retaining wall height depends on slope, soil, and setback.

Retaining wall design ideas are as follows:

- Incorporate a fountain or water feature into your retaining wall.
- Integrate an outdoor fireplace into your retaining wall design.
- Install landscape lighting in your retaining wall.
- Include built-in bench seating in your retaining wall.
- Don't forget steps if you plan to access the area above the wall.
- For a finished look, have wall caps installed.

7.9.3 Types of retaining walls

The commonly used retaining walls in drainage and landscaping works include the following.

7.9.3.1 Gravity retaining wall

This type of retaining wall uses its own weight to resist the lateral earth pressures. A gravity retaining wall typically has a base width of about 60% to 80% of the retained height. They are usually made of plain concrete or stone masonry. Gravity walls are used especially for both straight and curved wall designs, and they are typically under the height of 1.22 m, but dependent on product.

Gravity walls use their own weight to hold the soil behind them and are typically made with heavy materials such as stone, large concrete blocks, or cast-in-place concrete. They lean back toward the soil with interlocking edges and use their mass to resist pressure from behind. Gravity walls can be small (under the height of 1.22 m) or go upward to about 3 m without reinforcement. Municipalities usually require a developer to obtain a building permit for walls taller than 1.22 m.

In building SuDS, the gravity retaining walls are the commonest, especially those about 1.22-m tall. They help to hold back the soil in an infiltration system. SuDS practices are usually integrated with landscaping practices to make the site aesthetically appealing. In areas where the ground is sloping and needs to be leveled, gravity retaining walls may be used at the lower end so that the ground is leveled making it suitable for infiltration systems to function appropriately. Slopes greater than 20% would require leveling prior to the installation of infiltration systems. The three primary forces that act on the gravity retaining wall are as follows (Figure 7.9):

- Vertical forces from the weight of the wall.
- The lateral earth pressure acting on the back face of the wall.
- The seismic loads.

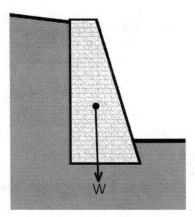

Figure 7.9 Gravity retaining wall.

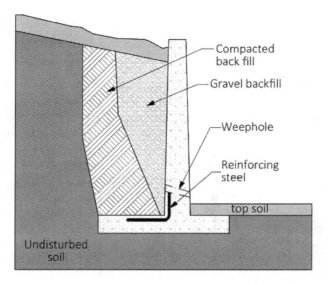

Figure 7.10 Cantilever retaining wall.

7.9.3.2 Cantilever retaining wall

A cantilever retaining wall is built using reinforced concrete, with an L-shaped, or inverted T-shaped foundation. A cantilever retain wall usually has a maximum height of not more than 5.5 m. The vertical stress behind the wall is transferred onto the foundation, preventing toppling due to lateral earth pressure from the same soil mass. A cantilever retaining wall works on the principle of leverage. The cantilever retaining walls are used in deep excavations with heights up to 5.5 m.

Cantilever retaining walls work on the principles of leverage and are often constructed in the shape of an inverted T with reinforced concrete or mortared masonry. Less building material is required for a cantilever wall than a gravity wall, and they can be poured on-site or manufactured at a precast concrete facility. They consist of a relatively thin stem and a base slab, which is divided into two parts: the heel and toe. The heel is the part of the base under the backfill while the toe is the other part. Rigid concrete footing is required for these walls that are usually under 7.62-m height (Figure 7.10).

7.9.3.3 Segmental retaining walls

Segmental retaining walls (SRWs) are designed to be used as a gravity wall either with or without reinforcement and can have heights in excess of 20 m. They are made in a variety of colors, sizes, and textures to meet the aesthetic requirements of your project. Some brands offer pins or clips to help create a continuous facing system. SRWs are modular blocks made from concrete that are often dry-stacked without mortar. The individual units interlock with each other to avoid overturning and sliding. Since they are manufactured in a plant, they meet industry standards and are uniform in weight, strength, and durability. SWRs are used for various commercial and residential applications, used for either straight or curved designs,

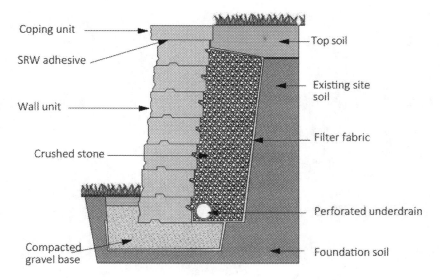

Figure 7.11 Segmental retaining walls.

and can be used to support SuDS applications. SRWs can be designed to conform to any shape while maximizing the site's usable space, particularly on high-sloped terrains. When reinforced, they typically have no height limitations (Figure 7.11).

7.9.3.4 Counterfort retaining walls

Counterfort retaining walls are similar to cantilever walls and require support along the backside of the wall (Figure 7.12). They use concrete webs usually known as counterforts. These counterforts are built at an angle to strengthen the stability of the wall. These webs are located at regular intervals along the length of the wall and reduce the natural pressures put on the wall from the soil while also increasing the weight of the wall. These are preferred over cantilever walls when the wall is taller than 7.62 m. Counterfort retaining walls are typically used for tall walls between 6 and 12 m.

7.9.3.5 Sheet or bored pile walls

Sheet pile walls are used in temporary deep excavations in tight spaces such as around marine locations, cofferdams, and seawalls, along with structural columns, pier shafts, and more. Sheet pile retaining walls are made out of precast concrete, steel, vinyl, or wood planks and are used in soft soil and tight spaces. The planks are driven into the ground by vibrating and hammering to ensure their stability and can be connected using a groove and tongue. Taller walls require an anchor of some sort that is tied to the wall and then built into the soil. They are good to use along waterfronts and can help with beach erosion, shoring, excavations, or cofferdams. Bored piles are often used when the vibrations from pile drivers are too strong for sheet piles to withstand. Anchors are sometimes needed to support the walls, but not always.

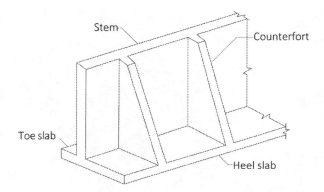

Figure 7.12 Counterfort retaining walls.

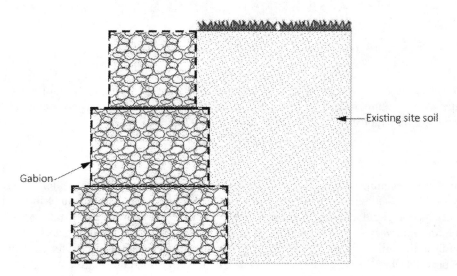

Figure 7.13 Gabion retaining walls.

7.9.3.6 Gabion mesh walls

Gabion retaining walls or gabion-style walls use wire mesh boxes filled with stone and rocks and then stacked together. The boxes are tied with wire and angled back toward the slope. Gabion retaining walls have a long life span as long as the wire used to hold them together is strong enough. High-quality wire is required because with time the wire will eventually corrode. Gabion mesh walls are used in stream or stormwater drainage applications where erosion is a concern. In hilly areas, the rivers usually have high gradient or velocity flows. There, it is not practicable to provide stone protection against bank erosion. For such locations, the most stable structural measure is the provision of Gabion (stone crate) retaining walls. Gabions are well suited for retaining walls because of their flexibility and also to make full use of readily available local stone. Gabion retaining walls are designed as gravity walls. The walls are usually well-extended landward into the bank line, at both ends, to avoid any outflanking during high floods. Gabion retaining walls can also be used in military applications to protect against artillery fire or in a modern residential or commercial setting where the wire mesh is filled with a recycled or aesthetically pleasing material (Figure 7.13).

7.10 CASE STUDY 7.1: COLLAPSED RETAINING WALL FOLLOWING FLASH FLOODS—CONTINUATION OF CASE STUDY (5.2): TYPICAL AUDIT AND DESIGN PROJECT (SECTION 5.7.2) (2021)

The wall (Figure 5.32) collapsed due to a combination of a number of factors, which included using poor quality backfill that did not meet shear strength parameters, poor workmanship, flash floods which fully saturated the poorly compacted backfill and thus increased the pore water pressures pushing the wall to collapse. The existing retaining wall therefore collapsed when flash floods found way to the estate. See Figure 7.14, interim response to the flash floods; temporal retaining wall. The wall had not been constructed to standard, and floods infilitrated the backfilled soil thereby fully saturating it. The increase in pore water pressures exerted enormous impact to the retaining wall pushing it to collapse. As a drainage engineer, the lessons to take from this case are as follows:

a. negligence on part of the developer, failing to build the retaining wall structure to the required standards. There were no approved drawings for the retaining wall and fence, thus garnering no compensation from the insurer.
b. Failing to provide adequate drainage for excess water caused excessive pore water pressures that pushed the wall to collapse.
c. Failing to compact the backfilled soil compromised the reliability of the retaining wall.
d. Poor workmanship affected the reliability of the retaining wall.
e. Drainage easements must be clearly delineated and widened to adapt to climate change consequences (see Section 12.1.3).

Figure 7.14 Collapased retaining wall due to excessive pore water pressure. Interim response to the flash floods; temporal retaining wall.

Water is the primary reason for the collapsing retaining walls. Therefore, retaining walls need good drainage, which is usually achieved by ensuring that no water builds up behind the wall. The designer needs to provide adequate drainage solutions for potential surface water sources ensuring drainage adjacent to the wall is properly considered. Excessive hydrostatic pressure building up behind the wall can cause the wall to collapse by bulging or cracking. Provision of subsurface drainage systems is recommended where retaining walls are likely to face excessive hydrostatic pressure. In order to minimize hydrostatic pressures, the groundwater could create, sometimes gravel is used as the backfill with perforated pipes draining away excess water to adjacent sewers or drains. Sometimes, weep holes are used to drain excess water. If landscaping or SuDS development behind the wall is required, a 150-mm layer of native soil should also be placed over the gravel fill. Weep holes are also used to allow water to drain through the wall and help to reduce the buoyancy and uplift on the structure and make it structurally strong and stable. Weep holes are provided at the bottom of the retaining wall allowing water to drain from the water lodge on the retaining wall. In the same way, waterproof membranes are used in basement walls; they are used to protect decorative veneer such as stucco, stone, and tile on retaining walls. Therefore, protecting the aesthetics the retaining wall faces is enhanced when waterproof membranes are applied on the back of the retaining wall. Suitable waterproof membranes for local climate and soil type are available from which the most effective can be selected.

7.11 CONCLUSION

Basic knowledge of soil mechanics is essential for drainage engineers and planners while planning and designing integrated drainage systems. Shear strength of soils used as backfill in trenches and retaining walls must be evaluated for short-term loading and long-term loading conditions. Also, the knowledge of subsurface drainage is critical. The permeability, soil types (gradation), and infiltration rates of soils are assessed in the design of SuDS. The drainage engineer must understand the techniques to deal with subsurface drainage.

References

1. Wopereis, M. C. S., Bouman, B. A. M., Kropff, M. J., ten Bergec, H. F. M., & Maligayab, A. R. (1994). Water use efficiency of flooded rice fields I. Validation of the soil-water balance model SAWAH. *Agricultural Water Management*, 26, 277–289.
2. Mackay city council. Engineering design guidelines, subsurface drainage design, planning scheme policy No. 15.04 Date Policy Took Effect: March 31, 2008.
3. MMSD, Partners for a cleaner environment. What is a Foundation Drain? | MMSD. www.mmsd.com/what-you-can-do/managing-water-on-your-property/foundation-drains
4. Tunnelling into trenchless: microtunnelling and pipe jacking explained – The Australian Pipeliner. Accessed 16 March 2016, 2:16 am. www.pipeliner.com.au/tunnelling-into-trenchless-microtunnelling-and-pipe-jacking-explained/
5. USAID Fact sheet. Flood control structures. Addressing climate change impacts on infrastructure: Preparing for change, 2012.

Further reading

1. Knappett, J. A., & Craig, R. F. *Craig's soil mechanics*, 8th ed. Spoon Press, an imprint of Taylor and Francis, ISBN 978-0-415-56126-6.

Chapter 8

Structural aspects of storm sewers

NOTATION

B_d : Width of the trench just below the top of the pipe
B_c : Pipe diameter
W : Load on the pipe per unit length
w : Weight of the fill material per unit volume
μ' : Coefficient of sliding friction between fill material and sides of the trench
μ : Coefficient of internal friction of the fill
K : Ratio of active lateral pressure to vertical pressure
C : Coefficient that depends on the depth of the trench, character of construction, and fill material
C_d : Coefficient that depends on the depth of the trench, character of construction, and fill material—normal (ordinary or narrow) trenches
C_c : Coefficient that depends on the depth of the trench, character of construction, and fill material—wide trenches/embankment cases (positive cases)
C_n : Coefficient that depends on the depth of the trench, character of construction, and fill material—negative projection cases (extension of narrow trenches)
σ : Normal stress
H : Depth of fill above the pipe
ϕ : Angle of internal friction

8.1 INTRODUCTION

Drainage sewer pipes experience loading conditions like other civil engineering structures. These loads include internal water load, traffic loads, temporary handling when loading, transporting, and offloading and installation, self-mass, and mass of backfill materials. These loads must be considered while planning for construction of drainage sewers so that they perform safely and reliably. When broken, drainage sewers cannot properly transport stormwater runoff. Several reasons lead to failure of buried stormwater drainage sewers including increased traffic on the ground, poor installation, shifting soil mass, settling, and leaking joints. The reasons why pipes fail are summarized below.

8.1.1 Differential settlement

This is a condition where pipes fail as a result of relative movement of the pipe sections arising primarily from the nature of soils beneath the pipe and surrounding the pipe causing circumstantial fracture. The forces leading to pipe circumstantial fracture are mobilized from either the excessive shear or excessive bending stresses. The manholes shown in Figure 8.1 might displace due to irregular soil support or weak soils leading to pipe fracture inside the pipe barrel or at the joints. In some instances, pipes may be built in a junction box which displaces as a result of differential settlement leading to relative movement of the pipe sections and pipe fracture. Differential settlement occurs when the soil beneath the structure expands, contracts, or shifts away. Some of the conditions leading to differential settlement include root systems from grown-up trees, poor drainage, broken water lines, poorly compacted soil fill, poorly compacted bedding material, waterlogged areas and weak soils, flooding, broken utility lines, and possibly vibrations from nearby construction sites. Cracking of pipes can also be due to bedding the pipe on expansive clay.

Poor compaction of soil fill and the pipe bedding material is very detrimental. Shear strength for the backfill, bedding material, and surrounding soil is enhanced when highly compactible soil is properly compacted to required standards to carry pipe loads. Good knowledge of soil mechanics specifically on shear strength and stiffness modulus is essential while taking decisions on foundation design, evaluation, and construction. Shear strength parameters are used to forecast behavior of soil under permanent deformation. Therefore, actions taken to modify/stabilize soil aim to improve soil shear strength.

Relative movement of pipes due to differential settlement can be avoided by using flexible joints in the sense that relative angular movement is not excessive whereby a 5° angular movement is considered safe for several types of joints. Once this limit is exceeded, a socket is likely to break by leverage.

8.1.2 Excessive pressure

Excessive pressure inside the pipe joint may cause the pipe to fail. That is the primary reason pipe surcharging must be avoided during design. Surcharging mainly occurs due to backwater effects as described in Chapter 2. When the pipe surcharges, joint material expands causing a bursting failure of the socket. This is one of the primary reasons pipes are designed not to surcharge. Storm sewer pipe rupture can be due to shifting soil, settling, increased traffic on the ground above, or use of heavy construction equipment aboveground.

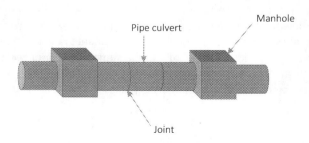

Figure 8.1 Storm sewer culvert with manholes.

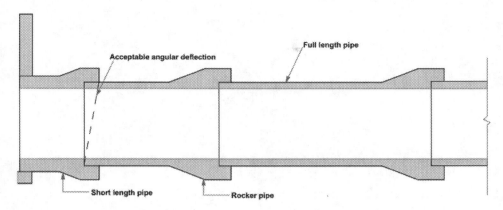

Figure 8.2 Socket and spigot pipes.

8.1.3 The expansion of the joints

In socket and spigot pipes (Figure 8.2), jointing material, which is usually a 1:2 cement mortar mixture of stiff consistency, may cause excessive pressure inside the joint when it expands causing a bursting failure of the socket. The excessive pressure inside the joint due to expansion of the jointing material causes a bursting failure of the socket. The joint usually filled with a 1:2 cement mortar mixture of stiff consistency, which is always rammed with a caulking tool. When storm sewer pipes are broken or ruptured, the stormwater cannot be able to properly drain through the system—leading to immediate and frequent backups. The leaking joints, where the seals between sections of pipe are broken, allow stormwater to escape.

8.1.4 Loading conditions

Different loading patterns may cause pipe fracture.

a. **Local overload** due to hard spots such as stones, hardcore, and uneven ground beneath the pipe causing diagonal tension cracks to spread from the point of overload.
b. **General overload in the vertical plane** causing fracture along the crown and invert and longitudinal fracture at springing level (Figure 8.3).

8.1.5 Restraining thermal expansion

This failure is mainly localized at the joints and the actual fracture is longitudinal due to circumferential tension. It primarily occurs due to restraint of thermal or moisture expansion at the joints. This type of failure is usually avoided through designing pipes allowing telescopic movement at the joints.

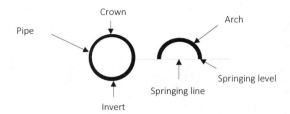

Figure 8.3 Springing line on a pipe.

8.1.6 Concrete shrinkage

Drying shrinkage of a concrete pipe and concrete boxes/manholes may cause longitudinal tension. This causes a circumferential fracture occurring especially in the barrel or the socket. This type of failure is usually avoided through designing pipes allowing telescopic movement at the joints.

8.1.7 Chemical attacks due to corrosion of storm sewers

Approximately 40% of the damage in concrete sewers can be attributed to biogenous sulphuric acid attack, which causes severe structural deterioration and ultimate structural collapse (Kong et al., 2011).[1] In partially combined systems, coagulation of grease at junctions of the water surface and the pipe happens in some storm sewers. When this happens especially in slowly flowing storm sewers, materials get deposited in sewers. The accumulation of materials, including those which are organic in storm sewers, gets slowly degraded by the bacteria in the runoff primarily due to prevalent conditions in sewers which is accompanied by the biological reduction of sulphates present in the flow. Corrosion of an older pipe, causing the pipe to break or collapse, is among the many causes of stormwater pipe failures. Chemical attacks are good at weakening concrete and clay pipes especially where the pipes are not properly manufactured to standards.

8.2 PIPE LOADING CONDITIONS/PATTERNS

Storm water sewers of whatever type are subjected to loads, which include traffic loads, self-mass, internal water load, mass of backfill, temporary handling, and construction. When a storm sewer pipe is broken or ruptured, it cannot be able to effectively convey storm water. The listed factors above, individually or collectively, influence the class/type of culvert and bedding to install.

The next sections outline the types of loads acting on storm sewers and theories on which pipe load calculations are based. Although design standards are available today that waive engineers from tedious calculations from first principles, it is always necessary for the planner/designer to have a sound knowledge and understanding of the theories from which loading calculations are derived or appreciate the assumptions paused and identify and analyze special causes encountered from time to time.

8.2.1 Types of loads that act on buried pipes

The following are the types of loads acting on buried pipes:

a. Internal water loads from the water within the pipe (only significant for larger-diameter pipes).
b. Loads due to earth overburden; effect of backfill load of soil acting on the pipe, including embankment.
c. Surcharge loads from small uniformly distributed areas such as those from pad foundations.
d. Imaginary external downward load.
e. Temporary and permanent superimposed loads, for example, vehicle wheels are surcharge loads acting on the pipe from concentrated loads such as traffic loads.
f. Surcharge loads acting from uniformly distributed loads of large extent such as material storage or changes in ground level.

Typically, the final loads in urban drainage design and development are derived from adding loads due to overburden (static loads) and the loads superimposed by vehicle wheels (dynamic loads). Loads due to overburden and vehicle wheels are the most commonly considered loads in urban stormwater conveyance hydraulic networks. These loads are then compared with the strength of the pipe and bedding classes.

Final load on pipe = earth overburden loads + surcharge and point loads

Where loads due to vehicle wheels are nonexistent, they are thus neglected and static loads considered to determine the pipe strength and bedding class.

8.2.2 Theories applicable in pipe load calculations

Theories applicable in pipe load calculations include the following:

a. **Marston's theory**—good at calculating backfill loading, earth overburden above the storm water pipeline. Marston developed a theory building on soil behavior, involving several assumptions.
b. **Boussinesq's theory**—for pressure within a semi-infinite elastic medium due to load at the surface.
c. **Rankine's equation**—for horizontal component of stress used in Marston's equation.

The present practice recommends that loads calculated from theories (a) and (b) above are added together to find the final loads acting on the pipe.

8.3 MATCHING IMPOSED LOADS TO PIPE STRENGTH

To prevent pipes from failing due to imposed loads, a number of factors are considered so that the imposed loads are matched with the pipe strength. The conditions of the ground might need additional support for the storm sewer pipe to ably carry loads in the vertical plane. In order to prevent longitudinal fracture and fracture along the crown and invert due to general overload in the vertical plane, a pipe which can withstand the loads together with additional support in form of haunching or concrete surroundings is selected. The primary comportment of the imposed loads are as follows:

a. **Earth above the pipe.** This refers to the embankment and backfill material above the pipe crown. Marston's theory is used to calculate imposed loads due to earth overburden on storm sewer pipelines.
b. **Superimposed point loads due to vehicle wheels**, cars, motorcycles, etc. Boussinesq's analysis of pressure within a semi-infinite elastic medium due to load at the surface is used to calculate point loading due to vehicle wheels.
c. **Internal water load.** This arises due to pipe wall stress as a result of the effect of the water contained in the pipe.
d. **Superimposed uniformly distributed permanent surcharge**, for example, loads due to building foundations and other structures.
e. **Superimposed uniformly distributed temporary surcharge** due to dumped material or heaped construction materials.

Important: The Boussinesq theory is not combined with the Marston's theory. The loads from the distinctly separate theoretical approaches are simply added together. At cover depths greater than 6.1 m or less than 1.2 m (roads) or 0.9 m (fields), pipes need to be completely surrounded with 150 mm of concrete to improve their strength. The earth load increases with depth of cover. The effect of wheel loads decreases with depth. Thus, the highest loads which pipes have to withstand are at shallow depths (due to wheel loads) or at large depths (due to earth load). At intermediate depths, the imposed loads are lower. In many circumstances, pipes are laid without external protection at these intermediate depths. This supports achieving sustainability goals going forward because of economizing concrete surroundings.

8.4 LOADS DUE TO EARTH OVERBURDEN

The famous theory used to compute loads due to earth overburden on storm sewer pipes is 'Marston-Spangler Load Analysis Theory'. The theory states that the load on the installed storm sewer pipe is equal to the weight of soil prism, which is termed as interior prism, on the pipe minus or plus the frictional shearing force transferred to the soil over the pipe by the trench wall side or exterior soil prisms on either side of interior prism.[2] *The direction and magnitude of the frictional force are dependent on the settlement of the prism over the pipe in relation to the neighboring soil prisms*. This is discussed further in the subsequent sections.

Assumptions

- The computed load equal to the load developed when maximum settlement is realized, and Rankine's theory is used to calculate lateral pressure which generates shearing force between the soil prism over the pipe and adjacent soil prisms.

Storm sewers experience several loading cases depending on how they are laid in the trench, the workmanship, the size and depth of the trench, and the orientation. The effect of soil and external loads is pretty an important consideration. The static load exerted on the buried sewer for normal conditions is calculated from **Marston's equation** below:

$$W = CwB^2 \tag{8.1}$$

Where:

W : Load on the pipe per unit length
C : Coefficient that depends on the depth of the trench, character of construction, and fill material
B : Width of the trench just below the top of the pipe
w : Weight of the fill material per unit volume

For normal ordinary trench construction, C can be calculated from

$$C = \frac{1 - e^{-2\mu' K H/B}}{2\mu' K} \tag{8.2}$$

Where:

H : Depth of fill above the pipe
B : Width of the trench just below the top of the pipe
K : Ratio of active lateral pressure to vertical pressure
μ' : Coefficient of sliding friction between fill material and sides of the trench

The value of $\mu' K$ ranges from 0.1 to 0.19 as shown in Table 8.1. Equation (8.2) is used to calculate loads for normal trenches. Other loading conditions manipulated the coefficients as shown in the preceding sections.

8.4.1 Derivation of Marston's equation

The derivation described herein is for the ordinary (**normal or narrow**) trenches loading conditions. Consider an element of the fill, ∂y on the horizontal plane X-X extending to the trench sides as shown in Figure 8.4. This is illustration based on normal trenches where the depth is relatively deep and the width is relatively narrow compared with the width of the pipe.

Assumptions

a. The vertical stress, σ is assumed to be uniform across the element.
b. A form of arching action is assumed to below **X-X**. However, in some practices, the fill adjacent to the pipe below horizontal plane **X-X** is assumed to carry part of the load although this is highly predetermined by the quality of workmanship.

The total load per unit length of the pipe can be calculated by taking the equilibrium of forces acting on the fill element.

Rankine's formula is used for horizontal component of stress:

$$K_a \sigma$$

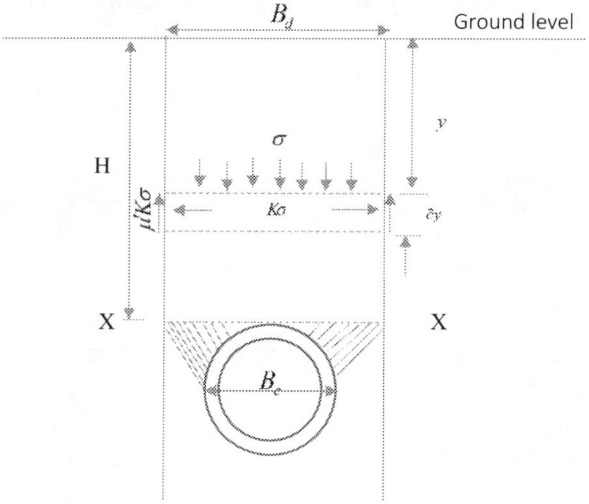

Figure 8.4 Derivation of Marston's equation.

Where:

$$K_a = \frac{(1-\sin\phi)}{(1+\sin\phi)} \tag{8.3}$$

The friction stress is $\mu'K\sigma$, where μ' is the *coefficient of friction between the fill and the trench sides*. Rankine's coefficient K is the ratio of active lateral pressure to vertical pressure. The coefficient of backfill/trench sliding friction μ' describes the friction between the backfill and the material making up the trench walls. These soil parameters can be determined from laboratory testing of the soils or can be estimated from generalized parameters. It is common to take the values of K and μ' together and treat them as an empirical function of the soil properties as it is unlikely that the individual values will be known in advance of the laying of stormwater drainage pipes. Table 8.1 provides the values of $\mu'K$.

For equilibrium, vertical forces per unit length on the element are resolved as follows:

$$B_d\sigma + wB_d\partial y = 2\mu'K\left(\sigma + \frac{\partial\sigma}{2}\right)\partial y + B_d(\sigma + \partial\sigma)$$

In which w is the weight of unit volume of fill and B_d is the width of the trench.
Integrating the equation between limits $y = 0$ and $y = H$ gives:

$$\sigma_H = wH\left(1 - e^{(-\alpha'_d)}\right)\alpha'_d$$

In which,

$$\alpha'_d = 2\mu'K(H/B_d)$$

Table 8.1 Value of the product $\mu'K$ and μK

Soil type	Maximum value of $\mu'K$ or μK.
Granular, cohesionless materials	0.192
Sand and gravel	0.165
Saturated top soil	0.150
Clay	0.130
Saturated clay	0.110

Table 8.2 Unit weight of backfill material

Material	Unit weight (kg/m³)
Dry sand	1600
Ordinary sand	1840
Wet sand	1920
Damp clay	1920
Saturated clay	2080
Saturated topsoil	1840
Sand and damp topsoil	1600

Load on the plane X-X per unit length of pipe = $\sigma H B_d$.

The above equation can be expressed as:

$$C_d w B_d^2 \tag{8.4}$$

In which,

$$C_d = \frac{1-e^{-\alpha'_d}}{2\mu'K} \tag{8.5}$$

Therefore, the equation (8.4) for backfill loads in narrow trenches is completed by the narrow trench fill load coefficient, which can be calculated using the equation (8.5). The unit weight of materials usually used as backfill are provided in Table 8.2.

Following Table 8.1, it is recommended that where detailed information of the nature of the soil is lacking, the designers uses a high value of ∞K for the wide-trench cases and a low value for the narrow-trench and negative projection cases in order to err on the safe side. Therefore, 0.19 is often used with C_c and 0.13 with C_d and C_n.

The narrow trench backfill load is in many cases taken from various charts and tables of values computed from typical soils and trench geometries. In several cases, the use of typical values is justified for soil properties commonly unknown at the time the design is undertaken.

8.4.2 Variation in the use of Marston's equation

The loading conditions of other cases manipulate Marston's equation (8.2) by adjusting loading coefficients. The variation in the use of Marston's equation considers how the pipe is buried, the size of the trench, and the embankment type, and projection from or onto the ground. Therefore, *fill load coefficients* for the different scenarios are derived for use in Marston's equation. This section is limited to analyzing complete and incomplete positive projection cases, incomplete and complete negative projection cases, and zero projection cases. The derivation of *fill load coefficients* is clearly elaborated addressing the theoretical approach. Further details on other loading cases including design charts and graphs can be obtained from design standards such as the BS 9295:2020 Guide to the structural design of buried pipes. Fill load coefficients such as $\mu'K$, μK, $r_{sd}p$, α, and C are provided graphically in several design manuals and standards. These parameters are used in several load factor expressions for different loading cases.

8.4.2.1 The complete embankment case or wide-trench case (positive projection)

When the pipe is installed in a wide trench or on the ground and an embankment installed over the top, it is called the positive projection condition, which can either be complete projection or incomplete projection. This refers to a situation where the pipe is laid on the original ground level and an embankment is placed on the top of it. It can also refer to a situation where the trench is wide and the pipe is considered to be remote from the trench walls. In that case, the load upon the pipe is usually greater than the weight of the fill immediately above it. Based on the *Iowa theory*, the *vertical planes tangential to the pipe sides* would form as shown in Figure 8.5. The additional load arises from friction stress on these planes mobilized by the settlement outside the shear planes in excess of that within the planes which is restricted by the presence of the pipe. In the wide-trench installation conditions, the frictional forces act downward and then effectively increase the load acting on the pipe. This is because the concrete pipe is much stiffer than the surrounding material which implies that the adjacent columns of soil settle more than the column directly above the pipe. This is the reason sandy soils exert more pressure on the concrete pipe due to their higher angle of internal friction than clayey soils in the wide-trench condition.

An analysis similar to the above (normal trench) but with ***friction stress reversed leads*** to:

$$C_c = \frac{e^{\alpha_c} + 1}{2\mu K} \tag{8.6}$$

Where,

$$\alpha_c = 2\mu K \left(H/B_c \right) \tag{8.7}$$

In this case, μ is the *coefficient of internal friction of the fill*. Table 8.1 provides the values of μK. Therefore, the load on the plane (X-X) per unit length for an embankment case (complete case) is given by

$$W_c = C_c w B_c^2 \tag{8.9}$$

Structural aspects of storm sewers 279

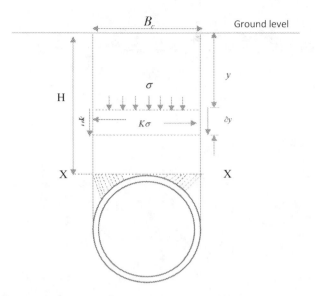

Figure 8.5 Derivation of wide-trench coefficient (complete embankment).

8.4.2.2 The incomplete wide-trench case (positive projection)

Figure 8.6 shows the derivation of wide-trench coefficient or embankment case (incomplete case). Two important scenarios are considered: deformation of the pipe and settlement of the fill. In this particular incomplete wide-trench case, when settlement occurs, the fill level *above* the pipe is always slightly higher than the fill level *alongside* the pipe. As the depth of the overburden increases, pipe deformation increases and the extent to which the fill over the pipe protrudes above the surrounding fill material decreases until a certain height will be no protrusion and shear at the ground surface will be *zero*.

When further fill is added, the shear planes do not extend into it and *the plane of equal settlement H_e at which the shear planes die out will fall slightly as the total depth of the overburden increases*. The height of the plane of equal settlement H_e is used to determine the backfill load for the incomplete projection case where the frictional slip planes cease to add to the backfill load within the depth of the backfill material. *The location of the plane of equal settlement therefore determines which condition is present and therefore which calculation is critical.*
Important:

a. If the plane of equal settlement is located above the embankment, then the installation is termed as complete trench condition or complete projection condition according to the direction of the shearing force. However, if the plane of equal settlement is within embankment as shown in Figure 8.6, then the installation is incomplete trench condition or incomplete projection condition.
b. When the embankment is sufficiently high, it is assumed that the shearing forces will terminate at some horizontal plane in the embankment fill; this plane is termed the plane of equal settlement.

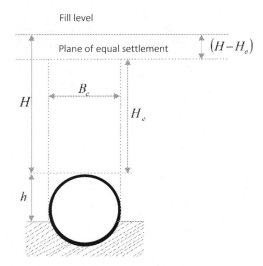

Figure 8.6 Derivation of wide-trench coefficient or embankment case (incomplete case).

The height of the plane of equal settlement above the top of the conduit, H_e, depends on the projection of the pipe above its foundation, h, and on the deflection of the pipe relative to the settlement of fill alongside over the depth h. This is measured by r_{sd}, the settlement-deflection ratio.

8.4.2.2.1 The settlement-deflection ratio r_{sd} and projection ratio p

To determine the height of the plane of equal settlement, H_e, requires determining two factors: the settlement-deflection ratio r_{sd} and the projection ratio p. The settlement-deflection ratio value describes the ratio between the settlement of the fill and the settlement of the ground underlying the pipe. **BS 9295:2020** Guide to the structural design of buried pipes provides a detailed account of the settlement-deflection equations and graphs for use in the design.

The derivation of r_{sd} is shown in Figure 8.7.

Figure 8.7 shows a pipe that has deformed due to the fill loading. At the same time the fill loading has settled. The left-hand half of this figure defines a 'critical plane' in relation to the hypothetical case when there is neither settlement of the fill nor pipe deformation. When there is pipe deformation or fill settlement, the 'critical plane' drops as shown to the right for Figure 8.7.

The right-hand half takes account of fill load settlement and pipe deformation. Therefore, in between the left-hand and right-hand halves, the following changes occur:

a. The depth of fill h in Figure 8.6 has decreased by s_m due to settlement.
b. The foundation alongside the pipe has yielded by an amount s_f.
c. The pipe diameter has deformed or deflected by an amount d_c.

Structural aspects of storm sewers 281

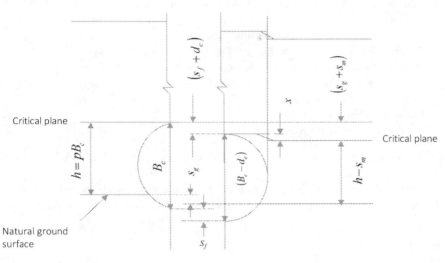

Figure 8.7 Settlement-deflection ratio for positive projection.

Therefore, the **critical plane** has dropped to the extent $(s_g + s_m)$ and the top of the pipe has dropped to the extent $(s_f + d_c)$.

The protuberance x of the top of the pipe above the critical plane is therefore given by Equation (8.10):

$$x = (s_g + s_m) - (s_f + d_c). \tag{8.10}$$

The ratio $x/s_m = r_{sd}$ is taken as the measure of the relative settlement and deflection termed as the settlement-deflection ratio.

Therefore, the settlement-deflection ratio is given by:

$$r_{sd} = \frac{x}{s_m} = \frac{(s_g + s_m) - (s_f + d_c)}{s_m} \tag{8.11}$$

Where:

r_{sd} : Settlement-deflection ratio (measure of the relative settlement and deflection). The ratio between the settlement of the fill and the settlement of the ground underlying the pipe.
p : Projection ratio = h/B_c. *Proportion of the pipe's outer diameter* which is above the bedding level or the natural ground level.
d_c : Deformation/deflection of pipe diameter.
x : Protuberance of the top of the pipe above the critical plane.
s_m : Decrease of fill height due to settlement (compression of soil columns).
s_g : Settlement of natural ground adjacent to the pipe.

s_f : Settlement at the bottom of the pipe (foundation alongside the pipe has yielded by an amount s_f).
h : Projection of the pipe above its foundation.
B_c : Taken to be the effective width of the trench (in wide trenches).
H_e : Height of the plane of equal settlement above the top of the pipe.
H : Fill level above the pipe. The height of the fill is the height between the top of the fill and the top of the pipe.
α_{ec} : For incomplete case projection.
α_e : For complete case projection.

The settlement-deflection ratio design values have been established as shown in Table 8.3.

The projection ratio p is the **proportion of the pipe's outer diameter** which is above the bedding level or the natural ground level. For an unyielding (rigid) foundation,

$$p = h/B_c \tag{8.12}$$

The product $r_{sd}p$ in which $p = h/B_c$ controls the extent of the shear planes and the height of the plane of equal settlement. *To cover different forms of bedding, $r_{sd}p$ is often regarded as a single entity.* The fill load exerted on the positive projecting pipe is affected by the multiplication of settlement ratio r_{sd} and the projection ratio p. Values commonly accepted for $r_{sd}p$ are 1.0 where bedding is Class A or B, that is, rigid and does not extend to the full trench width, and 0.5 where bedding is granular and for beddings such as Class B or concrete arch beddings.

The r_{sd} effect is usually not measured on-site before a drainage pipe is laid. The ratio varies typically between 0 and 1.0 with 0 being very soft soils and 1.0 being rock. The ratio can be negative in some cases. For example, where flexible pipes are used, it is possible for flexible pipes to be depressed below the critical plane so that r_{sd} becomes negative. Therefore, where negative r_{sd} is encountered, the fill above the pipe then moves downward in relation to the fill alongside and the shear stress is reversed. This leads to expressions algebraically similar to those of the negative projection cases except that B_c replaces B_d.

Complete:

$$2\mu KC_c = 1 - e^{-\alpha_c}$$

Table 8.3 Design value for settlement-deflection ratio

Storm sewer pipe type	Soil condition	Settlement ratio, r_{sd}
Rigid	Rock or unyielding foundation	+1
Rigid	Ordinary foundation	+0.5 to +0.8
Rigid	Yielding foundation	0 to +0.5
Rigid	Negative projecting installation	−0.3 to +0.5
Flexible	Poorly compacted fills on either side of pipe	−0.4 to 0
Flexible	Well compacted fill on either side of the pipe	0

Incomplete:

$$2\mu K C_c = 1 - e^{-\alpha_{ec}} + (\alpha_c - \alpha_{ec})e^{-\alpha_{ec}}$$
$$\alpha_c = 2\mu K (H/B_c)$$
$$\alpha_{ec} = 2\mu K (H_e/B_c)$$

Although $r_{sd}p$ is negative both for this particular case (where flexible pipes are used) and for negative projection (see Section 8.4.2.3), the meanings of r_{sd} are not truly analogous and the equations for determining H_e differ, as also do the straight line approximations to the relationship of C and α.

The value of the settlement-deflection ratio depends on the degree of compaction achieved in the fill material to the sides of the pipe. With good construction practices, a value of 0.7 is recommended for normal soils. This need only be amended for pipes bearing on rock or on particularly weak or compressible soils.

While this value could be measured on-site, use of standard bedding types allows the designer to determine a suitable projection ratio value for design. Bedding types D, F, and N have a value of 1.0. Bedding types B and S have a value of 0.7. See Figure 8.11.

In practice, the values for the settlement-deflection ratio and the projection ratio are often taken together as 0.7 for bedding types D, F, and N and 0.5 for bedding types B and S. During design we can assume a standard value based on the chosen bedding type though the designer can override these values if required. The derivation of the load-factor expression for this case needs to perform integration over the height of the plane of equal settlement.

Integration is thus performed over the interval H_e to obtain the load-factor expression. Therefore, the load at the plane of equal settlement being given by:

$$w(H - H_e)B_c \tag{8.13}$$

The result can be written as:

$$C_c = \left(\frac{e^{2\mu K(H_e/B_c)} - 1}{2\mu K}\right) + \left(\frac{H}{B_c} - \frac{H_e}{B_c}\right)e^{2\mu K(H_e/B_c)} \tag{8.14}$$

Or as:

$$2\mu K C_c = e^{\alpha_{ec}} - 1 + (\alpha_e - \alpha_{ec})e^{\alpha_{ec}}, \text{ in which } \alpha_{ec} = 2\mu K(H_e/B_c) \tag{8.15}$$

Because H_e depends on H as well as on μK and $r_{sd}p$, Equation (8.15) is inadequate when used alone. To compute (H_e) in Equation (8.15), the relationship between the settlement-deflection ratio must be considered. Therefore, Equation (8.15) is solved by first solving another complex Equation (8.16) involving $r_{sd}p$. Equation (8.16) was proposed by Spangler for obtaining H_e. This equation is solved by trial and error,[3] but the combined effect of the two Equations (8.15) and (8.16) is to produce a nearly linear relationship between C_c and α_{ec}. The plane of equal settlement H_e can be calculated directly using an iterative solution to the Equation (8.16), which includes a number of semiempirical coefficient choices which affect the outcome.

284 Integrated Drainage Systems Planning and Design for Municipal Engineers

$$r_{sd}p\frac{H}{B_c} = \left(\frac{e^{2\mu H(H_e/B_c)} - 1}{2\mu K}\right)\left(\frac{1}{2\mu K} + \left(\frac{H}{B_c} - \frac{H_e}{B_c}\right) + \frac{r_{sd}p}{3}\right)$$
$$+ \frac{1}{2}\left(\frac{H_e}{B_c}\right)^2 + \frac{r_{sd}p}{3}\left(\frac{H}{B_c} - \frac{H_e}{B_c}\right)e^{2\mu K(H_e/B_c)} - \frac{H_e}{2\mu K B_c} - \frac{H(H_e)}{B_c^2}$$

(8.16)

Several generic solutions for typical soil parameters and concrete pipe sizes have been developed over the years for use in the design process. These are used more commonly than necessarily needing to solve the complex Equation (8.16). Charts have been developed to estimate values for both C_c or W_c for typical conditions. These charts provide load coefficient relationships. Note: While these equations cover most design situations, for unusual situations where a precise value is required, interpolation between different equations is inaccurate.

BS 9295 proposed an approximate design equation (8.17) to solve for C_c. While the errors for the approximate equation (8.17) are between −1.4% and 2.8% for relevant conditions, Table 8.4 tabulates the errors allowing the equation to be adjusted accordingly. In order to determine the value of C_c for different design situations, the equation results are adjusted by interpolation of error Table 8.3. The value of (C_c) can be estimated for different values [(H/B_c) r_{sd}] and projection ratio (p) using charts.

$$C_c = \left(0.68(r_{sd}p)^{0.41} + 1\right)\frac{H}{B_c} - 0.1(r_{sd}p)^{0.57}$$

(8.17)

Therefore, regarding fill loads which the pipe is subjected to in positive projecting embankment condition, it may be estimated using the following expression:

$$W_c = C_c w B_c^2$$

(8.18)

The commonly used value of $r_{sd}p$ is 1.0 where the pipe bedding is rigid and does not extend to the full trench width, and 0.5 for beddings such as Class B (F.O.S = 1.9) (See

Table 8.4 Error table for Equation (8.17)

$r_{sd}p$	H/B_c						
	0.3	0.5	0.7	1	10	100	500
0	0	0	0	0	0	0	0
0.1	+1	+1.7	+2.1	+2.3	+2.8	+2.8	+2.8
0.3	+2	+1.9	+1.9	+1.9	+1.8	+1.8	+1.8
0.5	+1.6	+1.3	+1.1	+1	+0.8	+0.8	+0.8
0.7	+2	+1	+0.7	+0.4	−0.1	−0.2	−0.2
1	+4.4	+2.1	+1.2	+0.6	−0.5	−0.6	−0.6
2	+3.3	+1	+0.3	−0.3	−1.3	−1.4	−1.4
3	+2.1	+0.8	+0.3	0	−0.6	−0.6	−0.6
4	−1.6	−0.6	−0.2	0	+0.4	+0.5	+0.5
5	−4.2	−1.4	−0.4	+0.3	+1.4	+1.5	+1.6

Figure 8.11) or concrete arch beddings. In negative projection, $r_{sd} = -0.3$ has been tentatively adopted for rigid pipes.

8.4.2.3 Negative projection cases

This is pretty straightforward and applies where an embankment is placed above the pipe which is in a *narrow trench* below the original ground level. Therefore, it is an extension of a narrow-trench loading case as shown in Section 8.4.1. Both complete and incomplete cases arise similar to the wide-trench case because **a plane of equal settlement** occurs with large total depths of fill as shown in Figures 8.9 and 8.10.

By definition: *In the case of negative embankment condition, the pipe is placed in a trench which is narrow compared with pipe size and trench depth. In this installation condition, the top of the pipe is below the original ground surface and the fill material over the pipe exceeds the original ground level surface.*

8.4.2.3.1 Complete case negative projection

When interior soil prism over the pipe experiences settlement, an upward shearing force is created which decreases the load imposed on the pipe. As the backfill material compressibility and negative projection ratio increase, the interior soil prism settlement increases. That is why materials such as sawdust and other material with similar property are frequently added to the soil directly above the storm sewer pipe to increase interior prism settlement. See Figure 8.8. The load-factor expression for the negative projection cases (complete case) is given by Equation (8.19):

$$2\mu K C_n = 1 - e^{-\alpha_d}$$
$$\alpha_d = 2\mu K (H/B_d) \tag{8.19}$$

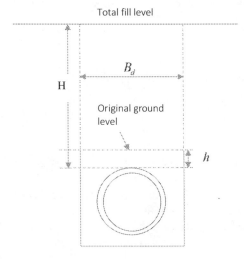

Figure 8.8 Complete negative projection case.

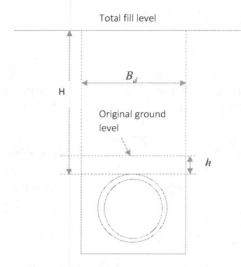

Figure 8.9 Incomplete negative projection case.

8.4.2.3.2 Incomplete case negative projection

If $H > H_e$ as shown in Figure 8.9, then the following formula is used to compute C_n:

$$2\mu K C_n = 1 - e^{-\alpha_{ed}} + \left(\alpha_d - \alpha_{ed}\right) e^{-\alpha_{ed}}$$
$$\alpha_{ed} = 2\mu K \left(H_e / B_d\right) \tag{8.20}$$

A complex equation connects H_e with H, $\propto K$, and $r_{sd}p$, but the incomplete case still gives a nearly linear relationship. Therefore, regarding fill loads which the pipe is subjected to in negative projecting embankment condition, it can be estimated using Equation (8.21):

$$W_n = C_n w B_d^2 \tag{8.21}$$

8.4.2.3.3 Settlement-deflection ratio for negative projection

The projection ratio is given by:

$$p' = h/B_d \tag{8.22}$$

Figure 8.10 shows how r_{sd} is derived for negative projection. The ratio r_{sd} has a negative sign and the datum surface from which h' is measured is the original ground level, instead of the pipe foundation. It is because h' is measured downward from datum level that the term negative projection is adopted (the wide-trench case may be termed positive projection). It should be noted that $r_{sd}p'$ is negative but the negative sign arises from the manner in which r_{sd} is defined, p' and h' being taken as positive.

Structural aspects of storm sewers 287

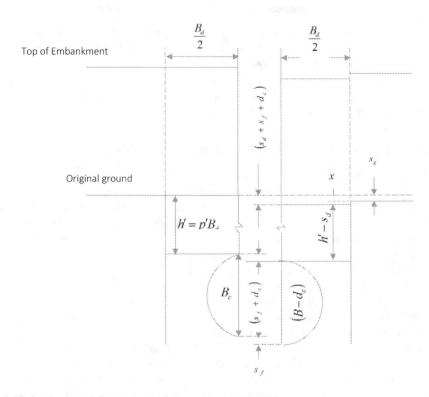

Figure 8.10 Settlement-deflection ratio for negative projection.

s_d is the compression within the fill for the height of $p'B_d$, where p' negative projection ratio, which is equal to the vertical distance from the top of the pipe to the original ground surface at the time of installation divided by trench width. If the natural ground surface is not leveled, then it is required to consider average vertical distance from the top of storm sewer pipe to both sides of the trench.

Similar to the positive projecting condition, it is required to specify the relationship between pipe deflection and relative settlement between interior soil prism and exterior soil prisms so as to evaluate H_e. The settlement-deflection ratio for negative projection is given by Equation (8.19):

$$r_{sd} = \frac{-x}{s_d} = \frac{s_g - (s_d + s_f + d_c)}{s_d} \qquad (8.23)$$

Table 8.3 provides recommended design values for settlement-deflection ratios for negative projecting condition. The value of C_n can be estimated for different values, r_{sd}, and projection ratio p value of 0.5, 1.0, 1.5, and 2. Values falling between the provided projection ratios, p, should be found by interpolation.

8.4.2.4 Zero projection case

This is the limiting case where positive projection or $r_{sd}p$ is zero and may be considered to occur where the top of the pipe is near the original ground level with an embankment as overburden. No shear planes are considered to form, and the load is simply the weight of overburden immediately above the pipe, $\gamma H B_c$, that is, $C = H/B_c$. In some circumstances, it may be arguable that the load should be taken over the trench width as $\gamma H B_d$ or $C = H/B_c$. Zero projection is then the limiting case for negative projection.

8.5 LOADS SUPERIMPOSED BY VEHICLE WHEELS

The objective of calculating vehicle wheels is to produce a load value which can be added to the fill loads. The famous Boussinesq formula is used for calculating the effects of loads superimposed by vehicles. The vertical stress component is calculated at any point below the surface of an elastic solid which extends infinitely from the surface as a result of a point load on the surface. Generally, the total vertical load on a horizontal plane beneath the surface requires the stress to be integrated over the area.

Since the effect of a point load extends indefinitely along the pipe, reducing as distance from the load increases, a specific length of pipeline must be chosen over which the load is to be calculated. A length of 0.9 m or less of an individual pipe is suitable. The variation of soil stress over the area and consequently the conversion factors depend on the depth of cover, on the breadth of the pipe, and on its form of support and bedding.

When it is desired to find the total vertical load on a horizontal area beneath the surface, stress has to be integrated over the area. Account has to be taken of the nonuniformity of stress over the area on which load is calculated. The load has to be divided by this length later to produce a load per unit length for addition to the fill load. The goal is to produce a load value which can be added to the fill loads. The load value is therefore taken as the uniform load which will give the same pipe stress as the calculated nonuniform load. Factors of safety have to be worked out to enable the loads obtained from Boussinesq–Newmark stage to be converted to equivalent uniform loads. Before the effective load is added to the fill load, a factor has to be applied to cover impact. Storm sewers under main traffic roads consider impact factor of 1.3. With the train of wheel loads in a chosen position, the loads from several wheels over a horizontal area tangential to the top of the pipe are assessed using *Newmark's influence values*. Holl Newmark developed methods to find total vertical load acting on a pipe. Several graphs have been developed covering different wheel loads for pipe-bedding classes which are used to doge extensive calculations.

Modern methods of calculating vehicle wheels have been developed over the years. For example, in the USA, the Design of Highway Loads often follows the American Association of State Highway and Transportation Officials (AASHTO) criteria. The AASHTO Load and Resistance Factor Design (LRFD) Bridge Design Specifications specifies the applicable highway loads and their distribution through the soil. This design data addresses the method of determining the live load pressure transmitted through unsurfaced roadways to circular, elliptical, and arch concrete pipe in accordance with the criteria of the AASHTO LRFD Bridge Design Specifications. The axles may be at right angles or parallel to the pipe axis. These methods are now used to estimate the vehicle loads acting on storm sewer pipes which are finally added to the backfill and other loads. The distribution of soil stress over the

Structural aspects of storm sewers 289

area at the top of the pipe differs according to the position of the wheels. The position giving the worst effect depends again on depth and breadth of pipe and on the type of bedding.

If any culvert or sewer pipe is within the heavy duty traffic highway right-of-way, but not under the pavement structure, then such pipe should be analyzed for the effect of live load transmission from an unsurfaced roadway because of the possibility of trucks leaving the pavement.

8.5.1 AASHTO LRFD design method[4]

The design method encompasses four steps:

1. Obtain the following project data:
   - Pipe shape, size, and wall thickness.
   - Height of cover over the concrete pipe and type of earth fill.
   - LRFD or other criteria.
2. Calculate the average pressure intensity of the wheel loads on the soil plane on the outside top of the pipe.
3. Calculate the total live load acting on the pipe.
4. Calculate the total live load acting on the pipe in kilonewtons per linear metre (kN/m).

According to the AASHTO method, the intermediate and thin thickness flexible pavements do not reduce the pressure transmitted from a wheel to the subgrade. Such pavements have no generally accepted theory for estimating load distribution effects and they are considered as unsurfaced roadways. However, thick and high-strength pavements designed for heavy truck traffic substantially reduce the pressure transmitted through a wheel to the subgrade and to the underlying storm sewer concrete pipe. The Portland Cement Association (PCA) (1944) developed a widely used manual for estimating the vertical pressure on the buried pipe due to wheel loads. This method is presented in the American Concrete Pipe Association (ACPA), 'Concrete Pipe Handbook', and 'Concrete Pipe Design Manual'. Over the years, there has been a number of recommendations drawn based on which the traffic loads are computed. Lightly trafficked areas differ from heavy trafficked areas. The process of calculation consists of several stages, some of them requiring extensive computations. The AASHTO LRFD criteria consider critical wheel loads for vehicle traveling perpendicular to pipe and vehicle traveling parallel to pipe. For more information on how to compute loads superimposed by vehicles' wheels on sewer pipes, BS 9295:2020 Guide to the structural design of buried pipes and AASHTO LRFD Bridge Design Specifications are good resources.

8.6 INTERNAL WATER LOAD

The water present inside a storm sewer pipe exerts a load on the pipe and supporting soils and bedding material. This internal water load has very often been neglected in traditional design charts and tables due to the relatively small impact except for particularly large pipes with small fill heights. The load due to water content is not equal to the total weight of the water. An equivalent water load factor dependent on the bedding type modifies the total weight of the water in the pipe as shown in the Equation (8.24).

$$W_w = C_w \rho g \pi \frac{d^2}{4} \qquad (8.24)$$

Where:

- W_w : Internal water load
- ρ : Density of water in the storm sewer taken as 1000 kg/m³
- d : Internal diameter of the pipe used to calculate the total volume of water
- g : Acceleration due to gravity = 9.81 m/s²
- C_w : Water load coefficient, which adjusts the total weight of the water to allow for the different actions on different bedding types. This has been taken on average as 0.75

8.7 PIPE STRENGTH AND BEDDING CLASSES

There are two factors considered to achieve the overall strength of the concrete pipe system.

   a. **Pipe strength.** Provision of the proper pipe strength to match imposed loads is essential. Concrete pipe classes are available from which a selection is made.
   b. **Bedding class.** A proper bedding class is needed so that the pipe performs reliably and safely. Several bedding classes exist following standards and codes of practice. The load carrying capacity of the concrete pipe is increased by the bedding class.

The designer usually selects a combination of bedding class with the correct concrete pipe class following reliability, durability, maintainability, functionality, and sustainability goals. Usually when all the loadings are determined, the pipe is then checked for strength. The crushing strength of the pipe is used to determine its strength. It is usually obtained from a three-edge crushing test. Previously, the bending moments and bending resistance of the pipe ring were calculated to determine the pipe strength to carry loads.

The components to consider while selecting the pipe to carry loads are bedding type and pipe crushing strength. However, the total allowable crushing load on the concrete pipe is determined by the pipe crushing strength multiplied by the bedding factor and divided by the factor of safety (FOS). In the UK and countries adopting British standards, the minimum crushing load is usually specified as a multiple of the nominal pipe diameter. The most commonly specified strength class is 120, so the crushing strength in kN/m will be 120 times the nominal pipe diameter in meters. BS 5911 provides the crushing values for 120 strength class pipes of standard sizes.

Choosing a FOS depends on a number of factors. The following factors are considered when choosing an appropriate FOS[3]:

   a. **Crushing strength of the concrete pipe.** Once concrete pipes are manufactured in accordance with the relevant standards, the desired crushing strength is achieved. Therefore, it is considered unlikely that the strength of concrete pipes manufactured in accordance with the relevant standards will drop significantly below the stated crushing strength.
   b. **Bedding factor.** Experimental data suggest that the bedding factors stated in standards are conservative. Therefore, a good contractor should be able to achieve the desired bedding factor in the vast majority of cases.

c. **Durability.** The pipes laid in aggressive ground may warrant a larger factor of safety if there are any risks that the pipe may deteriorate faster than generally expected due to chemical attack from either the ground or from effluent in the pipe.
d. **Loading equations.** The loading equations tend to be slightly conservative, particularly where soil material properties are common but unknown.

A factor of safety of 1.25 was commonly used where the concrete pipe proof load was high at 80% as per the old BS 5911. However, the new European Standard BS EN 1916 uses a value of 67% of the ultimate load for reinforced pipes. It is believed that this increases the risk of poor-quality concrete pipes being used on-site, therefore a FOS of 1.50 is sometimes used for reinforced pipes. BS EN 1295-1-1997 recommends a FOS of 1.25 for general use but a value of 1.5 for reinforced pipes with a 67% proof load. Usually, a conservative approach is to use a FOS of 1.25 for unreinforced pipes and 1.5 for reinforced pipes.

Usually pipes below 600-mm nominal diameter are unreinforced and larger pipes are reinforced. Sizes DN 225 to DN 600 inclusive are normally only manufactured unreinforced in the UK. Sizes DN 1000 and above are normally only manufactured reinforced in the UK. However, some manufacturers in the UK continue to test their reinforced pipes to 80% proof load as per the old BS 5911-100 standard, allowing a FOS of 1.25 to continue to be used. Whether the pipe will be reinforced or unreinforced depends on the manufacturer. BS EN 1295-1: 1998 recommends that the minimum value of FOS for the structural design of reinforced pipelines should be increased from the normal 1.25 to 1.5 if, as is the case of BS EN 1916:2002, the proof load is 67% of the minimum crushing load. The 2019 version of the same standard removed a specific recommended value and left the designer to use their engineering judgement based on the perceived risks of overloading and material and workmanship quality.

8.7.1 Concrete pipe design—conclusion

The structural strength of the concrete pipe and bedding system can be calculated as the concrete pipe crushing strength multiplied by the bedding factor divided by the FOS. This strength is checked against the total loads acting on the pipe to determine whether the pipe and bedding system proposed are adequate.

8.7.2 Worked example 8.1 for illustrative purposes (loading case: Ordinary [normal] trenches)

A 610-mm reinforced concrete pipe storm sewer is to be placed in an ordinary trench (narrow-trench case) 3.66-m deep and 1.22-m wide which will be filled with sandy gravel weighing 1680 kg/m^3. Determine the load on the pipe and the type of bedding required if the installation is to have a FOS of 1.5. Take the minimum crushing strength of a 610-mm reinforced concrete pipe as 72 kN/m. Neglect vehicle wheel loads and other loading categories/conditions.

8.7.2.1 Solution

From Equation (8.2),

CLASS A
Plain Concrete Cradle
BEDDING FACTOR = 2.6

NORMAL BACKFILL DEGREE OF COMPACTION DEPENDENT UPON SURFACE DESIGN REQUIREMENTS

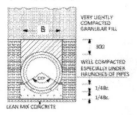

Class A concrete bedding, either plain or reinforced each with 120° cradle. Screed the formation level with lean mix concrete, place blocks on the screed to support pipes behind

Reinforced Concrete Cradle
BEDDING FACTOR = 3.4

NORMAL BACKFILL DEGREE OF COMPACTION DEPENDENT UPON SURFACE DESIGN REQUIREMENTS

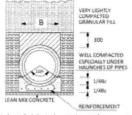

each socket. Lay pipes using packers on blocks to achieve the correct line and level. At pipe joints, form construction joints through the concrete bed to ensure flexibility of pipeline.

The minimum width of cradole is to be 1 1/4 Bc or Bc + 200mm. The minimum thickness is top be 1/4 Bc. Pour the cradle carefully form one side to prevent voids in the concrete. Backfill when the cradle has cured top the required strength.

KEY

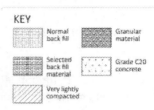

	Normal back fill		Granular material
	Selected back fill material		Grade C20 concrete
	Very lightly compacted		

CLASS D
Hand Trimmed Flat Bottom.
BEDDING FACTOR = 1.1

NORMAL BACKFILL DEGREE OF COMPACTION DEPENDENT UPON SURFACE DESIGN REQUIREMENTS

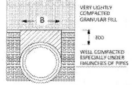

Suitable in fine grained soils and relatively dry conditions. Hand trim formation filling in any hollows. Form socket hollows as required with 50mm minimum clearance of sufficinet length to poermit jointing. Pipes are laid directly onto excavated trench base.

CLASS N
Flat Granular Type.
BEDDING FACTOR = 1.1

NORMAL BACKFILL DEGREE OF COMPACTION DEPENDENT UPON SURFACE DESIGN REQUIREMENTS

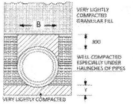

Lay pipes in a flat layer of selection material. Form socket hollows as required with 50mm minimum clearance of suffient length to permit jointing.

CLASS C
Hand Trimmed Flat Bottom.
BEDDING FACTOR = 1.5

NORMAL BACKFILL DEGREE OF COMPACTION DEPENDENT UPON SURFACE DESIGN REQUIREMENTS

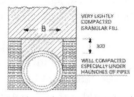

Suitable in uniform soils and relatively dry conditions. The bottom of the trench/ formation shall be profiled to fit the pipes over a width of 1/2 Bc with socket hollows as required with 50mm minimum clearance of sufficient lenth to permit jointing. Scarifying the formation level of the trench is generally adaquate in practice.

CLASS F
Granular Bedding
BEDDING FACTOR = 1.5

NORMAL BACKFILL DEGREE OF COMPACTION DEPENDENT UPON SURFACE DESIGN REQUIREMENTS

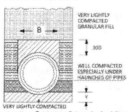

Lay pipes in a flat layer of granular bedding material on the formation level of the trench. Form socket hollows as required with 50mm minimum clearance of sufficient length to permit jointing. Pipes will settle slightly in to the bedding. Place side fill and compact well.

CLASS B
180° Granular Bedding
BEDDING FACTOR = 1.9

NORMAL BACKFILL DEGREE OF COMPACTION DEPENDENT UPON SURFACE DESIGN REQUIREMENTS

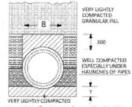

Lay pipes in a flat layer of granular bedding material on the formation level of the trench. Form socket hollows as required with 50mm minimum clearance of sufficient length to permit jointing. Compact the layers each side of the pipe upto the springing level taking care not to displace them.

CLASS S
360° Granular Bedding
BEDDING FACTOR = 2.2

NORMAL BACKFILL DEGREE OF COMPACTION DEPENDENT UPON SURFACE DESIGN REQUIREMENTS

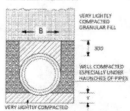

Lay, Joint and bed pipes as per Class B then place and well compact layers of the same bedding material at each side up to the crown level taking care not to displace the pipes. This is followed by a 300mm layer of the same granular bedding material but slightly compact directly over the pipe, after which ordinary backfilling is commenced.

Figure 8.11 Bedding classes.

$$C = \frac{1-e^{-2\mu'KH/B}}{2\mu'K} = \frac{1-e^{-2\times 0.165(3.66/1.22)}}{2\times 0.165} = 1.904$$

From Equation (8.1),

$$W = CwB^2 = 1.904 \times 1680 \times 1.22^2 = 4760.97 \ kg/m$$

Therefore, the load on the pipe is **4760.97 kg/m**. This result is compared with the minimum crushing strength of reinforced concrete pipes (RCPs). Thus, taking the minimum crushing strength of a 610-mm reinforced concrete pipe as 72 kN/m (7339.45 kg/m), a loading factor is obtained. To apply the FOS, we can either increase the applied load by a factor of 1.5 or reduce the crushing strength by a factor of 1.5. Therefore, the ratio of (load x safety factor)/strength becomes 0.97. From Figure 8.11 the Class D bedding would be adequate and cost-effective providing a load factor of 0.97.

Finding a cost-effective bedding class is important. Pipes are usually manufactured in different grades with standard strength and extra strength for different purposes to suit varying site conditions. Standard tables outlining the crushing strength of concrete pipes are available. Therefore, bedding classes usually differ highly influenced by the crushing strength of the pipes. The extra-strength pipe is more expensive than standard-strength pipe; the additional cost of concrete bedding would be much greater than the difference in pipe cost. A pipe of higher crushing strength might call for a relatively cheaper bedding class. An optimal solution in terms of cost and sustainability is arrived at by juggling crushing strength requirements and bedding class. This analysis is essential in conducting an environment lifecycle cost analysis in Section 9.4.1.

The embedded carbon (eCO2) from the bedding classes is compared with that of concrete pipes of different strengths to meet the sustainability goal. Thus, the most economical solution for the drainage pipeline would be derived by comparing alternative pipe strength and the corresponding bed classes arising from loading cases. The choice between bedding classes and corresponding pipe strengths may be left to the contractor with the condition that the contractor provides calculations on the most sustainable option during construction. For more information about structural design of buried pipes under various conditions of loading, it is good to consult BS 9295:2020 Guide to the structural design of buried pipes

8.8 BS 9295:2020 GUIDE TO THE STRUCTURAL DESIGN OF BURIED PIPES

BS 9295: 2000 issued by the British Standards Institute (BSI) describes the UK established method of structural design of buried pipes under various loading conditions.[5] The theoretical approach for different loading cases described in this chapter is well expounded in the standard. The standard explains the procedures and provides loading tables for situations where general assumptions can be made. It covers concepts used for structural design, classification and calculations of loads, rigid, flexible, and semirigid pipe design, and looks at other design considerations. The standard superseded BS 9295:2010 and came into effect on February 29, 2020. For further guidance, the designer is advised to consult BS 9295: 2000. It is complementary to BS EN 1295-1:2019 and PD CEN/TR 1295-2:2005.

8.9 STORM SEWER CONSTRUCTION MATERIALS

Several manufacturers produce pipes and other materials used for storm sewers following national standards. The prices of different pipe materials change from time to time which is significantly influenced by location and complexity of the stormwater drainage scheme. It is important for the engineer to support the developer to understand the pros and cons with the selection of drainage material in construction. It is good practice to install relatively inexpensive storm sewer materials because storm sewers rarely flow under pressure. The developer who is considering cost as the prime factor in construction of their building, unplasticized polyvinyl chloride (uPVC) pipes will most likely be the option, especially for general multifamily apartment construction, where ownership will likely change frequently, and where upfront cost is essential. However, for developers in development of condo, hotel, hospitals, and educational and office construction, selection of cast iron over uPVC pipes should be carefully considered due to durability, reliability, and noise reduction, which may relate to possible occupant complaints. In large-scale municipal drainage works, concrete pipes, high-density polyethylene (HDPE) pipes, and cast iron pipes are the best candidates. Before, a decision is made to specify or procure a standard storm sewer pipe, applicable standards are cross-checked for conformity. Storm sewer pipes come in different diameters, lengths, and joints. Several pipe types and other materials are used for the construction of storm sewers. The pipes include the following.

8.9.1 Reinforced concrete pipes

RCPs come in various sizes ranging from 300- to 2500 mm. the commonest sizes range from 450-to 1800 mm. The large-diameter pipes are usually reinforced although the small ones usually made of unreinforced concrete, that is, below 450 mm. However, reinforced pipes of smaller diameter are also available. All concrete pipes made in sizes greater than 600 mm are usually reinforced. In situ conduit construction (not necessarily circular in cross-section) competes with pipes of diameters greater than 1.5 to 1.8 m. Concrete pipes have two types: socket and spigot type and a plain-end concrete pipe. Socket and spigot pipes with rubber rings are commonly used although their use has slowly declined in the recent past.

8.9.2 Corrugated steel pipes

According to the National Corrugated Steel Pipe Association (NCSPA) (2021), corrugated steel pipe (CSP) has been used successfully in storm sewer applications since the early 1900s.[6] Countless miles of CSP provide reliable service throughout the highway system and in all sizes of cities and municipalities. As the need for cost-effective, durable infrastructure has grown over the years, the corrugated pipe industry has responded. Large-diameter steel pipes are most commonly used in pressure pipes for storm water conveyance. Steel pipes have applications similar to those of cast iron pipes. Since they are lighter, longer, and ductile, there are situations where steel is preferable to cast iron, but it is always necessary to thoroughly protect steel pipes from corrosion if they are to have the intended life span. In instances where shock loads are likely to occur either internally (water hammer) or externally, steel pipes should be chosen, but steps should be taken to see that the possibility of an internal vacuum is avoided. CSP storm sewers maintain the advantage of the strength of steel with a variety of coatings providing long service life. With allowable fill heights over 30 m and minimum covers as low as 300 mm, CSP can be matched to the strength required. Established long-term material properties ensure product performance throughout the design life. Steel pipes are about 8-m long but are not manufactured to precise standard lengths. The range of diameter and wall thickness are wide and will cover any conceivable pipe-sewer application.

Corrugated metal pipes (CMP)/Armco pipes (metallic pipes). Large storm sewers constantly apply CMP or arches. CMP is good at resisting differential settlement. In waterlogged areas, well-coated CMP pipes are preferred to prevent corrosion. Other considerations to use CMP may include the quantity of flows, nature of soils, subsurface conditions, and hydraulic efficiency requirements.

8.9.3 Unplasticized polyvinyl chloride

uPVC is the world's third most widely produced synthetic plastic polymer after polyethylene and polypropylene. uPVC sewer pipes are widely used in drainage systems for housing and light commercial applications and serve as an excellent alternative to clay and concrete pipes. They are light in weight and highly resistant both to chemicals and erosion.

8.9.4 Vitrified clay drainage (clayware) pipes

In the UK, vitrified clay plain-end pipes and fittings are manufactured in accordance with the requirements of BS EN295-1. The standard lengths of pipes are convenient for handling and laying and allow for flexible joints at sufficiently frequent intervals to enable the pipeline to withstand settlement or other ground movement after installation. Clayware pipes manufactured from clay and are slowly losing popularity due to increasing need to use environmentally sustainable, economical, and durable products for drainage works. Clayware pipes are known for being so ductile and brittle. Thus, their strength is very low. They are used extensively in diameters up to 0.45 m and commonly have spigot-and-socket joints. Clayware pipes are manufactured in two shapes: the bell-and-spigot (spigot-and-socket) ends and in a plain-end configuration.

8.9.5 Brickwork

These were commonly used in the past up to diameters of about 900 mm. Today, they are restricted to very large sewers because concrete pipes of larger diameters are now available. Brickwork is still used to build sewer appurtenances especially in aggressive conditions. It is also used as the lining for large reinforced concrete sewers constructed in situ.

8.9.6 High-density polyethylene pipes

HDPE pipes are much lighter in weight compared with ductile iron or concrete alternatives, which make transportation and installation significantly easier and safer. HDPE joints are watertight as the pipe and joints are fusion welded together, needing no extra couplers, grout, or other sealants to install. Uninterrupted water flow in HDPE pipes makes it cost-effective. HDPE pipes are resistant to chemical, biological, and abrasive damage, promoting a smooth surface, hence better and perhaps more importantly, *sustained* Manning's roughness coefficients, throughout the lifetime of the pipe.

HDPE pipes are now fast becoming the choice to build storm sewers over RCPs for a number of reasons. For example, repairing HDPE pipes is not very expensive compared to RCPs. It is easier to renovate HDPE pipes and takes little time compared with RCPs. HDPE pipes have been proven environmentally friendly. HDPE pipes support trenchless installation methods such as horizontal directional drilling (HDD) if pipes must be installed near an ecologically sensitive place.

HDPE pipes have long life spans and adopt well to custom designs and layout. HDPE pipes' strength, low weight, and flexibility support unique designs and sizes. They are leak free and significantly reduce construction time. Prefabricated joints and fittings help to reduce on-site construction time. HDPE joints are watertight as the pipe and joints are fusion welded together. Thus, they need no extra couplers, grout, or other sealants to install. HDPE pipes are durable and can withstand seismic activity, thus resisting stresses and cracking deformation much better, and it is easier to handle during construction.

8.9.7 Ductile iron pipes

Ductile iron pies are proving competitive in sustainable drainage systems.[7] They are used in tree trenches, creating root spaces in pipe trenches. Solutions to improve growth conditions of city trees must be found in underground space. The creation of root spaces in pipe trenches can only be successful if root-resistant pipe systems are used that at the same time allow the use of pore space-rich backfill materials for rainwater storage in the pipe trench. Ductile iron pipe systems with cement mortar coatings offer these degrees of freedom. If robust and root-resistant ductile iron pipes with cement mortar coatings are used as the storm water transport pipes, there are degrees of freedom in the choice of backfill materials in the pipe trenches. The use of robust ductile iron pipes increases the amount of freedom in backfill material. This enables the sponge city principle to be adequately implemented in the street space.

8.9.8 Cast iron pipes

Cast iron has the advantage that it lasts longer than, for example, uPVC and is resistant to fire. Cast iron is noncombustible and does not require complex and expensive fire-stopping systems. Cast iron often lasts significantly longer than PVC piping when utilized in drainage systems. Decades ago, the debate on drainage pipe material selection was oftentimes not a debate at all. Cast iron was the primary drainage material utilized for almost all construction, however, today's sustainability goals require the designer to optimize the design. That means pipe section would look at the associated carbon and cost. It is, therefore, not a straightforward process.

8.10 CONCLUSION

BS 9295:2020 Guide to the structural design of buried pipes is an essential standard in the structural design of storm sewers. AASHTO LRFD Bridge Design Specifications are also very helpful in the structural design of storm sewers. This AASHTO LRFD design data addresses the method of determining the live load pressure transmitted through unsurfaced roadways to circular, elliptical, and arch concrete pipe in accordance with the criteria. Several other standards and codes of practice are used to calculate bedding classes. The structural design of storm sewers is governed by loading cases: vehicle wheels, backfill loads, and internal water load. The pipes are designed to withstand these loads and to resist failure due to several factors.

Marston's theory is used for different loading cases due to earth overburden. The Bousnessq theory is used to calculate loads due to vehicle wheels. Different charts are used to estimate loading cases. In urban areas, pipes are laid to intermediate depth and concrete surroundings, and haunching is economized to support sustainability goals.

References

1. Tee, Kong Fah, Qing Li, C., & Mahmoodian, M. (2011). Prediction of time-variant probability of failure for concrete sewer pipes. *International Conference on Durability and Building Materials and Components*, PORTO – PORTUGAL, April 12–15, 2011.
2. Marston-Spangler load analysis for sewer sanitary pipes installation (theconstructor.org). Accessed August 4, 2022. https://theconstructor.org/geotechnical/marston-spangler-load-analysis-sewer-sanitary-pipes/17981/
3. Concrete Pipe Design – CivilWeb Spreadsheets (civilweb-spreadsheets.com). Accessed August 4, 2022. https://civilweb-spreadsheets.com/drainage-design-spreadsheets/buried-pipe-design-spreadsheet/concrete-pipe-design/
4. Highway Live Loads on Concrete Pipe. American concrete pipe association. American Concrete Pipe Association, 2009. www.concretepipe.org/wp-content/uploads/DD_1.pdf
5. BS 9295:2020 Guide to the structural design of buried pipes.
6. Storm Sewers: National Corrugated Steel Pipe Association (ncspa.org). https://ncspa.org/steel-product/storm-sewers/
7. European Association for ductile iron pipe systems. Stormwater management with ductile iron pipes. Drainage and climate concepts: Storage of stormwater Sponge city principle for stormwater management, Soil-pipe system as a solution. https://eadips.org/wp-content/downloads/data-facts-en/EADIPS-FGR-DATA-FACTS-Stormwater-Management.pdf
8. 'Vertical pressure on culverts under wheel loads on concrete pavement slabs'. Portland Cement Association, 1944.

Further reading

1. BS 9295:2020 Guide to the structural design of buried pipes.
2. American Concrete Pipe Association. Highway Live Loads on Concrete Pipe. www.concretepipe.org/wp-content/uploads/DD_1.pdf
3. AASHTO LRFD Bridge Design Specifications.
4. British Standard EN 1295-1-1997.

Chapter 9
Evaluation of integrated drainage system designs

NOTATION

P	:	Present worth
A	:	Amount
i	:	Interest rate
I	:	Inflation rate
$\dfrac{P}{A}$	:	Present worth factor
n	:	Time in years
LCA	:	Least cost (lifecycle) analysis
C	:	Original cost
S	:	Residual value
M	:	Maintenance cost
N	:	Rehabilitation cost
R	:	Replacement cost

9.1 INTRODUCTION

The primary objective of today's drainage planning is to plan, design, and develop a well-engineered and sustainable drainage systems for municipals and to provide advice on alternative stormwater best management practices (BMPs), referred to as 'SuDS' in the UK. As a design engineer, environmental conservations measures must be extensively incorporated in the design. This means a number of things must be tackled including highlighting on the receiving waters, for example, river restoration programs where necessary. A highlight of receiving waters such as the river's past, present, and future flows performance, if necessary, may be conducted especially when much runoff generated in the municipal ends there. The main reason for doing the river performance analysis is to investigate the likelihood of river banks getting eroded by the stormwater runoff generated in the municipality, for example, using the extreme value analysis (EVA) theory. Environmental conservation is a good practice to promote the ecosystem indefinitely. That is the overarching goal of every developmental project in the 21st century.

Therefore, evaluating drainage improvement program effectiveness against stormwater runoff volume reduction program is a crucial step in planning integrated sustainable drainage systems (SuDS). In several cases, both program projects are equally essential to the council.

This is because prevention is better than cure. SuDS significantly reduce the stormwater runoff generated in the municipality hence minimizing the risk of drainage structures flowing to capacity. This enables the structures to function adequately meeting the intended life span. Reliability is built in the drainage system by juggling the lifecycle costs, hydraulics, and structural analysis. Also, assessing each program's costs and benefits is another vital step so that the planned structures and systems fit within the financial constraints. Therefore, an optimal solution should weigh costs and benefits accruing from the implementation of different programs, that is, cost-effectiveness approach to stormwater management.

A stormwater drainage network legal policy framework is required to highlight on who is responsible for what. The municipal council left to maintain council drains, storm sewers, streams, ponds, rivers, lakes, and municipal stormwater management assets. Individuals left to manage systems on their property such as roof gutters, downpipes, and pipelines, which connect to council drainage systems. This must be done to organize the responsibilities of the stakeholders to maintain drainage infrastructure without shifting blames to unconcerned parties. Therefore, a legal point of stormwater discharge (LPSD) is the first step going forward in the direction of sustainable development. The municipals need to enforce the policy on stormwater management and to distribute the risk within stakeholders and make the community own the entire stormwater management program.

In order to fit within the municipal financial constraints as you know stormwater projects do not come with revenue streams, specifying materials is also a crucial component in the design and development of drainage solutions. In some areas, stone-pitched channels can work better while in other areas only concrete, high-density polyethylene (HDPE), cast iron, and unplasticized polyvinyl chloride (uPVC) pipes may be feasible. However, for long-life drainage structures, the structural component favors concrete-made channels than stone-pitched channels on a big scale. The hydraulic component also shows a higher capacity for concrete channels than stone-pitched channels because frictional resistance is lower in concrete channels. Additionally, from a global perspective, by significantly reducing the steel and cement amounts required in drainage network construction, the designer enormously contributes to the achievement of sustainable development goals because cement and steel production is a high energy-intensive process.

Integrated drainage planning and design is based on the hydrologic and hydraulic models of different scenarios. This influences the lifecycle costs, system reliability, and carbon emissions. By studying the hydrologic and hydraulic model of drainage schemes, the designer can evaluate the amount of carbon offset from one scheme to another. The designer can also compare the lifecycle costs.

9.2 ADVANTAGES AND POTENTIAL DISADVANTAGES OF SUDS

9.2.1 Advantages of SuDS

9.2.1.1 Reduction in stormwater runoff quantities

SuDS reduce stormwater quantity that ends up in the gray drainage systems (the standard pipe networks). This helps the standard pipe networks not to be overworked. This is due to the reduction in the impervious surfaces. The minimization of impervious surfaces promotes infiltration and reduction of the volume and the impact of stormwater runoff. Therefore, properly integrated SuDS can significantly reduce stormwater peak flows by reducing of the volume of stormwater generated, along with managing and releasing stormwater as close to the runoff source as possible. In a case where the drainage system is an integrated system that has both the SuDS and gray infrastructure, SuDS help to minimize the runoffs by intercepting the flows and absorb or infiltrate the flows, for example, infiltration systems.

9.2.1.2 Reduction in runoff velocities

Steep terrains greatly warranty the need to have SuDS in the stormwater management program. The SuDS always intercept the runoffs thereby decelerating the flow—this deceleration causes a reduction in the time of concentration for a specific point downstream. Therefore, in case of flooding, reducing the travel time of flows helps to greatly reduce the impact of floods. In reducing flood impacts, the preservation and restoration of the natural flood-carrying capacity of streams and floodplains is enhanced. Through a reduction of runoff flow quantities and decelerating runoff flow movement, SuDS greatly reduce the frequency and impacts of flooding affecting public safety, property, and infrastructure damage as well as public services in an urban setting.

9.2.1.3 Improve aesthetics

The beauty of the area is greatly improved with SuDS. For example, planting trees in a city, rain gardens and flowers, improves the aesthetical appearance of the area and it becomes more attractive, improving the well-being of the dwellers. SuDS that include attractive vegetation can improve property aesthetics translating into increased property values. A developed area designed with SuDS looks attractive, livelier, and always has a relaxed environment. Such places are good for living stress-free moments, vacations, and can contribute to the treatment plans for depressed and stressed people.

9.2.1.4 Recreational activities

SuDS support recreational activities by creating leisure parks hence improving the well-being of the city dwellers. This can be associated with cycling and walking lanes which are a sustainable mode of transport. Public health and well-being can be significantly improved in the presence of large-scale green spaces leading to increased recreational opportunities in society. Larger-scale SuDS facilities that include public access, such as constructed wetlands, green parks, micro-parks, offer top-notch recreational opportunities.

9.2.1.5 Reduce greenhouse gases

This is one of the overarching goals of the sustainability campaign, globally. SuDS can sequester carbon thus contributing to a carbon sink. Any activity that reduces GHGs, for example, CO_2, water vapor, etc., from the atmosphere is advocated for globally. Tree planting is an exercise that contributes enormously to the going green campaign as one tree can absorb about 21 kg of CO_2 from the atmosphere, annually. For example, tree trenches adjacent to a subsurface infiltration beds can serve multiple benefits. Tress and other vegetation can reduce urban heat island effect, which reduces energy use and the incidence and severity of heat-related illnesses. Integrated SuDS developments in an area such as trees, bio-swales, bioretention gardens, and planter boxes can significantly improve the air quality and reduce climate change.

Often referred to as 'lungs', trees are vitally important to our ecosystem because they are able to absorb and store carbon and produce additional oxygen to our planet. Trees provide a cool temperature range and reduce humidity levels, stabilize soils, reduce flooding, and improve water quality. It is most likely that trees are what makes us ably live on this planet, without them it would be impossible to sustain lives on this planet.

9.2.1.6 Provide shades for the city dwellers

Trees can provide shades for the city dwellers. It is good practice to holistically plan for every person. Some people will always find shades provided in the city very helpful.

9.2.1.7 Regulate temperatures

The temperatures for a city without trees are quite high more than a city with trees. Trees regulate the temperatures when they absorb carbon dioxide and release oxygen. And of course, carbon dioxide is responsible for increased heating up of the cities.

9.2.1.8 Enormously contribute to climate action policy

SuDS are an excellent approach to managing stormwater because they present a multitude of benefits. The development of SuDS contributes to the effective implementation of climate action policy and other policies on disaster risk management, clean water, agriculture, and health. Applying SuDS to greenfield and urban infill and redevelopment sites leads to the shift toward a green economy while developing cost-effective infrastructure that delivers better environmental and social outcomes for society.

9.2.1.9 Educational opportunities

Learning about plant species and animals is one of the benefits of incorporating SuDS while planning our cities/municipalities. Long ago, people lived in undisturbed environments and they could possibly learn more about the different species of plants and animals quite easily. Things have since changed; there are quite many children born and raised in cities who hardly learn about many such kind of rare species. It is, therefore, good practice to innovate SuDS that incorporate these species for educational purposes. Signposts can be installed on specific SuDS to label the species, species' history and origin and how they live for case of animals, etc. The visible nature of green infrastructure and SuDS offers enhanced public education opportunities to teach the community about mitigating the adverse environmental impacts of our built environment. Therefore, signage can be used to inform viewers of the features and functions of the various types of facilities. In that case, SuDS can be very beneficial to educators especially when integrated in school campuses and other public places.

9.2.1.10 Recharges groundwater

SuDS increase infiltration and natural groundwater recharge to protect groundwater supplies and stream baseflows. Recharge is the primary method through which water enters an aquifer.

Properly developed SuDS facilitate the movement of stormwater downward from surface water to groundwater. SuDS that incorporate locally adapted or native plants promote water conservation.

9.2.1.11 Harvesting rainwater for domestic/commercial use

In some areas, rainwater harvesting (RWH) cannot be ignored. The RWH can *waive* some money spent on portable water. This water can be used for irrigating crops and for feeding animals. Rain barrels and cisterns could reduce water consumption and associated costs, reduce demand for potable water, increase available water supply for other uses, and significantly reduce stormwater discharges from roofs.

9.2.1.12 Biodiversity and resilience

Through SuDS, air quality is improved enabling climate change adaptation, helping to absorb flooding, and enhance biodiversity. The drainage schemes' ecological impact on habitats and species or natural features is significantly reduced through an integration of SuDS in drainage practices worldwide. In many cases, communities which are more resilient plan for integrated/holistic approaches. In a case of mitigating and controlling floods, such communities that integrate SuDS with gray infrastructure will always bounce back to normal more quickly than those that depend entirely on the collection, treatment, and conveyance of stormwater through the standard pipework solutions. Therefore, SuDS enhance biodiversity and community resilience hence establishing a unique sense of place especially when featuring native plants, for example, bioretention gardens. On the other hand, SuDS can appeal to funders' desires to achieve public health benefits, community revitalization, and habitat creation, among other benefits, as compared to traditional standard pipework drainage solutions.

SuDS are an essential element of climate change resiliency planning due to its ability to adapt to environmental and climatic fluctuations compared with standard pipework drainage solutions. That allows pipework drainage systems not to be overworked. SuDS are vitally important because they provide food and habitat for several birds and pollinators. They connect between larger natural heritage features such as wetlands and forests. With SuDS integrated in city and municipal operations, the impact of heavy rainfalls on properties is significantly reduced. SuDS can provide a multitude of services allowing communities to recover following extreme weather events and reducing the impact of extreme heat days on standard pipework drainage systems and human health.

9.2.1.13 Preserving ecological sites

SuDS are good at protecting and conserving heritage sites with historical significance. When planning for improvement of sacred sites, one cannot do so without incorporating SuDS. Properly landscaped areas integrated with SuDS preserve the beauty of ecological and cultural heritage sites hence supporting community resilience and sustainability.

9.2.1.14 Reduce pollutants

SuDs capture and reduce urban air pollutants, for example, dust, O_3, SO_2, NO_2, carbon monoxide (CO), etc. For example, the dust settles in the green spaces. The quality of air in the city is improved by the implementation of SuDS. Through a reduction of stormwater pollutant loads, the quality of ground and surface water is protected.

9.2.1.15 Enhancing conservation practices for sustainable development

SuDS promote energy conservation by limiting damage due to erosion and sedimentation. Thus, SuDS prevent erosion, scour, and sedimentation of stream channels. SuDS vegetation helps control erosion, provides structural stability, promotes infiltration, and removes pollutants from stormwater runoff. It can also enhance the appearance of the SuDS and help them blend into the landscape.

Also, they protect natural resources, along with the aquatic habitats and species within them—greatly improving aquatic and wildlife habitat. In that case, SuDS reduce infrastructure requirements and maintenance costs for associated stormwater handling facilities protecting natural resources. SuDS such as green spaces can increase soil porosity, reduce stormwater runoff volume, reduce peak stormwater flows, and help reduce the risk of flooding.

SuDS such as permeable pavers reduce stormwater runoff and standing water, promote infiltration and groundwater recharge, improve the longevity of infrastructure, and may be easier to maintain than standard pavement. SuDS such as bio-swales and rain gardens can improve property and neighborhood aesthetics, reduce localized flooding, promote infiltration and groundwater recharge, enhance pedestrian safety when used in traffic calming applications, etc.

9.2.1.16 Ease of working with underground utility services

Permeable pavers can quickly be removed and brought back after constructing new underground utility service lines and during utility service repair and maintenance activities. This is a great advantage compared with asphalt pavement which requires to break the ground and after construct the pavement afresh. Permeable pavers are quicker to install and take less time to remove and bring back.

9.2.1.17 SuDS foster collaboration with NGOs

Public authorities in association with nongovernmental organizations (NGOs) could start an awareness campaign to sensitize the general public about the role of SuDS such as RWH projects. In emerging economies, RWH-jars, cisterns, and tanks can be sold to each household, at a subsidized fee, and encourage them to utilize them appropriately. RWH projects are anticipated to have an annual cost savings due to reuse from flushing or area irrigation. Ferro-cement tanks can equally be utilized especially in poverty-ridden areas globally. The advantage of ferro-cement tanks is that the community can be trained on how to build them which is more sustainable integrating women and youth and forming water user committees to operate and maintain the ferro-cement tanks. The operation and maintenance of the ferro-cement tanks is enhanced through water user committee trainings on sanitation and hygiene, roles and responsibilities, collection of funds, and repairing the RWH system.

9.2.1.18 Ease of funding compared with traditional gray infrastructure

SuDS can easily attract funding compared with standard pipework drainage solutions because of the multiple benefits they bring to the community. Building resilient and ecologically sensitive communities, aesthetically appealing built environments, community revitalization, habitat creation and sustainability, and exploiting public health benefits are associative attributes for SuDS which are very attractive to funders compared with the standard sewer systems known for collecting, treating, and conveying stormwater runoffs.

9.2.1.19 Economic benefits

SuDS provide employment opportunities to community through design, construction, and maintenance of SuDS, rejuvenating downtowns and streetscapes and retrofitting areas. SuDS also improve local waterway conditions and aquatic species health supplying high-quality groundwater. Through outdoor recreational activities and attracting new businesses, services, and new area residents, SuDS can become economically beneficial and inspiring. An integrated mix of these activities boosts the economic prospects of the region leading to prosperity.

9.2.2 Potential Disadvantages of SuDS

9.2.2.1 Tropical climates

In tropical countries, SuDS can be breeding sites for mosquitoes, that is, detention ponds. This can be very detrimental. In such cases, they have to be thoughtfully done by specifying the most appropriate SuDS. For example, going for specific trees species as part of SuDS, as advised by an expert, is not such a bad idea. The development of bioretention gardens, detention ponds, bio-swales, trees, and planter boxes in tropics must be conducted following expert advice in some locations to avoid outbreak of diseases (e.g., malaria) from mosquitoes.

9.2.2.2 Engineering, art, and continuing education

Some SuDS require skills (engineering and art) to implement them and in some locations this may not be available. That means, it may call for an extra cost to hire experts to design and implement SuDS. Sometimes the development of SuDS blends knowledge and skills with site-specific work of art which requires integrated skills from engineers and land artists. This mix of skills may be costly for some regions where talented and skillful personnel is unavailable.

In some cases, continuing education is necessary for engineers, planners, and operation and maintenance teams, in order to keep up-to-date with new and advancing stormwater management technologies, byelaws, regulations, and ordinances. Also, since the shift of managing stormwater is now placed at the source, residential owners constantly need educative information regarding landscape practices, and pollutant prevention techniques. This would be an enforcement work of municipal authorities to ensure that effective continuing education is fostered in the interest of sustainable development—and this calls for cost planning and running software activities in the financial year.

9.2.2.3 Heavy initial cost

SuDS come at a cost and not all cities/municipals may afford what it calls for. Although SuDS may reduce costs over time, some require higher initial capital costs for design and construction.

9.2.2.4 Large space requirements

Space constraints dictate the type of SuDS to develop in the area. Many SuDS require enough space. In some instances, available easements/land may not be sufficient to support the development of SuDS. For example, surface detention ponds and retention ponds may require large spaces/sufficient land which may not be readily available.

9.2.2.5 Compliance with regulatory frameworks

Several authorities issue guidance manuals on how to develop SuDS. Developers are required to comply and conform to the standards and codes of practice. Thus, management plans are required in some cases to ensure effective operation and maintenance of the SuDS. In some cases, SuDS require application reviews, inspections, and enforcement during and after construction. Sometimes, permanent protection of SuDS from alteration is also required to ensure longevity, reliability, maintainability, and operability. Organizations and individuals responsible for operating and maintaining SuDS are required to be part of SuDS planning right from inception stages, and developers must verify and confirm that maintenance funds will be readily available. In cases, where operators are used, they must submit evidence of availability of funds to properly manage the SuDS.

9.3 INTEGRATED DRAINAGE SYSTEM RELIABILITY ANALYSIS

9.3.1 Introduction

The study of reliability and sustainability is essential in all civil engineering branches. Operational and functional reliability can be modeled as independent variables to inform sustainability. A drainage scheme can only be reliable when its functionality is uphold and when it is operable and maintainable. During drainage systems planning and design, sustainability goals must be met and at the same time the desired reliability uphold. That means economizing concrete use in gray drainage infrastructure and maximizing the use of SuDS as much as possible and as applicable. Therefore, the drainage master plan of the area is viewed as a system of interconnected subsystems such as infiltration systems, bioretention gardens, and storm sewers, among others. For purposes of sustainability, an environmental lifecycle assessment (LCA) is always conducted for integrated drainage systems.

The study of reliability is important in all engineering disciplines. From an engineering context, reliability is the ability of a device, system, or structure to perform its intended function or a closely related function in a stated period of time. Therefore, such device, system, or structure becomes less reliable or unreliable when it no longer performs the intended purpose before the expiry of its design life span. When this happens, the device, system, or structure is said to have failed to meet the reliability goal and therefore significant costs are borne to the owner of the asset.

Improving the reliability of a system means continuously studying the system failures. Basing on the failure data, fundamental approaches can be laid to improve on the reliability of the system. For example, designing maintainable systems, structures, devices, or equipment is an approach to improving reliability. Similarly, improving on availability of systems by increasing the uptime of the device, system, or structure is also a measure to improving reliability.

The reliability analysis of a civil engineering structure like storm sewer system entails the analysis of the constitutive systems, their parts, and their interactions through dependences and interdependences. Because civil engineering structures are large, they cannot be assessed for reliability through full-scale testing as it happens in the case of small mechanical and electrical devices. Models built in software platforms such as Autodesk Storm and Sanitary Analysis evaluate the effectiveness of SuDS for reducing wet weather pollutant loadings. Predominant gray infrastructure uses steel and concrete. Therefore, minimizing discharges by integrating SuDS in the draiange system leads to cost reduction because of reduction in storm sewer sizes and this means reducing embedded carbon too!

Usually reliability model testing collects failure data from which the reliability formulations are conducted. Therefore, deciding whether a product or process would perform satisfactorily in its design lifetime is basically evaluated from the collected failure data and using statistical and probability distribution functions such as the Weibull distribution to analyze.

The system reliability is measured as the ability of the proposed drainage system to perform its intended function or a closely related function in a stated period of time, that is, the design period, say 25 years. This design life span is derived by significantly reducing down the probability of occurrence of a storm that strains the designed drainage system. The likelihood of a 25-year flood occurring at least once in the next 25 years is 0.64 while that for 10 years is 0.34. **Note** also that the likelihood of a 25-year flood occurring in any given year is 0.04. Therefore, it is synonymous to speak of a 4% storm as a 25-year storm. The hydraulic design life is much lower than the structural design life for concrete storm sewers. On the other hand, in determining how the system will perform efficiently and effectively, we analyze the hydrologic and hydraulics model parameters/inputs and the structural aspects. In that case, we look at a lot of what-ifs, ranging from cost, technology, and human resources.

9.3.2 Drainage system model reliability

Several major drainage master plans for municipalities are designed to accommodate flows of average recurrence interval (ARI) of 25-year storm and checks done with the 50-year ARI storm. If such typical model is simulated for 25-year ARI storm, and it passes the 50-year ARI check, then the model passes the preliminary step for reliability checks. Design manuals specify the checks to be conducted on an ARI less or bigger than 50 years.

Several software platforms enable SuDS to be integrated into the models such as infiltration systems, bioretention gardens, detention ponds, and vegetated swales. Thus, water quality analysis can be modeled. The second stage would be to evaluate the strength of storm sewer concrete pipelines by specifying the crushing strength for concrete pipes and the corresponding bedding classes. The reliability goal should be borne in mind at the earliest start of the design, that is, the system should be able to withstand a storm of ARI 25 years and a storm sewer concrete crushing strength of 120 kN/m. With a factor of safety incorporated in the evaluation of pipeline designs, the reliability is enhanced. To ascertain that the storm sewer concrete pipes pass the desired test, Weibull++ software can be used to analyze failure rates in concrete pipes and associated systems. With Weibull++, life data analysis results can support decisions—inputting multiple types of life data in the software, and use all major lifetime distributions to get results quicker with flexible plots and reports. The software can also be used to ensure reliability within a specific time frame and confidence by demonstrating how the product's reliability goals can be met. It can be used to design effective reliability and demonstration tests with appropriate sample size and test duration—applicable on drainage infrastructure materials, for example, storm sewer pipes, concrete, and other SuDS materials.

During construction, the crushing strength of concrete pipes is tested to conform to the required standards before laying in the trench. The American Society for Testing Materials has developed and published a standard of practice for least cost LCA of concrete culvert, storm sewer, and sanitary sewer systems.

9.3.3 Sensitivity analysis

Sensitivity analysis gives you an insight in how the optimal solution changes when you change the coefficients of the model. After the obtaining the optimal solution, the modeler can create a **sensitivity report**. The main goal of sensitivity analysis is to gain insight into which assumptions are critical, that is, which assumptions affect choice. The process involves various ways of changing input values of the model to see the effect on the output value. In some situations, the modeler can use a single model to investigate several alternatives.

Sensitivity analysis can best be defined as the study of how uncertainty in the output of a model can be attributed to different sources of uncertainty in the model input. This implies a technique used to determine how different values of an independent variable impact a particular dependent variable under a given set of assumptions. A basic sensitivity analysis for the project hydraulics and hydrologic models is highlighted below using the what-if analysis. Deterministic models are developed in software platforms from which reliability analysis can be computed. See Section 6.9.

9.3.3.1 Model inputs variables for uncertainty checks

- Imperviousness (runoff coefficients and curve numbers).
- Rainfall data.
- Manning's roughness factors or Chezy coefficients.
- Subcatchment area delineation, for example, $Q = 0.278CiA$ by rational method for $A \leq 0.8ha$.
- Entrance and exit coefficients for pipes/drains.
- Subcatchment slopes.
- Channel and drain slopes.
- Froude number.
- Hydrologic soil groups (HSGs), etc.

9.3.3.2 Model output variables

- Nodes and links hydrographs.
- Time of concentration.
- Design intensities, frequencies, and recurrence intervals for design storms.
- 90th percentile precipitation depth of the cumulative curve for storms.
- Subcatchment peak flows and volumes.
- Energy losses due to pipe end wall structures.
- Channel and drain runoff velocities.
- Channel and drain capacities.

- Surcharging of nodes, channels, and drains.
- Flooding of nodes, channels, drains, etc.

9.3.3.3 Using what-if analysis to model system reliability

Using the what-if analysis, the following questions can be set among others:

- What if there exists uncertainties in the rainfall data?
- What if there exists uncertainties in the physical plan leading to uncertainties in the runoff coefficients?
- What if there exist uncertainties in subcatchment area delineation?
- What if there exist uncertainties in subcatchment basin slopes and link slopes?
- What if there exist uncertainties in head loss coefficients taken/considered contrary to site conditions?

9.3.4 Environmental lifecycle assessment

9.3.4.1 Introduction

LCA, also called environmental LCA (ELCA) or lifecycle cost analysis (LCCA), is a systematic, standardized approach to quantifying the potential environmental impacts of a product or process that occurs from raw materials extraction to end of life.[1] LCCA is an approach used to assess the total cost of owning a facility or running a project. LCCA considers all the costs associated with obtaining, owning, and disposing of an investment project. The LCCA for both gray and SuDS infrastructure is prepared to compare alternatives on the assumption that reducing cost implies reducing carbon too. The materials or technologies with the least lifecycle cost is deemed to have little carbon emissions too. LCCA is helpful where a project comes with several alternatives and all of them meet performance necessities but differ with regard to the initial, as well as the operating, cost. In this case, the alternatives are compared to find one that can maximize savings. For example, LCCA helps to determine which of the two alternatives, that is, gray infrastructure assets and SuDS, could potentially raise the initial cost but would reduce the operating cost. The most expensive drainage infrastructure asset may provide superior performance and quality but may require a significant amount of maintenance. Yet, a relatively cheaper material or asset or technology may require less regular maintenance, but its overall cost is significantly lower.

In many cases, SuDS outperform traditional drainage systems in environmental and social terms. However, high maintenance costs and shorter life expectancy hinder SuDS economic feasibility. An LCCA is based on a consistent methodology applied across all products and processes and at all stages of their production, transport, energy use, maintenance, and disposal or recycling at end of life. A full LCCA includes the impacts of energy use and associated emissions during the life of the product or structure, not just construction. Figure 9.1 shows the lifecycle costs. Carbon emissions are synonymous with capital expenditure (CAPEX) and operational expenditure (OPEX), that is, capital carbon and operational carbon. This carbon is generated throughout the project lifecycle stages.

The International Standards Organization (ISO) established a standardized LCA (least cost) method beginning in 1996. The ISO 14040 series standards outline the major procedural steps of LCA most commonly followed for comparisons of process and product alternatives

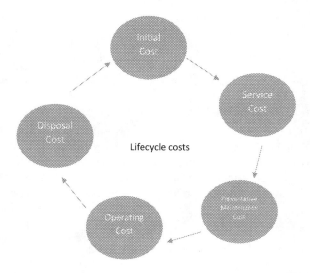

Figure 9.1 Lifecycle costs.

Table 9.1 ISO 14040:2006 environmental management—lifecycle assessment—principles and framework

ISO 14040:2006 environmental management—lifecycle assessment—principles and framework	
Goal and scope definition	Establish objectives; methods; temporal, spatial, and technical system boundaries; functional unit and criteria (impact categories); map all relevant human and natural material and energy flows.
Life cycle inventory (LCI)	Catalog all resources used and emissions for each process, product, or activity. Process flow and/or input/output modeling may be necessary to calculate material and energy flows.
Lifecycle impact assessment (LCIA)	Determine the environmental consequences. Classify the inventory into predefined impact categories and characterize the magnitude of each element's contribution to the impact categories. (Further aggregation, normalization, and valuation are optional).
Interpretation	Determine the dominance, sensitivity, and uncertainty of results. (Interpretation can be conducted with without LCIA).

and documentation of industry-wide 'eco-profiles'. According to the ISO 14040 series, LCA is structured in four phases (Table 9.1).

Several standards draw reference from the ISO 14040:2006 Environmental management—LCA—principles and framework. In drainage schemes, the most widely adopted approach to LCA (least cost) for drainage projects is the ASTM procedure (ASTM C-1113). This procedure was developed by the American Society for Testing and Materials (ASTM), Committee C-13 on concrete pipes, in 2007.

The ASTM procedure provides LCA techniques to evaluate alternative pipeline materials, structures, or systems that satisfy the same functional requirement. The LCA technique evaluates the present value constant costs to install and maintain alternative drainage systems including planning, engineering, construction, maintenance, rehabilitation, and replacement and cost deductions for any residual value at the end of the proposed project design life. The decision maker, using the results of the LCA can then readily identify the alternative with the lowest total cost based on the present value of all initial and future costs. Thus, the procedure can be applied to both SuDS and gray infrastructure.

When comparing construction alternatives, an ELCA provides a level-playing field. An LCA is based on a consistent methodology applied across all products and at all stages of their production, transport, energy use, maintenance, and disposal or recycling at end of life. A full LCA includes the impacts of energy use and associated emissions during the life of the product or structure, not just construction.

9.3.4.2 Factors considered in LCA for drainage schemes

As outlined throughout this book, several interrelated fields are applicable in the design and development of sustainable drainage projects. These include project planning, specifications' identification, hydrology, hydraulics, structures' analysis, installation, durability, and economics. Now that the world is fast-tracking sustainable drainage systems, drainage planning has become an interesting avenue where LCA (least cost) must be a part of the process to design and develop nonstructural and structural drainage systems to solve a flooding problem.

Several years ago, the reliability of drainage projects was not well evaluated. The aspects of durability and economics were less integrated in the planning, design, and development of drainage infrastructure. The initial capital investment was the main factor considered to decide whether the drainage infrastructure is affordable or not. However, lower capital cost does not necessarily mean the most economical product or system. To determine the most economical choice, the principles of economics and reliability analysis must be applied through an LCA (least cost), focusing on cost-effectiveness. New researches have since shown that LCA (least cost) is vitally important as several drainage schemes have failed prematurely, an indicator of lack of considering economic indicators and reliability metrics throughout the project lifecycle phases.

ASTM produced results about LCA for drainage schemes after several years of dedicated research. The ASTM standard practice includes the following factors: project design life, material service life, capital cost, interest (discount) rate, inflation rate, maintenance cost, rehabilitation cost, replacement cost, and residual value.[2]

9.3.4.2.1 Capital cost

Capital cost is the original cost incurred during planning, designing, and constructing a project. This includes the direct and indirect costs such as removal and disposal of existing materials, structures or systems, mobilization, administration, clearing and grubbing, excavation, pipe material and placement, bedding and backfilling, surface restoration, traffic maintenance, engineering, and contingencies. The real tender prices can be used for many of the capital cost items.

9.3.4.2.2 Project design life

Project design life is the *number of years of relatively maintenance-free performance*. Design life is the forecast life expectancy of products based on their design. Based on the American research findings where a review of all published culvert surveys, and current (USA) state

312 Integrated Drainage Systems Planning and Design for Municipal Engineers

practices were published in the National Cooperative Highway Research Program Synthesis of Highway Practice titled *Durability of Drainage Pipe*, up to 50 years of relatively maintenance-free performance should be required for culverts on secondary road facilities and up to 100 years for higher-type facilities, such as primary and interstate highways and all storm and sanitary sewers.

9.3.4.2.3 Material service life

Service life is the number of years of service that a material, system, or structure will provide before rehabilitation or replacement is required. Service life is the forecast life expectancy of products based on real-world results. Once the 'project design life' is established, the proven 'service life' of the pipe material or system must be evaluated. Many survey reports have been prepared in the recent past by several impartial specifying agencies worldwide, such as the U.S. Federal Highway Administration, U.S. Soil Conservation Service, U.S. Bureau of Reclamation, U.S. Corp of Engineers, and several state U.S. Departments of Transport. These report include predictive equations and charts for a given set of environmental conditions that accurately predict service life. Therefore, following such reports, evidence can be adduced to adequately predict the service life. The project design life and service life must be established by the principal or owner during the drainage project feasibility and evaluation process.

According to the U.S. Army Corps of engineers:

- A concrete pipe has a service life of 70–100 years.
- A corrugated metal pipe (CMP) may obtain up to a 50-year service life with the use of coatings.
- HDPE pipe is lightweight and flexible. Thus, the designer should not expect a material service life greater than 50 years for any plastic pipe. Their service life greatly depends on the installation and surrounding soil of the embankment, which will add to the initial cost of the pipe. Other factors that affect the service life of HDPE pipe include the flammability of polyethylene and ultraviolet (UV) sensitivity.

9.3.4.2.4 Inflation/interest factor

The general expression for the compounding of interest or the inflating of a future cost is:

$$A = P(1+i)^n, \quad \text{or} \quad A = P(1+I)^n \tag{9.1}$$

Where:

A : Future sum of money or cost
P : Present sum of money or cost (present worth)
i : Interest rate
I : Inflation rate
n : Number of periods

Present worth is the reciprocal of the compound amount and is given by the equation:

$$P = \frac{A}{(1+i)^n}, \quad \text{or} \quad P = \frac{A}{(1+I)^n} \tag{9.2}$$

Evaluation of cash flows over a period of time should consider inflation. The amount of money that should be planned to cover a future expenditure is affected by both interest rates and inflation rates. The effects of interest and inflation rates tend to offset each other and the net effect on the lifecycle cost is primarily due to the difference in these two rates. **Note:** Interest may be earned on the money set aside, but inflation will increase the amount of the final expenditure. Interest rates or inflation rates in the future over a 20-, 50-, or 100-year period should be based on substantial historical data not imaginary data. The difference between inflation rates and interest rates remains relatively constant over years.

The use of the inflation/interest factor to simplify lifecycle cost estimation was first proposed by the Jet Propulsion Laboratory of California Institute of Technology under a contract with the National Aeronautic and Space Administration.

Considering an annual inflation rate 'I', a current expenditure of P will be at n years cost: $P(1+I)^n$. With an annual interest rate of i as shown in Equation (9.2), the discounted value of this cost at the present is:

$$P(1+I)^n \times \frac{1}{(1+i)^n} = P\left(\frac{1+I}{1+i}\right)^n \qquad (9.3)$$

The term within the brackets is the inflation/interest factor F. The interest rate of a time period is always greater than the inflation rate, usually by at least 1 or 2 percentage points. Therefore, the inflation/interest factor is always less than 1.

9.3.4.2.5 Maintenance cost

Maintenance expenses are costs incurred when performing routine and periodic actions to keep a drainage asset in its original condition. The inflation/interest factor to the 'nth' power is used as a multiplier to inflate future maintenance, rehabilitation, and replacement costs and then discount these future costs back to present constant money values. The 'n' term is the number of years in the future at which the costs are incurred.

9.3.4.2.6 Rehabilitation cost

Rehabilitation cost means any repair to or replacement of any portion of the drainage system. Replacement and/or rehabilitation may include, but not limited to, all reasonable engineering, design, inspection, and other consultant costs such as geotechnical engineers, contractor costs, material costs, labor costs, and equipment costs.

9.3.4.2.7 Replacement cost

These are costs incurred to replace the drainage asset.

9.3.4.2.8 Residual value

In case a material, system, or structure has a *service life greater than the project design life*, then it would have a residual future current money value, which should be discounted back to a present constant money value utilizing the inflation/interest factor and subtracted from the original cost.

9.3.4.3 The ASTM procedure

The most widely adopted approach to LCA (least cost) for drainage projects is the ASTM procedure (ASTM C-1113). This procedure was developed by the ASTM, Committee C-13 on concrete pipe in 2007.[3] ASTM developed and published ASTM standard of practice C-1131-95 (2007) for least cost LCA of concrete culvert, storm sewer, and sanitary sewer systems. The practice covers procedures for using LCA techniques to evaluate alternative pipeline materials, structures, or systems that satisfy the same functional requirement. Thus all drainage systems including SuDS can be analyzed by the ASTM C-1113. Since 2007, the application of least cost LCA to road and drainage projects has increased dramatically.

In several instances, engineers and executive officers need to repair and replace integral sections of drainage infrastructure that have experienced premature failures. Therefore, local and state governments worldwide have increasingly included some type of analysis in their material selection process. The world has now shifted from taking decisions solely based on the lowest initial capital costs to LCA philosophy that involves prediction of interest and inflation rates for the years ahead. The ASTM Standard of Practice C1131 adopts a five-step procedure.

9.3.4.3.1 Identify objectives, alternatives, and constraints

It is important that the specific objectives be established to enable alternative means of accomplishing them to be identified. For example, alternatives for a road drainage system may include a pipe culvert, box culvert, or a bridge. Constraints may include head and tail-water levels, maximum and minimum grades, access requirements, etc.

9.3.4.3.2 Establish basic criteria

As discussed earlier the criteria include:

a. Project design life.
b. Material, system, or structure service life.
c. First or capital cost.
d. Maintenance, rehabilitation, and replacement costs.
e. Residual costs.

Consideration should also be given to the comprehensiveness of the LCCA evaluation.

9.3.4.3.3 Compile data

The necessary data to calculate the LCA of the potential alternative must be collected.

9.3.4.3.4 Compute LCA for each material, system, or structure

Cost categories to be considered include:

a. Capital cost.
b. Maintenance and operating cost.

c. Rehabilitation or repair cost.
d. Replacement cost.

If there is a residual value at the end of the project design life, this value should be discounted back to a present value and subtracted from the original cost. The present value of all future costs is determined by multiplying each cost by the appropriate inflation/interest factor.

9.3.4.3.5 Evaluate results

The final stage is to evaluate the lifecycle cost results.

9.3.4.4 ASTMC-1131 formula

The ASTMC-1131 formula is given in Equation (9.4):

$$LCA = C - S + (M + N + R) \tag{9.4}$$

Where:

LCA	:	Least cost (lifecycle) analysis
C	:	Original cost
S	:	Residual value
M	:	Maintenance cost
N	:	Rehabilitation cost
R	:	Replacement cost

9.3.4.4.1 Residual value

Using straight-line depreciation, present value of the residual value can be defined as:

$$S = CF^{n_p}\left(\frac{n_s}{n}\right) \tag{9.5}$$

Where:

S	:	Residual value
C	:	Present constant money factor
F	:	Inflation/interest factor
n_s	:	Number of years of service life exceeds design life
n	:	Service life
n_p	:	Project design life

9.3.4.4.2 Inflation factor

$$F = \left(\frac{1+I}{1+i}\right) \tag{9.6}$$

Where:

F : Inflation/interest rate factor
I : Inflation rate
i : Interest rate

9.3.4.4.3 Maintenance cost

The present value of maintenance costs can be determined by applying the inflation/interest factor to each cost occurrence and summing all values. If maintenance costs are established on an annual basis, the following equation can be used with a nominal discount rate.

$$M = C_m \left(\frac{1-(F)^n}{\left(\frac{1}{F}\right)-1} \right) \qquad (9.7)$$

Where:

M : Maintenance cost
C_m : Annual maintenance cost
i : Nominal discount rate
I : Inflation rate
F : Inflation/interest factor
n : Number of years requiring annual maintenance

9.3.4.5 Worked example—for illustrative purposes

An 80-year design life has been assigned to a storm sewer project to be constructed for a private subdivision. Two alternative pipe with different wall thicknesses are included in the bid documents.

Material A with a project bid price of $350,000 has been assigned a 70-year service life with an annual maintenance cost of $8,000/year. To meet the project design life, an $85,000 rehabilitation cost will have to be incurred at the end of the 70-year service life.

Material B has an 'in ground' cost of $395,000 with a 100-year projected service life. The annual maintenance cost has been estimated at $7000/year.

Planning and design costs applicable to all alternatives are $250,000. Based on historical data, a 4% inflation rate and 8% interest (discount) rate are appropriate for this project.

Solution

Material A	Description and calculations	Material B
$250,000	Planning and design cost	$250,000
$350,000	Bid price	$395,000

Evaluation of integrated drainage system designs 317

$184,640	Maintenance cost $8,000 $7,000	$161,560
	$$\left(\frac{1-(F)^n}{\left(\frac{1}{F}\right)-1}\right) = \left(\frac{1-(0.96)^{80}}{\left(\frac{1}{0.96}\right)-1}\right) = 23.08$$	
$4879	Rehabilitation cost $(F)^n C_m = (0.96)^{70} \times 85{,}000$	0
0	Residual value	($3,015)
	$$S = CF^{n_p}\left(\frac{n_s}{n}\right) = 395{,}000(0.96)^{80}\left(\frac{20}{100}\right)$$	
$789,519	**Total cost**	**$803,545**

Result: Material A is more cost-effective since the LCA is $ 14,026 less than material B.

The discount rate is quite sensitive relative to the inflation rate. Imagine the discount rate is increased from 8% to 10% in the above example, resulting in an unreasonably large difference of 6% between discount rate and inflation rate. Then the inflation/interest factor F = 1.04/1.10 = 0.95. By increasing the discount–inflation differential from a realistic 4% to an artificial high 6%, the LCA changes dramatically. *In some cases, LCA results can reverse when the inflation differential is so high and the more cost-effective material changes from A to B.* This implies that evaluating interest (discount) rates relative to inflation rates is essential. The determination of these two rates should be based on historical data of appropriate economic indicators rather than arbitrary assumptions.

Material A	*Description and calculations*	*Material B*
$250,000	Planning and design cost	$250,000
$350,000	Bid price	$395,000
$149,520	Maintenance cost $8000, $7000	$130,830
	$$\left(\frac{1-(F)^n}{\left(\frac{1}{F}\right)-1}\right) = \left(\frac{1-(0.95)^{80}}{\left(\frac{1}{0.95}\right)-1}\right) = 18.69$$	
$2344	Rehabilitation cost $(F)^n C_m = (0.95)^{70} \times 85{,}000$	0
0	Residual value	($1304)
	$$S = CF^{n_p}\left(\frac{n_s}{n}\right) = 395{,}000(0.95)^{80}\left(\frac{20}{100}\right)$$	
$789,519 **$751,864**	**Total cost** **Total cost**	**$803,545** **$774,526**

Result: Material A is still more cost-effective since the LCA is $ 22,662 less than material B. However, the difference between LCA for A and B has changed drastically.

9.3.4.6 Analyzing reliability goals—typical example for illustrative purposes

The reliability goal was set by the municipal authority that the operation and maintenance budget for *all* sustainable drainage systems integrated in the drainage system should be less than or equal to 70% of the total operation and maintenance budget per month. That was a management decision. Integrated drainage system in division A integrates SuDS of *different types* with gray infrastructure and the same applies to drainage system in division B. Over ten years, the maintenance costs for SuDS have been as shown in Table 9.2 for both divisions. The bid prices for SuDS maintenance have fluctuated over the years. Going by municipal infrastructure maintenance goal, all yearly expenditure on SuDS should be less than or equal to 70% of the total maintenance budget allocated to integrated drainage system in divisions A and B for the municipal authority. The remaining 30% is for maintaining gray infrastructure.

Going by average, the maintenance cost is **US$ 7000/year** for both schemes in divisions A and B. The municipal authority wishes to reduce the budget going by average to develop more typical integrated drainage systems in other municipal divisions and wants to go by average so that funds are spread in all old and new systems. The reliability goal set is that the gray operation and maintenance budget should not exceed 30% of the total municipal operation and maintenance budget allocated for each drainage infrastructure in a given division. This can be communicated mathematically as $R(7,000)_{SuDS} \leq 0.3$. However, would the **US$ 7000**/year allocated for SuDS be enough if we are planning to establish *typical* integrated drainage systems of either system A or B in other divisions? Is it logical to consider the average?

Solution

The technique is to assume the monthly expenditures follow the Weibull distribution. Tables 9.3 and 9.4 analyze the total SuDS maintenance expenditure in divisions A and B, respectively. Figures 9.2 and 9.3 show the ln(ln(1/(1-Median Rank))) versus ln(maintenance cost) for divisions A and B, respectively.

Table 9.2 SuDS maintenance expenditure in divisions A and B

Year	Total SuDS maintenance expenditure in division A (US$)	Total SuDS maintenance expenditure in division B (US$)
2012	8000	4800
2013	6300	9000
2014	8500	8800
2015	8900	7200
2016	7000	6600
2017	5500	5000
2018	8500	4900
2019	4800	7000
2020	4000	8900
2021	8500	7800
Total	**70,000**	**70,000**

Table 9.3 Weibull analysis for SuDS maintenance expenditure in division A

Year	Maintenance cost per year Amount (x)	Rank (r)	Median ranks	1/(1-Median rank)	ln(1/(1-Median rank))	ln(ln(1/(1-Median rank)))	ln(maintenance cost)
2020	4000	1	0.067308	1.072165	0.06968	−2.66384	8.29405
2019	4800	2	0.163462	1.195402	0.178483	−1.72326	8.476371
2017	5500	3	0.259615	1.350649	0.300585	−1.20202	8.612503
2013	6300	4	0.355769	1.552239	0.439698	−0.82167	8.748305
2016	7000	5	0.451923	1.824561	0.60134	−0.5086	8.853665
2012	8000	6	0.548077	2.212766	0.794243	−0.23037	8.987197
2014	8500	7*	0.644231	2.810811	1.033473	0.032925	9.047821
2018	8500	8*	0.740385	3.851852	1.348554	0.299033	9.047821
2021	8500	9*	0.836538	6.117647	1.811178	0.593977	9.047821
2015	8900	10	0.932692	14.85714	2.69848	0.992689	9.093807

*Note: Duplicate values (maintenance cost) for years 2014, 2018, and 2021 would have the same score/rank. However, in this case, the tie-break situation was resolved and gave a unique rank to each number.

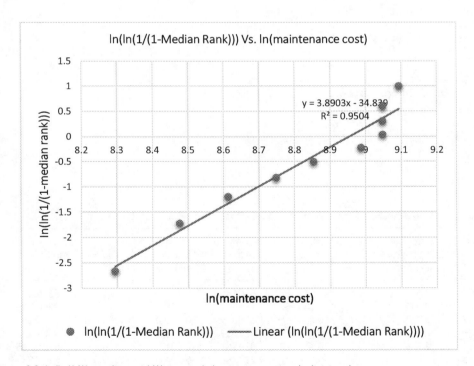

Figure 9.2 ln(ln(1/(1-median rank))) versus ln(maintenance cost), division A.

From Figure 9.2 and from the Weibull cumulative distribution (4.31),

$\beta = 3.8903$

Hence,
From Figure 9.2 and from the Weibull cumulative distribution (4.31),

$\beta = 4.2293$

Hence,

$-\beta \ln \alpha = -37.853$
$\alpha = \$7,709.286$

Interpretation of the results

The Weibull shape parameter, β, indicates whether the failure rate is increasing, constant, or decreasing, that is, failure to keep within the proposed budget of US$ 7000. $\beta < 1.0$ indicates a decreasing failure rate. $\beta = 1.0$ indicates a constant failure rate. $\beta > 1.0$ indicates an increasing failure rate. With all this, still it does not reveal whether SuDS in drainage system A or B meet the reliability goal of $R(7,000) \leq 0.3$. For this, we need to know the formula for reliability assuming a Weibull distribution:

$$R(x) = e^{-\left(\frac{x}{\alpha}\right)^{\beta}} \qquad (9.8)$$

SuDS in drainage system A

$$R(7,000)_{Design\ A} = e^{-\left(\frac{7,000}{7,749.243}\right)^{3.8939}} = 0.510$$

Table 9.4 Weibull analysis for SuDS maintenance expenditure in division B

Year	Maintenance cost per year Amount (x)	Rank (j)	Median ranks	1/(1-Median rank)	ln(1/(1-Median rank))	ln(ln(1/(1-Median rank)))	ln(maintenance cost)
2012	4800	1	0.067308	1.072165	0.06968	−2.66384	8.476371
2018	4900	2	0.163462	1.195402	0.178483	−1.72326	8.49699
2017	5000	3	0.259615	1.350649	0.300585	−1.20202	8.517193
2016	6600	4	0.355769	1.552239	0.439698	−0.82167	8.794825
2019	7000	5	0.451923	1.824561	0.60134	−0.5086	8.853665
2015	7200	6	0.548077	2.212766	0.794243	−0.23037	8.881836
2021	7800	7	0.644231	2.810811	1.033473	0.032925	8.961879
2014	8800	8	0.740385	3.851852	1.348554	0.299033	9.082507
2020	8900	9	0.836538	6.117647	1.811178	0.593977	9.093807
2013	9000	10	0.932692	14.85714	2.698481	0.992689	9.10498

Evaluation of integrated drainage system designs 321

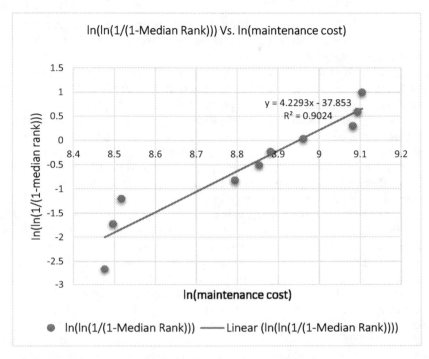

Figure 9.3 ln(ln(1/(1-median rank))) versus ln(maintenance cost), division B.

SuDS in drainage system B

$$R(7{,}000)_{Design\ B} = e^{-\left(\frac{7{,}000}{7{,}709.286}\right)^{4.229}} = 0.514$$

*Reliability	A	B
$P(x \geq 7{,}000)$	0.510	0.514
$P(x \geq 7{,}500)$	0.415	0.411
$P(x \geq 8{,}000)$	0.322	0.311
$P(x \geq 8{,}060)$	0.312	0.299
$P(x \geq 8{,}500)$	0.238	0.221
$P(x \geq 9{,}000)$	0.167	0.146
$P(x \geq 9{,}500)$	0.110	0.090
$P(x \geq 10{,}000)$	0.07	0.050

*Reliability = probability to fail.

Both systems indicate an increasing failure rate, that is, failure to keep within the target goal. However, system A is more reliable than system B as far as meeting the desired goal is concerned. Budgets for both systems do not meet the reliability goal, but the budget is more reliable for system A than system B.

$R(8,060)_{System\ A} = 0.312$ and $R(8,060)_{System\ B} = 0.299$, which are very close to the desired reliability goal. The goal of the public authority is to minimize the SuDS maintenance budget to as low as US$7000 per scheme per year. Therefore, setting US$8060 as the maintenance budget for the SuDS per year is not such a bad idea.

When US$7000 is allocated for maintaining SuDS, both systems will be severely constrained, and the operational reliability would be terribly affected. It is not enough. This information is important for designers and planners wishing to integrate a variety of SuDS in the locale. It clearly demonstrates that although the average SuDS expenditure over ten years for both systems is US$ 7000, it is not enough to benchmark on it to improve the drainage infrastructure of the municipality. If it is fixed as the yearly maintenance expenditure for SuDS in each division when a replica of either system A or B is done in other divisions, it will severely affect the operational and functional reliability of the drainage system, consequently impairing the overall functionality of the drainage infrastructure. This is because inefficiency and ineffectiveness in integrated drainage infrastructure induced from allocating little funds for maintenance would have a big impact on the entire system. Local problems in the system can induce problems elsewhere and it becomes a global problem for the whole drainage system. Therefore, the decision to go by average should be dropped and the SuDS maintenance budget increased reasonably to meet the reliability goal.

This approach to reliability analysis provides a quick insight about how several other scenarios can be modeled to arrive at the most reliable estimates for either capital costs, operation and maintenance costs, rehabilitation costs, or replacement costs, among several constraints, emphasizing that averages taken from historical data are not logical. Therefore, considering inflation and interest rates, the estimation departments analyze the combination of budgets that make the system reliable throughout the lifecycle phases, upholding its functionality. Enhancing reliability requires reliability metrics to be incorporated throughout the lifecycle phases of the drainage project. That means maintenance personnel must be part of the design and planning teams during early stages of project development.

9.3.4.7 Design life and service life used as reliability indicators

Design life is the forecast life expectancy of products based on their design. Service life is the forecast life expectancy of products based on real-world results. Reliability is connected to the development of the product. It is a function of serviceability and maintainability. Therefore, the service and maintenance costs must be well-evaluated to uphold the desired operability and functionality. For example, coatings with longer service lives are better as they reduce maintenance costs in pipes.

Service life is the duration throughout which a product is used economically. That means if it is no longer economical to service a system or product, then it is no longer sustainable because it is more likely to associate with more carbon emissions to maintain, service, or operate it. Service life is the measure of the durability of assets and manufactured products. During the service life, a coating in a storm sewer pipe has the ability to protect a substrate and maintain its acceptable appearance with no need for maintenance apart from cleaning. The elapse of service life can be determined by the ineffectiveness of the product that results from aging, frequent failures, and increased repair expenses. The factors that can determine the service life of a product include the following:

a. Quality of the manufacturer.
b. Materials used.
c. Flexibility in use.
d. Intensity of use.
e. Operating/environmental conditions.
f. Care in distribution and use.
g. Built in obsolescence.
h. Maintenance and repairs.

In case a material, system, or structure has a *service life greater than the project design life*, then it would have a residual future current money value, which should be discounted back to a present constant money value utilizing the inflation/interest factor and subtracted from the original cost. After the designer has done hydraulics and hydrology properly, then they do the structural integrity analysis for loading cases (see Section 8.4). Section 2.7 requires designers to consider the durability and maintainability of materials while planning.

Assuming the design life of a storm sewer is equated to average recurrence interval, but which is not always the case, so that the storm sewer can carry the peak flows. The design life of the pipe material is not always equated to the ARI. Note that when the design life is equated to ARI, it influences the cost and the routing method also influences cost because the Froude number choice would definitely impact the size of the hydraulic structure. The selection of Manning roughness factor would be influenced by the cost after the feasibility analysis is completed. The feasibility analysis and lifecycle cost analysis inform one another. Depending on the size, projected risk, and complexity of the project, an ARI of *25 years, 50 years*, and *100 years* would be selected. This would translate to the design life of the hydraulic structure. This is an assumption. Reliability is built in the system by juggling the lifecycle costs, hydraulics, and structural analysis. For example, according to the U.S. Army Corps of engineers, concrete pipe has a service life of 70–100 years. If the designer opts for a 50-year ARI design storm, which means a concrete pipe would ably withstand the service life. That again means the design life can be assumed to be 50 years with a residual value equivalent to a cost of about 20 years.

9.4 ACHIEVING THE SAFE NET-ZERO GOAL

Safe net-zero is the internationally agreed upon goal for mitigating global warming in the second half of the century, and the Intergovernmental Panel on Climate Change (IPCC) concluded the need for net-zero CO_2 by 2050 to remain consistent with 1.5°C temperature rise goal[4].

Embodied (capital) carbon comes from the consumption of embodied energy consumed to extract, refine, process, transport, and fabricate a material or product (including buildings). Operational carbon comes from all energy sources used to keep our infrastructure such as buildings warm, cool, ventilated, lighted, and powered. When you talk of carbon emissions, you talk about lifetime emissions (embedded + operational), these emissions occur throughout the lifecycle of the drainage system. They come from the energy consumed in the process.

Inception, planning, design, and implementation	Operation, maintenance, and decommissioning
Embedded carbon	Operational carbon

9.4.1 Estimating carbon offsets for integrated drainage systems

The governing hierarchy for reducing environmental impact while designing and developing the built environment is in the order:

Reduce—Reuse—Recycle—specify green.

However, in designing and developing integrated drainage systems aiming to achieve sustainability goals, the above hierarchy is not usually straightforward and the main pillars are to 'reduce and specify green'. These influence each other as shown in Figure 9.4.

The primary goal of specifying green is to reduce runoff as much as possible so that the risk of flooding is significantly reduced. This means the sewer capacity is reduced significantly for a design storm of given average recurrence interval considered when appraising the drainage sewerage system when SuDS are specified. This leads to reduction in raw material use. The use of recyclable materials such as recyclable aggregates and reusing materials in drainage schemes would always come after site-specific SuDS have been prioritized. In order to significantly reduce materials used to build sewers and drains, recyclable materials would be considered.

Specifying green means incorporating SuDS in the drainage system such as RWH, infiltration systems, bioretention gardens, tree planting, etc. The primary goal of SuDS is to collect stormwater, infiltrate runoff into soil, capture pollutants, groundwater recharge, reduce runoff velocity, act as a carbon sink (absorb GHGs), etc. Sustainable practices are usually recommended in order to reduce stormwater volume entering the gray drainage system (minimize runoff hence reducing flooding risk). A sustainable system/process/product is economical, socially beneficial, and environmentally friendly. For example, US$660, 10,000 L RWH–HDPE tank, 30-year life span are adequate for 464.51 m² plot, waives about $1444 in 30 years, that is, $0.96/m³, typical in Uganda, East Africa. According to Viessmann, a UK based company, a fully grown typical tree can absorb around 21 kg of CO_2/year (www.viessmann.co.uk).[5] Trees control temperatures creating a cool temperature range. Therefore, estimating how much carbon is associated with a drainage project development requires determining the total capital and operational carbon for both gray infrastructure and SuDS. This is weighed against the carbon that SuDS would take out of the environment through the trees and plants.

Capital carbon (Extract raw materials, transport, manufacture materials, transport to site, and construct).	Carbon embedded in the production of materials, e.g., concrete, and carbon from vehicles and plants used in construction, e.g., front-end loaders, backhoes, trucks, etc.

Figure 9.4 Reduce and specify green.

Operational carbon
(Fuel from vehicles conducting routine maintenance—cleaning litter, mowing, and removing sediment, retrofitting SuDS, and minor repairs).

Carbon from vehicles used to maintain SuDS and gray infrastructure, e.g., mowers used to keep turf grass at an appropriate height. Heavy equipment such as front-end loaders and backhoes are usually employed where huge amounts of sediment loads accumulate and where SuDS need replacement.

The drainage system total lifecycle carbon emission would be calculated from operational carbon plus embodied carbon (eCO_2). Table 9.5 provides the amount of carbon emissions per unit consumption of fuel. Table 9.6 provides the amount of carbon emissions per vehicle type. Table 9.7 provides UK Environmental agency transport emissions (gCO_2/t km) These are beneficial when calculating the operational and embedded carbon associated with activities surrounding drainage system construction and maintenance. Concrete is one of the main contributors to the eCO_2 footprint of most infrastructure assets. According to Circular Ecology, eCO_2 of concrete wildly varies depending on the details of the concrete mixture, in terms of the material constituents per cubic meter.[6]

Table 9.5 Calculation of CO_2 emissions[7]

Fuel type	Kg of CO_2 per unit of consumption
Grid electricity	43 per kWh
Natural gas	3142 per tonne
Diesel fuel	2.68 per liter
Petrol	2.31 per liter
Coal	2419 per tonne
LPG	1.51 per liter

Table 9.6 Transport conversion table

Vehicle type	Kg CO_2 per liter
Small petrol car 1.4-L engine	0.17/km
Medium car (1.4–2.1 L)	0.22/km
Large car	0.27/km
Average petrol car	0.20/km
Small diesel car (>2 L)	0.12/km
Large car	0.14/km
Average diesel car	0.12/km
Articulated lorry, diesel engine	2.68/km (0.35 L fuel per km)
Rail	0.06 per person per km
Air, short haul (500 km)	0.18 per person per km
Air, long haul	0.11
Shipping	0.01 per tonne per km

Table 9.7 UK Environmental agency (transport emissions (gCO$_2$/t km))

Transport	UK Environmental agency (transport emissions (gCO$_2$/t km))
Road	317
Rail	41
Water	9

9.4.1.1 Embodied carbon

9.4.1.1.1 Introduction

Carbon footprint is used to calculate the amount of damage caused by an individual, household, institution, or business to the environment through harmful carbon dioxide emissions.[8] The chief construction materials for drainage infrastructure assets is concrete. We strive to reduce the use of concrete as much as we can.

eCO$_2$ is defined as the CO$_2$ produced over a defined part of the lifecycle of the product. The CO$_2$ is primarily associated with the consumption of energy over the relevant part of the lifecycle but can also include emissions which occur directly as a result of the production process. The production of concrete as the main construction material produces enormous amount of embedded carbon. When comparing eCO$_2$ of concrete 0.12t/t with cement 0.88t/t or with other building materials (e.g., steels) 0.82t/t, the eCO$_2$ of concrete is relatively not that high.

The sources of eCO$_2$ for reinforced concrete comes from:

a. Constituent materials.
b. The fuel and process used to manufacture Portland cement.
c. The amount of Portland cement replacement.
d. The strength class of the concrete and the resulting mix composition.
e. The level of steel reinforcement.
f. Transport impacts of the aggregate and cement.

According to the Portland Cement Association,[8] the manufacture of cement produces about 0.9 kg of CO$_2$ for every 1 kg of cement. Since cement is only a fraction of the constituents in concrete, manufacturing a cubic yard of concrete about 1769.01 kg is responsible for emitting about 181.437 kg of CO$_2$. The release of 181.437 kg of CO$_2$ is approximately equivalent to:

a. The CO$_2$ associated with using an average tank of gas in a car.
b. The CO$_2$ associated with using a home computer for a year.
c. The CO$_2$ associated with using a microwave oven in a home for a year.
d. The CO$_2$ saved each year by replacing night light bulbs in an average house with compact fluorescent light bulb.

9.4.1.1.2 Embedded carbon from concrete constituents

9.4.1.1.2.1 PORTLAND CEMENT

- Portland cement has very high embodied CO$_2$ content.
- Cement manufacture is energy intensive.

9.4.1.1.2.2 AGGREGATES

- Aggregates have a very low eCO_2 compared to Portland cement and contribute only 3% to the total for reinforced concrete.
- Transportation of the aggregate to the batching plant and to site accounts for most CO_2 emission.

9.4.1.1.2.3 OTHER MAJOR CONSTITUENTS

- The eCO_2 figure presented is based on fair estimate in UK industry (Table 9.8).

eCO_2 source data are available for Finland, UK, and USA. Although there is variation between the source data and Pulverized Fly Ash (PFA), Ground Granulated Blast-Furnace Slag (GGBS) has a significantly lower eCO_2 value than Portland cement (Table 9.9).

9.4.1.1.3 Embedded carbon from concrete constituents, C32/40 mix

A typical C32/40 mix would produce embedded carbon as shown in Figure 9.5. The eCO_2 contributed by each of the concrete constituents is as shown in Table 9.8.

9.4.1.1.4 Variations of eCO2 in concrete strength and mix design

eCO_2 also varies by strength class (i.e., the amount of cement content in the mix) is as given in Table 9.10.

9.4.1.1.5 Embedded carbon in reinforcement

eCO_2 varies only slightly with steel reinforcement. Therefore, reducing cement content will have greater effect than reducing the weight of reinforcement. See Table 9.11.

Table 9.8 Embedded carbon (eCO_2) figure presented is based on fair estimate in UK industry

Material	eCO_2 (tCO_2/t)
CEM I (Portland cement)	0.822
Coarse aggregate	0.008
Fine aggregate	0.0053
Water	0.000000249
Reinforcement steel	0.97 (predominantly recycled)

Table 9.9 Embedded carbon by cement product for UK, USA, and Finland

Material	Finnish eCO_2[11]	UK eCO_2 (tCO_2/t)	US eCO_2 (tCO/t)
Portland cement	0.670	0.822[4]	0.900[1]
Fly ash	0.0053	0.011[a]–0.025[12]	0.002[13]
GGBS	0.026	0.050–0.070[15]	0.028[15]

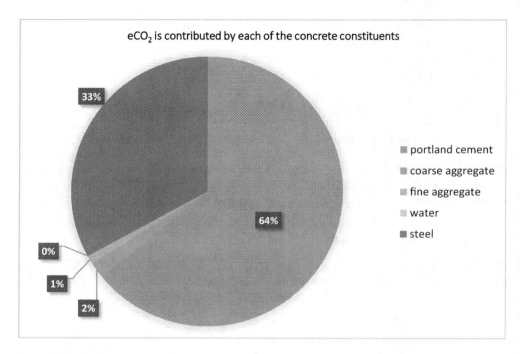

Figure 9.5 Embedded carbon (eCO$_2$) is contributed by each of the concrete constituents.

Table 9.10 Embedded (eCO$_2$) varies by strength class

Class	Cement (kg/m^3)	Coarse aggregate (kg/m^3)	Fine aggregate (kg/m^3)	Water (kg/m^3)	eCO$_2$ (tCO$_2$/t)
C8/10	180	1025	912	189	0.07
C25/30	290	1053	826	176	0.11
C28/35	320	1059	801	174	0.12
C32/40	350	1066	773	174	0.13
C40/50	430	1000	796	174	0.15
C50/60	450	1175	600	149	0.16

Table 9.11 How embodied carbon varies with steel reinforcement

Kg/m^3 of reinforcement	100	125	150	175	200
tCO$_2$/t of reinforced C32/40 concrete	0.16	0.17	0.18	0.19	0.194

Table 9.12 Dimensions of reinforced concrete pipe

Internal diameter (mm)	Wall thickness (mm)		
	Wall A	Wall B	Wall C
310	44	51	
380	47	57	
460	51	63	
530	57	70	
610	63	76	95
690	66	83	101
760	70	89	108
840	73	95	114
910	76	101	120
1070	89	114	130
1220	101	127	146
1370	114	140	159
1520	127	152	171
1680	140	165	184
1830	152	178	197
1980	165	190	209
2130	178	203	222
2290	190	216	235
2440	203	229	248
2590	216	241	260
2740	229	254	273
2900	241		
3050	254		
3200	267		
3350	279		
3500	292		
3650	305		
3800	318		
3960	330		
4110	343		
4270	356		
4420	368		
4570	381		

Concrete, as a material, can never be truly sustainable, but we can reduce its impact on the environment. All concrete pipes made in sizes larger than 610 mm are usually reinforced. Table 9.12 provides the dimensions of reinforced concrete pipes.

9.4.1.2 What to do in drainage schemes to reduce carbon emissions?

Possible alternatives include:

a. Specify green, use SuDS to minimize concrete usage.
b. Government policy is paramount to support the overarching goal of sustainable development.

c. Reduce CO$_2$ emissions using other types of cement manufacturing process.
   d. Reduce Portland cement consumption.
   e. Reduce primary aggregate usage.

9.4.1.3 Operational carbon in drainage systems

Operational carbon is carbon that is associated with the operation and maintenance of drainage schemes. It comes from trucks and vehicles used to clean or desilt storm sewers, mowing of SuDS, repairing drainage assets, rehabilitation, and replacing drainage assets. The operation carbon can be calculated from fuel. See Tables 9.1 and 9.2 for carbon produced from vehicles and/or fuels.

9.5 EMBODIED CARBON WORKED EXAMPLE FOR ILLUSTRATIVE PURPOSES

Figure 2.7 shows the postdevelopment site condition integrated with sustainable drainage systems; 100% of the proposed site condition will have impervious surfaces. A 910-mm reinforced cement concrete storm sewer pipeline is meant to carry the flows from the fully impervious postdevelopment site condition and is to be laid on a slope of $S_0 = 0.008$. The developer wishes to exploit SuDS benefits and to reduce stormwater runoff that ends into the storm sewer using a bioretention garden(s) designed to meet water quality volume, WQ_v, requirements only. The bioretention will absorb a considerable fraction of the stormwater runoff and infiltrate it into the ground. The reinforced cement concrete storm sewer will be reduced from 910- to 610 mm, to reduce embedded carbon and the overall construction cost of the storm sewer pipeline. The 90th percentile precipitation depth of the cumulative curve for storms for the area is $P = 22.86$ mm, and the total area contributing to the bioretention is $A = 4.5$ acres.

   a. Calculate the amount of capital carbon saved in reducing the storm sewer from 910- to 610-mm diameter, where the total length of the storm sewer is 500 m. Take 0.11 tCO$_2$/t of eCO$_2$ for a C25/30 concrete mix and 0.18 tCO$_2$/t of eCO$_2$ due to the steel reinforcement and neglect emissions from other constituents, like mortar for sealing joints.
   b. What should be the capacity of the bioretention garden(s)? Use Driscoll's equation to find R_v (using i in decimal) and calculate the water quality volume.

Solution

Part (a)

- Estimate the amount of steel reinforcement in the RCC pipe in kilograms.
- Calculate the embedded carbon of steel reinforcement.
- What is the density of C25/30 concrete mix?
- What is the thickness of the concrete culvert? For wall thickness C, see Table 9.12. For 610-mm diameter, (t = 95 mm) and for 910-mm diameter, (t = 120 mm).

A 610 mm of wall thickness C, that is, 95 mm would have a volume of:

$$0.3142 \times 0.61 \times 500 \times 0.095 = 9.10\,m^3$$

A 910 mm of wall thickness C, that is, 120 mm, would have a volume of:

$$0.3142 \times 0.61 \times 500 \times 0.12 = 11.50\,m^3$$

The difference saved $= 11.50 - 9.10 = 2.4 m^3$

A mix of 1:2:2 of concrete is used for making reinforced concrete pipes with longitudinal reinforcement equal to at least 0.25% of the cross-sectional area of concrete.

Thus, steel reinforcement occupies 0.25% of the reinforced concrete by volume.
Concrete $= 0.9975 \times 2.4 = 2.394 m^3$
Reinforcement steel $= 2.4 - 2.394 = 0.006 m^3$
Take density of concrete in this problem $= 2400\,kg/m^3$
Take density of reinforcement steel in this problem $= 7850\,kg/m^3$
Mass of concrete $= 2400 \times 2.394 = 5745.6\,kg$; 5.75 tons
Mass of reinforcement steel $= 7850 \times 0.006 = 47.1\,kg$; 0.0471 tons
Taking 0.11 tCO_2/t of eCO_2 for a C25/30 concrete mix and 0.18 tCO_2/t of eCO_2 for reinforcement steel provides:
Embedded carbon in concrete $= 0.11 \times 5.75 = 0.6325\,tCO_2$
Embedded carbon in concrete $= 0.18 \times 0.0471 = 0.008478\,tCO_2$
Total embedded carbon saved from being pumped into the atmosphere $= 0.6325 + 0.008478 = 0.640978$ tons (640.978 kg).

Part (b)
Using Driscoll's equation to find R_v (using i in decimal), thus,

$$R_v = 0.05 + (0.9) \times (1.0) = 0.95$$

And

$$WQ_V = 4.047 PAR_v = 4.047 \times 22.86 \times 4.5 \times 0.95 = 395.50\,m^3$$

Therefore, the capacity of the bioretention garden is 395.50 m^3.

9.5.1 Analysis of example (Section 9.5)

From the nomograms (Figure 5.8), laying a storm sewer of 610 mm implies that the sewer would only be able to accommodate and conserve a peak discharge of about 33.5 m^3/min laid on a slope $S_0 = 0.008$. However, laying a 910-mm storm sewer would accommodate about 100 m^3/min peak discharge. Considering a variety of hydrologic methods, the designer would adopt a back-and-forth approach to derive a proper drainage solution.

The primary goal of the designer is to scale down gray infrastructure size in the drainage system as much as they could. The use of SuDS such as the bioretention gardens comes with advantages of mimicking the natural environment and scaling down the runoff volumes and peak flows. Through mimicking the natural environment, SuDS such as trees and other vegetation planted in the SuDS absorb carbon monoxide and other pollutants from the environment, hence greatly contributing to sustainability goals going forward.

Concrete is one of the main contributors to the embodied carbon footprint of most drainage infrastructure assets. According to Circular Ecology, eCO_2 of concrete wildly varies depending on the details of the concrete mixture, in terms of the material constituents per cubic meter.[9]

In the example in Section 9.5, several constraints may be juggled before the decision is taken to drop a 910- for 610-mm diameter pipeline. An ELCA would provide a level-playing field when comparing construction alternatives, that is, installing bioretention garden(s) to scale down the size of the pipeline. Thus, all stages of the lifecycle for SuDS and standard pipework should be evaluated. That includes material production, transport, energy use, maintenance, and disposal or recycling at end of life. The associated carbon emissions must be evaluated throughout the lifecycle phases and an optimal solution considered because SuDS come with heavy initial costs but score highly on environmental and social benefits. Therefore, the cost of embedded carbon from gray infrastructure must be weighed with the cost of installing SuDS as a carbon sink, in the long run.

9.6 CONCLUSION

Offsetting carbon is a back and forward approach that requires careful evaluation of the steps where SuDS are integrated to reduce runoff quantities, thereby reducing runoff volumes that dictate the capacity of sewers. When the capacity of sewers is reduced significantly, eCO_2 is offset in gray infrastructure because construction materials such as concrete are reduced significantly. However, operational carbon requires to evaluate maintenance practices.

Gray infrastructure requires trucks that usually clean the sewers, collect garbage and sediment. The amount of fuel they burn would be calculated from the routine maintenance charts. And depending on the quantity of sediment loads sewers carry in any given storm, the frequency of maintenance is determined.

SuDS do not require sophisticated equipment to maintain them. In many cases, SuDS require heavy initial capital investments. However, when designed and constructed properly, they can have a long life span. Therefore, the SuDS benefits outcompete standard pipework solutions at this time when climate action is an essential topic of the century.

References

1. Life cycle cost analysis: What is life cycle cost analysis? CFI Team. Accessed March 5, 2022. https://corporatefinanceinstitute.com/resources/knowledge/finance/life-cycle-cost-analysis/
2. Life cycle cost analysis in drainage projects: Technical brief. Concrete Pipe Association of Australasia. www.cpaa.org.au/
3. Design Data 25: Life cycle cost analysis. American Concrete Pipe Association. (2012). www.concrete-pipe.org
4. Intergovernmental Panel on Climate Change (IPCC). Accessed August 5, 2022. www.ipcc.ch/sr15/
5. Viessman. How much CO_2 does a tree absorb? Accessed July 29, 2022. www.viessmann.co.uk/heating-advice/how-much-co2-does-tree-absorb
6. Concrete embodied carbon footprint calculator – Circular Ecology. Accessed July 20, 2022. https://circularecology.com/concrete-embodied-carbon-footprint-calculator.html

7. University of Exeter. Accessed July 29, 2022. https://people.exeter.ac.uk/TWDavies/energy_conversion/Calculation%20of%20CO2%20emissions%20from%20fuels.htm
8. Carbon dioxide (CO_2) emissions of concrete. A paper presented by Ir. Stephen Leung, Arup materials technology, February 27, 2009, SCCT 2009. Accessed July 29, 2022. www.devb.gov.hk/filemanager/en/content_680/6_mr_stephen_leung_carbon_dioxide_emissions_of_concrete.pdf
9. Carbon footprint: Integrated paving solutions. 0020-11-105. PCA – The Portland Cement Association – America's Cement Manufacturer. Accessed July 29, 2022.www.cement.org/docs/default-source/th-paving-pdfs/sustainability/carbon-foot-print.pdf

Further reading

1. Reliasoft. Weibull++ – Life data analysis software – ReliaSoft. Accessed July 29, 2022. www.reliasoft.com/Weibull/

Chapter 10

Construction and management

10.1 INTRODUCTION

Drainage construction requires careful planning because the operational and functional reliability of a drainage scheme primarily depends on proper design and construction. This means skilled and experienced personnel should be employed to design and develop drainage solutions. In that case, construction and management solutions must be cost-effective and conforming to municipal ordinances, byelaws, and national statutes governed by the overarching goal of sustainable development.

Environmental conservation practices must be highly considered throughout the design and construction phases conforming to local stormwater ordinances and byelaws geared toward regulating materials discharged to the drainage system, employing skillful and experienced personnel to build drainage schemes or inspect drainage lines and to make connections. Before connecting to the public sewer, the municipal may require scheduled inspections to ascertain that the work has been done satisfactorily and to the required standards.

Social and environmental management during construction and postconstruction phases must be guided by an environmental and social management plan (ESMP) so that the environment is not severely destroyed, and where there is a likelihood of causing severe harm to the environment, mitigation measures are proposed or scheme routes are changed appropriately. Extensive ground investigations may need to be considered in some areas, especially where hazards are spotted, for example, contaminated land, waterlogged areas, and cultural heritage sites. ESMP is provided in accordance with standards and guidelines such as ISO 14001. In designing and developing drainage solutions, safety considerations must be incorporated through the design and construction phases meeting standards such as the UK CDM regulations (2015) so that staff and workers and the general public are protected from harm.

10.2 PROCEDURAL MANAGEMENT IN DRAINAGE CONSTRUCTION

The governing principles in drainage construction are safety and sustainability. The sustainability metric is based on the cost and environmental and social management plan. The cost, time, standards, and codes of practice guide the final decisions when drawing a work plan for construction. As a basis of quality control and management, contract documents are relied upon and any expected variations reported. The sewer plans and profiles are used to construct the sewer lines accurately and cost-effectively.

DOI: 10.1201/9781003255550-10

As a result of climate concerns, work plans are drafted bearing in mind inclement weather and as a rule of thumb weather forecasts must be incorporated in the method statement and work plans. The effect of not planning effectively to account for inclement weather can bring about unprecedented flooding especially in congested urban areas. In many cases, drainage development is upgraded from existing drainage systems in urban centers and because drainage construction comes with excavations, sometimes with blockage of existing storm sewers and diversions, flooding may result because the area land use becomes very disturbed hence distorting the original flow path of the runoff. In doing so, the land use patterns rapidly change especially where excavations are conducted and blocking of existing sewers is inevitable as drainage works commence and progress. The distortion of drainage corridors as drainage works commence and progress can affect the runoff process or flow path. Therefore, drainage construction in urban centers must be sensitive about flood-prone zones and work plans must be prepared taking into account inclement weather, distortion in drainage corridors, and the overall safety of workers and nearby communities.

Trench excavations are conducted in various ways depending on the scope and complexity of the project. Several techniques may be deployed such as the use of mechanical equipment or hand tools. Sometimes explosives are used to cut and open more difficult areas. However, safety must be clearly observed, and where explosives are to be used, safety codes must be conformed to including the use of mats to contain debris.

10.3 EXCAVATION OF TRENCHES FOR STORM SEWERS

Storm sewers come with enormous amounts of excavations. The excavations require procedures to safely dig the trenches and sometimes to provide embankments depending on the nature of drainage scheme because some cases may require pipes to be laid on the original ground and the trench sides are considered remote from the pipe. In modern sewer plans, storm sewers are represented by links and nodes on plans. Links refer to pipes or drains while nodes could be manholes, mini-manholes, junction boxes, outlets, outfalls, inlets, etc. The specifics of the nodes would be provided on separate drawings and the flow movement through the sewers and drains provided on a single master plan drawing for a catchment basin clearly indicating the flow direction through the links.

For quality control purposes, the contractor follows the links properly. Each link has an inlet node and outlet node. These enable the contractor to trace the direction of flow of the stormwater runoff. Each node has the maximum rim elevation which is the ground elevation. Sometimes, the rim elevation of a node may be aboveground depending on what site constraints dictate influencing the design. The inverts of nodes and links are usually obtained taking into consideration the minimum cover required. Prior to construction, an offset line is established where it will not be disturbed or covered, then measure from the offset line to lay out the trench on the ground. Storm sewer trenches are mainly excavated with a mechanical excavator rigged as a back-acting shovel except where constraints cannot allow. The production rate for mechanical excavators is high and produces a great amount of work in a short period hence catching up with the construction timelines. Excavation by hand is greatly warranted where the ground has several underground utilities, poor and waterlogged soils, and very close to buildings. Therefore, continuous sheeting may be required. Spoil is placed well away from the side of the trench.

Where there are buildings and other structures, that is, walls, roads, and bridges, next to where excavations will be dug, it may be appropriate to prop and/or underpin the structures to prevent the excavation causing their collapse. For safety purposes, trench excavation usually proceeds in the upstream direction coming from the nodes downstream to the nodes upstream in accordance with the plans and specifications. This allows runoff to safely discharge off to streams in case it rains during construction. In waterlogged areas, a sump may be created by channeling on the side of the trench bottom so that excess water is pumped out. Excavated trenches are usually fenced off to keep people away from the construction area avoiding accidents. Trench shoring machines are also used to eliminate the hazard of workers falling in trenches hence improving safety measures. Trenches in unstable materials require sheeting and bracing to prevent collapse of sidewalls. The commonly used supports consist of corrugated, perforated steel sheets held apart with timber or adjustable steel props. Trench supports are fixed as the excavation proceeds.

During construction, the **line and grade** of each sewer must be carefully established and maintained so that self-cleansing velocities will be obtained. The depth is gauged by means of a boning rod and sight rails, see Figure 10.1. The following steps outline the procedure to accurately locate the sewer line and grade:

a. Excavate the trench close to its final grade, and place **batter boards** across it at intervals of 10 to 15 m as shown in Figure 10.1.
b. Establish the centerline of the sewer on the batter boards by measuring from the offset line and nailing an upright cleat so that one edge is on the centerline.
c. Establish the elevation of each cleat and mark thereon an elevation which is an identical distance above the finished grade of the sewer at each batter board.
d. Drive a nail into each cleat at the grade mark and run a string from nail to nail. This string is at the slope of the sewer and directly above its centerline.

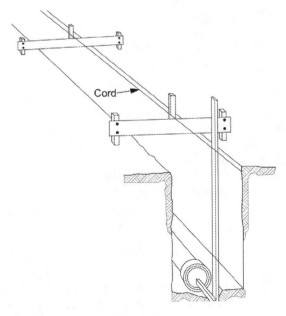

Figure 10.1 Establishment of line and grade of a sewer.

e. Establish a line by dropping a plumb bob from the string, and grade is checked with an ell-shaped gauge marked at the distance equal to the vertical displacement between the string and the invert of the sewer.
f. When the gauge is installed in the sewer as shown, the mark should match the string. The grade is checked in this method on each length of pipe.

In large-diameter storm sewers, the above technique differs slightly. The modern method of locating sewer line and grade is by using a laser. Pipe lasers are essential for underground trenching and pipe work and when aligns a pipe quickly. This pipe level tool can be set in place with brackets that will allow you to align sewer pipes and storm drains with each other.

10.3.1 Dewatering of excavations

Dewatering of the trenches is a key component in lowering the water table or removing excess water. Therefore, the control of surface and subsurface water is required so that dry conditions are provided during excavation and pipe laying. Sometimes, thorough ground investigations must be conducted and groundwater conditions obtained before they are encountered during the course of excavation. Usually, the dewatering process is done by pumping or evaporation. It is usually done before excavation and helps to lower the water table that might cause problems during excavations. Dewatering can also refer to the process of removing water from the soil by wet classification.

The water can be in the form of rain, snow, and sleet directly into excavations, especially when they are uncovered. The runoff following the downpours can also run in. Water can laterally enter through the sides depending on the height of the surrounding water table, and proximity of the watercourses (rivers, streams, lakes, etc.) can cause excess water.

The contractor is advised to follow the links (conduits and drains) properly on the drawing plans and profiles. Each link has an inlet node and outlet node. These enable the contractor to trace the direction of flow of the stormwater. Each node has the maximum elevation which is the ground elevation. This node could be a manhole or meeting location/junction for links. For safety proposes, the contractor starts constructing the drainage system from the outfalls, so that when it rains, the runoff can safely route off without flooding the working area and excavated trenches.

To protect the stability of the sides of the excavation, drainage channels can be cut around the excavation and the water can be channeled away to sumps where it can be pumped away. The water table is then lowered below the level of the excavation, in extreme cases, watercourses might be redirected. Excavation sides can be strengthened by using higher shoring and sandbags on the outside. However, the continued flow of water from the area surrounding the excavation may cause settlement difficulties.

When excavations extend below the groundwater table, water will flow into the opening. If the subsurface strata are sufficiently permeable, the velocity of flow may be sufficiently great to fluidize the soil, creating a quick condition. This can undermine the sheeting and cause failure of the trench.

The quick conditions in the trench can be avoided by lowering the groundwater table with well points. Well point is a line of relatively small driven wells parallel to the trench, which are pumped at a rate large enough to depress the groundwater table below the bottom of the opening. Well points may be placed along either or both sides of the trench, about 2 m from the centerline, 1 m apart, and extending well below the bottom of the trench. The individual wells are attached in series to a header which is connected to a common pump. Well pointing or deep wells can be considered where sandy or silty soil exists. Ground freezing or providing an impermeable barrier by injecting cement, bentonite, or a chemical is an option, but can be expensive and cause disruption to the ground. The disposal of any water should be discussed with the appropriate environmental agency.

When water is encountered unexpectedly, the trench bottom may be stabilized temporarily with gravel, rock, or rubble. Material of this kind causes arching between the grains of the soil and will prevent its fluidization. The flow of water will not be halted by this technique. If the flow is not much to cause fluidization, it may be removed by letting it run along the trench bottom to a sump, from which it is then pumped. In the construction of large sewers, an open-jointed tile underdrain may be placed in the bottom of the trench below the location of the major structure to provide more adequate drainage and a dry trench bottom for construction. The drain is left in place when the sewer is completed but should be plugged so that it does not result in permanent drainage of the soil.

Accidental damage to nearby tanks, water mains, or other water supply pipework in the environs of the excavation can also lead to an ingress of water. All exposed pipes should be identified and supported.

10.3.2 Excavation techniques

A trench, which typically refers to a narrow excavation of 10-m wide or less, requires a combination of the right technology, techniques, and safety measures to serve its purpose. The most important factors in the trenching process are the competence of the worker forming the trench and the equipment they use. In shallow trenches with favorable soil conditions, the trench may be dug at a rate of up to 10 m/min. If the contractor chooses to do so, trenches can also be excavated with standard construction equipment such as backhoes, clamshells, or draglines. Excavation for sewers should be done with a mechanical equipment wherever possible. Specialized equipment for trench excavation employs continuous chain drives carrying buckets that cut the soil, carry it to the surface, and discharge it to a conveyor which carries it to the size of the trenches up to 1.3-m wide and 9 to 10-m deep in a single pass. The main methods of the execution of trenches are conventional methods: excavators +/− rock breakers, drill and blast, and trenching. Although the most appropriate method depends on the terrain, lithology, etc., trenching turns out to be the most efficient system in most cases:

10.3.2.1 Drill and blast application

- In very strong, abrasive, and massive rock. Excavation in rock should be carried below the bottom of the pipe to a depth equal to one-fourth the pipe diameter or 100 mm, whichever is greater. The space between the rock and the pipe is then filled with appropriate bedding material.
- Not suitable for rock masses composed of small, loose blocks.
- Relatively easy to set up and execute.
- Small amounts of rock may be excavated by drilling and hammering with either hand tools or machinery similar to pavement breakers.
- Backhoes and dump trucks required to remove blasted material.

- Particularly useful on steep slopes.
- When explosives are used, the technique should conform to modern safety codes—including the use of mats to hold the debris. Once the rock has been broken, it can be removed with ordinary construction equipment.

10.3.2.2 Conventional methods: Excavators +/− rock breakers application

- In rock masses composed of small, loose blocks.
- Not suitable in very strong, abrasive, and massive rock.
- Easy to set up but can be very time consuming.
- Useful for tight curves.
- Useful where the ground changes from soil to rock over short distances.

10.3.2.3 Trenching application (hand excavation)

- In weak to strong rocks.
- Generally, not economic in very strong, abrasive, and massive rock.
- Unsuitable for rock masses containing loose cobbles and boulders or those containing pockets of wet clay.
- When bell-and-spigot pipe is used, hand excavation of the bedding material is required at each bell.
- Larger machines require considerable lateral working space and operate most efficiently over long distances in relatively homogeneous rocks.
- Hand excavation is also required in the vicinity of services of other subsurface utilities in order to ensure that they are not damaged. Except for such locations, machinery should be employed.

The type of excavations should be indicated on the drawings and plans and may be divided into the following classes:

a. Where quicksand is known to occur, it should also be included as a separate item.
b. Solid rock includes solid rock in its original bed, or well-defined ledges removable only by blasting, and boulders over 0.25 m^3 in volume.
c. Hardpan includes materials such as disintegrated limestone, shale, soapstone, slate, fireclay, cemented gravel, and boulders less than 0.25 m^3 in volume. This type of material can be excavated by hand or machinery with minor difficulty.

10.3.3 Excavation limitations

Trench width and depth are important considerations in excavations. Trench grades should be constantly checked against the elevations which are established in the sewer profile. The hydraulic efficiency of the sewer pipeline is influenced by the quality of trenching. Where excavations are poorly conducted leading to low or high spots in the line, the hydraulic efficiency of the sewer line may be adversely affected, which may necessitate correction reworks and additional maintenance once the sewer line construction is completed.

From Section 8.4, the backfill loads and pipe strength requirements are seen to be a function of the trench width. Therefore, maximum trench widths are established in the plans or standard drawings and where maximum trench widths are not indicated in any of the construction contract documents, trench widths should be as narrow as possible, with side clearance adequate to ensure proper compaction of backfill material at the sides of the pipe.

The backfill load transmitted to the pipe depends on the trench width. Therefore, in evaluating the backfill load, the designer or planner assumes a certain trench width and then selects a pipe strength capable of withstanding this load as shown in Section 8.6. If the constructed trench width exceeds the width assumed in design, the pipe could be overloaded and possibly structurally distressed.

10.4 EXCAVATION SUPPORTS

Trenches which are excavated in unstable ground require bracing and sheeting in order to prevent collapsing of the sidewalls. The consequences of the failure of soils, which may include heavy pecuniary losses due to civil cases arising out negligence acts, justify all kinds of expenses necessary to prevent such occurrences. Therefore, recommendations with regard to supporting open excavations are primarily based on geotechnical evaluation of the site conditions. Excavation supports are provided to safeguard trench side soil against collapse. During deep excavation, the sides of the trench collapse if the soil is weak. The side supports prevent the side of the trenches against collapse. The importance of excavation supports is as follows:

a. For the safety of excavation workers.
b. For the safety of the surrounding property.
c. For the safety of passersby and vehicles.
d. For the safety of public service property such as telephone cable, water pipes, and electric cable.

Trenches more than 1.5-m deep and 2.5 m in length should be held by **shoring or bracing** or alternatively sloped to the angle of repose of the soil. The trenches are excavated by sloping the sides or are usually dug vertically at the bottom so that the soil load on the pipe is reduced. Three terminologies to take note of are as follows:

a. **Sheeting** includes the support materials in contact with the walls of the excavation.
b. **Bracing** refers to crosspieces extending from one side to the other.
c. **Rangers** are structural members which transfer the load from sheeting to braces.
d. **Shoring** refers to a temporary structure made from wood, metal, or other structural materials to support the work space, purposely installed to ensure the safety of workers and the site.
e. **Sheeting** refers to the process of holding a polling board or plank together or using a sheet instead of a plank.
f. **Poling board** is a flat wooden plank which is in direct contact with the soil of the trench and is arranged in a vertical position on the side is called a polling board or planking.

The methods of holding excavations include stay bracing, box sheeting, vertical sheeting, runners, and sheet piling.

a. **Stay bracing.** This type of timbering is used when soil is moderately firm and excavation depth is not more than 2 m. It consists of vertical boards placed against opposite walls of the trench and supported by two cross braces. This type of support is not very reliable and should be used only in shallow trenches in fairly stable materials. The spacing between vertical members depends on the soil. The vertical members should be at least 50 × 100 mm and the cross braces at least 100 × 100 mm. This type of excavation support is used when soil is moderately firm and excavation depth is not more than 2 m. Polling boards are placed on the sides of the trench and it is kept in position with one or two struts rows. These polling boards are placed at a distance of 3 to 4 m and are up to the full depth of the excavation. Width of excavation polling board is about 250 mm. Thickness of excavation polling board is about 40 to 50 mm.
b. **Polling boards** are placed on the sides of the trench and it is kept in position with one or two struts rows. These are short pieces of board placed against the walls of the trench and supported by rangers and cross bracing. The system may be constructed of random lengths of timber and it is not continuous over the wall. Poling boards are used in materials which will stand without support at depths of 1 to 1.5 m. The sheeting can thus be installed after the trench has been partially excavated and need not be driven ahead of the excavation. Sheeting in this method is usually 50-mm thick, the rangers 100 × 150 or 100 × 200 mm, and the braces 150 × 150 mm or 200 × 200 mm, depending on the width of the trench.[1]
c. **Vertical sheeting** is the strongest and most elaborate method employed in trench construction. It is commonly used in soft soils and in those where groundwater may be encountered. It can be made of different support material such as steel or timber/wood sheeting. The trench is first excavated as far as possible without the danger of bank failure. The first set of horizontal rangers is then placed about 300 mm below the ground surface against three planks which bear against the squared and trimmed sides of the trench at the ends and midpoint of the ranger. Rangers on opposite sides of the trench are cross braced and the sheeting is then driven vertically between the rangers and the wall of the trench. Figure 10.2 shows shoring the sides of an excavation by 'close vertical sheeting'.
d. **Box sheeting.** This employs both the horizontal sheeting and horizontal rangers. It is a system of ready-constructed units which is easily and quickly installed into the excavation using an excavator or a similar machine, providing reliable shoring. They can be moved along as work progresses. It is suitable for unconsolidated soils which are excavated in stages equal to the height of an individual board which is then placed and temporarily braced. It is used in loose soil and when depth of excavation is not more than 4 m. It is a type of loose soil timbering. A box-like structure is made using sheeting, wales, struts, and bracing. When three or four boards have been installed, the rangers and permanent bracing are put in position. This method may also be used for the upper 1 to 1.5 m of an excavation which employs vertical sheeting at lower levels.
e. **Runners.** Trench walls are prone to collapse in very soft or hard soil. A special type of shoe made of iron should be placed at the bottom of the polling board used in such circumstances and 0.3 to 0.4 m above the bottom surface during excavation; the planks are lowered to a greater depth with the help of shoe. These types of polling boards or sheets are called runners.

Construction and management 343

Figure 10.2 Shoring the sides of an excavation by 'close vertical sheeting'.

  f. **Sheet piling.** When excavating over a large area. Excavating soil and its surrounding soil loose and soft. Water comes during excavation of soil. When excavation depth is more than 10 m. Greater width of trench.

10.5 PIPE LAYING AND JOINTING

Pipe inspection is a must to ensure that the pipe has no leaks, sound enough and undamaged. Before the pipe is lowered into the trench, the grade of the bedding material must be checked with levels. The grade of the sewer should be held at least within 10 mm of that shown on the plans. For effective pipe laying, slings or rolling hitches are used to ensure that as the pipe is picked up and lowered into the trench, neither the pipe nor its gasketing material is damaged. Pipe sections are then placed on line and grade in the bottom of the dewatered trench and are pressed together with a hand lever or a winch. The gasketing material should be lubricated according to the manufacturer's instructions before the sections are joined. If the sections are to be joined by a ring, for example, with plain-end clay pipe, plastic pipe, asbestos, cement pipe, the ring may be installed on one end before the assembly is lowered into the trench. In case of the use of pipes with bell-mouth, the belled end faces upstream. The pipes are fitted and matched so that when laid in work, they form a culvert with a smooth uniform invert. Any pipe found defective or damaged during laying is removed.

    In laying pipes, the foundation must be cured and approved by the supervising engineer. The pipe should be inspected to ensure that it is sound and that the ends are undamaged. In places where two or more pipes are to be laid adjacent to each other, the pipes are usually separated by a distance of a minimum 450 mm or equal to half the diameter of the pipe. Proper care is taken while lifting, loading, unloading, and lowering of concrete pipes at a factory or site so that the pipes do not suffer any undue structural strain, any damage due to fall or impact. The lowering and positioning of pipe are done with the use of a tripod–pulley arrangement or simply by manual labor to minimize damage to the pipe. The laying of pipes is commenced from the outlet and proceed toward the inlet and be completed to the specified lines and grades. During construction, the contractor follows the links properly from the drawings and plans provided by the designer. It is always advisable that the contractor starts building the drainage sewer or drains from the outfalls so that when it rains, the runoff quickly routes to the discharge points without flooding the area. Each link has an inlet node and outlet node. These enable the contractor to trace the direction of flow of the stormwater. Each node has the maximum elevation which is the ground elevation.

Future household connections may be required and therefore, wye or tee sections are provided in the sewer network. These should be properly plugged and mortared shut. The trench should be filled as soon as the pipe has been placed and the installation inspected. In the case where class A (concrete) bedding is required, the backfill is delayed until the concrete has set up adequately to support its weight. Draining of the trench must be continued until the backfill is compacted.

In some cases, the installation of sewers below airfields, highways, railroads, etc., may be conducted without interference with surface traffic by jacking and boring. In this method, pipes are driven by hydraulic jacks mounted in a pit at one side of the obstacle. The jack and bore method, also known as horizontal auger bore method, has always been one of the most economical and reliable ways to bore under objects such as established highways and railroads. The jack and bore makes the task much simpler, while reducing labor costs and increasing productivity. The jack and bore or auger boring is executed with an auger boring machine by jacking a casing pipe through the earth while at the same time removing earth spoil from the casing by means of rotating auger inside the casing.

A cutting ring may be installed on the first section and lubrication fittings placed on subsequent sections. A cutter head operating ahead of the pipe opens the hole into which the pipe is jacked, while an auger operating on the same shaft as the cutter draws the excavated material through the pipe to the jacking pit. Jacking is generally done on a slight upgrade to keep the cutting face and pipe dry. The pipe should be kept in motion once the operation is begun. If the pipe remains stationary for any time, particularly under surfaces subject to vibratory loads, the soil may consolidate around the pipe, freezing it in position. It may then be necessary to jack and bore and bore from the other side to meet the frozen pipe.

The pipes could be joined by two methods, namely, collar joint or by flush joint. In collar joint, the collars can be of 150- to 200-mm wide RCC and have the same strength as the pipes to be jointed. The caulking space is in the range of 13 and 20 mm depending on the diameter of the pipe. The caulking material can be of 1:2 cement mortar, which is rammed with caulking irons. In the flush joint, the flush joint may be an external flush joint or an internal flush joint. In both cases, the ends of the pipes are specially shaped to form a self-centering joint with a jointing space 13-mm wide. The jointing space can be filled with 1:2 cement mortar, mixed sufficiently dry to remain in position when forced with a trowel or rammer. The pipes are jointed that the bitumen ring of one pipe shall set into the recess of the next pipe. The ring is thoroughly compressed by jacking or by any other suitable method. After finishing, the joint is kept covered and damp for at least four days.

Fill beneath the streets or other construction must be placed carefully, at optimum moisture content (OMC), with appropriate compaction until the ground surface is reached. In easements where the surface will be unused for traffic or other load-bearing purposes, the backfill above a level 600 mm above the crown need not be compacted. Fill material must be free of brush, debris, frozen material, and large rocks. No rock should be placed within 900 mm of the top of the pipe nor within 400 mm of the ground surface. The fill must be carefully placed in layers not more than 150-mm thick and be tamped under, around, and over the pipe to a height of 600 mm above the crown. Until this level has been reached, the earth should be placed very carefully. Backfilling with a bulldozer on a bare pipe will almost certainly displace it and may break it.

10.6 CONSTRUCTION AND ENVIRONMENTAL MANAGEMENT

Construction of drainage systems comes with lots of excavations and vegetation loss. The planning of drainage construction must account for environmental management, health and safety of workers, social aspects, and must be constructed to perform efficiently and effectively. To meet the overarching sustainability goal, the drainage system must be developed in an economic sense to avoid unnecessary reworks and waste.

10.6.1 Environment and social impact assessment for drainage systems

10.6.1.1 Introduction

In drainage systems development, an environment and social impact assessment (ESIA) is always conducted to:

a. Describe and provide a baseline of the drainage system for the city, municipal, or town and immediate environs, including the biophysical, social, and cultural aspects.
b. Identify all likely positive and negative environmental impacts due to the construction and future operations of drainage system.
c. Identify and evaluate all significant negative and positive environmental impacts and propose appropriate mitigation and reinforcement measures for incorporation into the contract documents and consequently into construction and operational phases.
d. Compile an environmental impact statement incorporating an environmental and social management plan for all aspects of the development plan for the drainage system.

10.6.1.2 Potential project benefits of drainage schemes

The opportunities (benefits) experienced from the proposed project are of economic, social, and environmental in nature and include, among others, the following:

a. Provision of employment opportunities during the construction phase.
b. Improved access boosting trade and commerce.
c. Reducing floods.
d. Reducing stormwater volumes through sustainable drainage systems (SuDS)
e. Absorbing greenhouse gases.

10.6.1.3 Potential project negative impacts

The potential impacts during the mobilization, construction, demobilization, and maintenance phases are as follows.

10.6.1.3.1 Potential negative impacts of mobilization phase

This phase entails mobilization of labor and equipment, acquisition of land for the contractor, and the engineer campsites and quarry sites.

10.6.1.3.1.1 LOSS OF VEGETATION
Clearance of the camp site areas if on a Greenfield may lead to loss of vegetation due to the activities for construction of the camp.

10.6.1.3.1.2 NUISANCE FROM DUST AND NOISE
The camp clearance and additional construction traffic in the area will lead to dust and noise pollution.

10.6.1.3.1.3 INCREASED INCIDENCE OF DISEASES
The community in surrounding area may be more prone to contracting diseases as they start to interact with some of the migrants who come in looking for jobs. The incident of communicable diseases such as HIV/AIDS may increase, especially in emerging economies.

10.6.1.3.2 Potential negative impacts of construction phase

The major construction activities include extraction and transportation of materials (e.g., gravel, sand, and water) and clearing the vegetation for the right-of-way. Other activities include formation of the excavations and construction of drainage structures. The potential negative impacts include, but are not limited to, the following.

10.6.1.3.2.1 TRAFFIC FLOW INTERRUPTION
Traffic flow cannot be wholly avoided in urban centers when drainage schemes are under construction. There is always interruption to traffic and vehicles diverted to use alternative routes. When a drainage scheme is expected to interrupt traffic, traffic calming measures must be provided throughout the town when drainage construction is progressing. Traffic/vehicle accidents involving human beings and animals along access roads are also possible.

10.6.1.3.2.2 AIR AND NOISE POLLUTION
During construction, the roads leading to the site might have a traffic growth, which is likely to affect existing levels of air pollutants, associated with traffic emissions. Therefore, following environmental good practices and guidelines to minimize adverse impacts is essential. The guidelines should be harmonized with various national legislative instruments currently in use and/or being developed.

Increased dust and noise generation from the construction traffic along the drainage network may lead to increased levels of respiratory diseases and hearing difficulties, respectively. Thus, the health and safety of the nearby communities and laborers may be impaired if proper mitigation measures are not put in place. Noise and dust disturbance may lead to vibrations affecting built-up structures by creating cracks in walls as well as affecting the sick. Lack of communication prior to project activities such as blasting close to settlements could lead to risks of heart attacks and shock-related problems. Exhaust fumes from stationary equipment as well as moving construction machinery and equipment can emit SO_2, NOx, CO, and CO_2.

Noise is another threat to the quality of the environment within the drainage corridors which may have residential buildings as well as sensitive receptors such as residential areas/ homesteads along the route. It is expected that there will be an increase of vehicles such as buses and cars moving on the roads both during the day and at night. These vehicles can generate high noise in the college areas. It is, therefore, important to include noise control, mitigation, and management measures in the ESMPs, which are fully harmonized with various legislative instruments that are currently in use and/or being developed. The air quality and noise and impacts can be minimized by careful consideration and decision-making and putting in place noise, dust, and waste management mitigation measures.

10.6.1.3.2.3 EROSION

Mass movements due to cuts during the excavations to create way for drainage structures and creation of access to material sites may cause sheet erosion. This may cause sedimentation of nearby watercourses.

10.6.1.3.2.4 CONTAMINATION OF WATER SOURCES (WATER POLLUTION)

Several communities, especially from emerging economies, rely on shallow wells and surface water pools for domestic water supply. Drainage construction activities could lead to siltation/ sedimentation of these water sources in areas where the main source of water is surface flows.

Drainage construction activities such as clearance of vegetation and excavation loosen soils that may be eroded as sediments fall into drains and gardens within the project site. Filling activities also cause sediments to end up in water. These sediments may alter the flow of water in these habitats with offsite impacts such as ending into Lake Victoria. This can pollute certain parts of water which may result into loss of plants/animals in those parts of the water and death of microorganisms that are water dependent.

Sediments into water systems resulting from construction activities including construction hydrocarbons (petrol, diesel, oils-lubricants) and their improper disposal into watercourses may result into their degradation. This may bring about diseases in the people depending on this water. Construction lubricants in water also reduce the amount of oxygen reaching these waters and consequently affect oxygen-dependent organisms.

10.6.1.3.2.5 INCREASED DISEASE PREVALENCE IN QUARRY AREAS

New quarry areas can be established while existing ones reopened and/or expanded. New quarry areas may affect current land use and vegetation and with long-term negative influence on the environment. The quarry areas, if not reinstated properly, can become potential sites for mosquitos and snail breeding posing a risk of increased incidents of malaria and bilharzia, respectively.

10.6.1.3.2.6 HIV/AIDS AND OTHER SEXUALLY TRANSMITTED DISEASES

During drainage construction, an influx of people into the project area could increase the spread of HIV/AIDS and other sexually transmitted diseases (STDs). It is good practice for the project to implement HIV/AIDS sensitization, gender awareness, and other social programs to take good health care of the workers including the adjacent communities around the project area to enhance safety and health.

10.6.1.3.2.7 INCREASED COMPETITION FOR WATER SOURCES
The construction activities may exert pressure on water sources for domestic and livestock consumption. Scarcities of water can create a conducive environment for some diseases.

10.6.1.3.2.8 FLOODING POTENTIAL
Due to blocking existing storm sewers, diverting runoff, and land use changes, creating diversions, and cofferdams, flooding may be inevitable. This primarily due to the high likelihood of distorting the natural drainage paths during construction.

10.6.1.3.3 Presence of environmentally sensitive areas

Environmentally/ecologically sensitive areas are prone to disturbance and are easily degraded in the course of drainage infrastructure development. In such areas, extra precautions must be taken to avoid significant environmental impacts. In many cases, this means extra investment in mitigation measures, while in some cases these areas may have to be avoided. Environmentally sensitive areas include gazetted forest reserves, wetlands (including swamps, open waters, peat land as defined by the Ramsar Convention on wetlands of international importance), desert habitats, national parks, and game reserves; cultural and historical sites and densely populated urban centers are also environmentally sensitive areas. It is essential to evaluate drainage routes to ensure that they don't severely destroy ecologically sensitive areas (swamps, streams, forests, etc.) that could easily be degraded if the drainage system is built or expanded.

10.6.1.3.3.1 IMPACTS DUE TO WASTE
Different types of wastes may be generated, both solid and liquid wastes. These must be properly managed to prevent environmental pollution, promote good sanitation practices, and prevent the spread of diseases.

10.6.1.3.3.2 LOSS OF FLORA AND FAUNA
Drainage construction needs land and the use of land can have direct impacts in terms of destruction of habitats and more subtle effects on biodiversity such as disturbance and fragmentation.

10.6.1.3.4 Resettlement of people

In some cases, drainage schemes may require resettling people from land so that drainage routes are given adequate land to achieve the desired hydraulic structure capacity. This may come with enormous social and environmental impacts such as loss of livelihoods and inadequate compensation in some areas. It is good practice for the engineers and planners to devise mechanisms to minimize resettlements as a result of drainage scheme developments. Loss of livelihoods of nearby residents should be minimized.

10.6.1.4 Environmental management system

Drainage systems are developed to have a minimum impact on the environment. They are always a part of the project developments such as buildings and roads or highways. The purpose of integrating SuDS into the overall drainage systems is to mimic the natural environment as much as possible so that the hydrology of the area is kept in balance. In that case, the area experiences minimum disturbance as a result of drainage activities. In the bid to have efficient and effective operation and maintenance plans, public authorities always prepare an environmental management system (EMS) which conforms to national and international standards. An EMS assists the organization in managing its significant environmental impact that arises from drainage and other activities. An example of an EMS is the ISO 14001.[2] This is issued by the International Organization for Standardization (ISO).

The ISO 14001 provides guidelines that any organization must follow to formulate a policy that takes into account the legislative requirements and information about environmental impacts. The standard is limited to environmental aspects that are directly under the administration of the organization and over which it can be expected to have an influence. The ISO 14001 Standard helps to reduce environmental impact and establish a sustainable growth path. An ISO 14001 certificate officially acknowledges the organization's commitment to environmental best practices. The certification can be used to instruct parties, clients, investors, or public authorities. The ISO 14001 does not explicitly provide environmental performance criteria, but an organization would be expected to adopt procedures considered as best practice. There are a number of ISO standards associated with ISO 14001 which give guidance on environmental management and audit:

a. **ISO 14004 Environmental management systems**—General guidelines on implementation give practicable help on the design, development, and maintenance of an EMS. It provides guidance for an organization on the establishment, implementation, maintenance, and improvement of a robust, credible, and reliable EMS. The guidance provided is intended for an organization seeking to manage its environmental responsibilities in a systematic manner that contributes to the environmental pillar of sustainability.
b. **ISO 19011 Guidelines for auditing management systems**. This document provides guidance on auditing management systems, including the principles of auditing, managing an audit program, and conducting management system audits, as well as guidance on the evaluation of competence of individuals involved in the audit process.[3] These activities include the individual(s) managing the audit program, auditors, and audit teams.

　　It is applicable to all organizations that need to plan and conduct internal or external audits of management systems or manage an audit program.

　　The application of this document to other types of audits is possible, provided that special consideration is given to the specific competence needed.
c. **ISO 14020 Environmental labels and declarations**—General principles.
d. **ISO 14021 Environmental labels and declarations**—Self-declared environmental claims (Type II environmental labeling).
e. **ISO 14024 Environmental labels and declarations**—Type I environmental labeling principles and procedures.
f. **ISO 14025 Environmental labels and declarations**—Type III environmental declarations —Principles and procedures.
g. **ISO 14031 Environmental management**—These are environmental performance evaluation guidelines that explain how to identify suitable environmental indicators for measuring performance against policies, objectives, and targets. This document

gives guidelines for the design and use of environmental performance evaluation (EPE) within an organization. It is applicable to all organizations, regardless of type, size, location, and complexity. This document does not establish environmental performance levels. It is not intended for use for the establishment of any other EMS conformity requirements. The guidance in this document can be used to support an organization's own approach to EPE including its commitments to compliance with legal and other requirements, the prevention of pollution and continual improvement, among others.

h. **ISO 14040 Environmental management**—Lifecycle assessment—Principles and framework.
i. **ISO 14044 Environmental management—Lifecycle assessment**—Requirements and guidelines.
j. **ISO/TR 14049 Environmental management—Lifecycle assessment**—Illustrative examples on how to apply ISO 14044 to goal and scope definition and inventory analysis.

10.6.1.5 Environment and social impact assessment

An Environment and social impact assessment (ESIA) allows the impact of a proposed development on all aspects of the environment and society to be investigated and recorded. This includes visual and noise impact, air pollution, economic implications for the location, and the effect on areas of land protected because they are in some way sensitive or enjoy special protection. The development of drainage systems has shortcomings on the environment. An ESIA must always be carried out to determine the likely outcomes of implementing the proposed drainage system.

In Uganda, a country that follows the UK standards, a scoping report is always submitted to the National Environment Management Authority (NEMA) prior to commencement of the ESIA assessment. The scoping report usually outlines the methodology and terms of reference that guided the ESIA team to conduct a thorough assessment and thus recommend the possible measures to conserve the environment through designing an ESMP for the entire project lifecycle, that is, from design to construction and operation and maintenance of the drainage system. Therefore, the project for drainage development or improvement must be consistent with government policy on rationalization and harmonization of public agencies and enhancing social and environmental sustainability.

In Uganda the environmental assessment must be guided by the applicable legal framework. In Uganda, in particular this is provided for by the National Environment Management Policy, 2018; the National Environment Act, No. 5 of 2019; and the Constitution of the Republic of Uganda, 1995. Also, this environmental and social assessment enables achievement of the goals of international treaties including the UN Convention on Biological Diversity (UNCBD), 1992 and contributes to the achievement of the UN sustainable development goals (UN SDGs) and Uganda's Green Growth Development Strategy.

The ESIA usually covers:

a. Project description and justification.
b. Policy legal and administrative framework.
c. Description of the project environment.
d. Project alternatives.
e. Potential impacts and mitigation/enhancement measures.
f. Environmental and social management plan.
g. Monitoring program.

h. Public consultations and disclosure.
i. Conclusion and recommendations.

10.6.1.5.1 Scope of environmental activities

The scope of implementing drainage systems warrants the following activities:

a. Removal of structures and vegetation located in the drainage corridor to commodate the proposed drainage structures and associated elements.
b. Construction of longitudinal drains and sewers over most of the length requiring excavation and also construction of cross drainage structures like culverts.
c. Provision of temporary crossings and traffic diversions during construction.
d. Construction of temporary facilities for the workers and equipment.

The above works facilitate attaining the project goals and on the other hand impact on the physical, biological, and social environments of the project area.

10.6.1.6 Environmental screening

A mandatory review of all proposed investment subprojects undertaken with the purpose of categorizing them by expected environmental risks and impacts, filtering out proposals grossly detrimental for the environment, and determining appropriate extent and type of environmental assessment to be applied to the investment subprojects accepted for further processing.

In order to ensure sustainability of the drainage systems, the initial environmental assessment (screening) is always done. The screening covers issues including environmental unknowns and recommend appropriate mitigation measures. The screening and environmental impact process determines how and when a specific activity will trigger a given environmental safeguard or policy, and what mitigation measures need to be put in place to guide the implementation of the project.

The screening process ensures that the project's activities with potentially significant impacts will require further full environmental impact study (EIA) whose scale will be appropriate to the planned activities. The detailed EIA process follows guidelines for EIA. The latter guidelines outline sector-specific EIA needs, categorizing infrastructure projects into those that may require full and mandatory EIA to be conducted before implementation.

10.6.1.6.1 Overall impacts assessment

The ESIA is part of the project, based on the project design, in order to anticipate possible environmental conflicts, and undertake mitigation measures for avoiding permanent negative impacts of the project on the environment. There are several potential impacts associated with activities undertaken during implementation and operation of the drainage systems. The potential impacts are classified into temporary or transient and permanent impacts depending on the influence period of the impact. The impacts are also examined under two categories: negative and positive environmental impacts. The various impacts in these two categories are then examined in order of their level of importance and significance. They are also examined in categories of their time of occurrence. Table 10.1 shows the *potential* environmental and social impacts due to project location, design process, equipment, during

Table 10.1 Potential environmental and social impacts due to project location, design process, equipment, during construction, operation and maintenance in reference to drainage and stormwater works

#	Project activities	Project phase	Nature of impact Direct	Nature of impact Indirect	Permanency of impact Reversible	Permanency of impact Irreversible	Magnitude of the impact H	Magnitude of the impact M	Magnitude of the impact L	Duration of impact N	Duration of impact L	Duration of impact M	Duration of impact S	Overall Impact	Mitigation approach
Environmental and social impacts due to location of the project site
3	Generation of spoil	Construction	Direct		Reversible			M				M		X	Usually included in works Contract
5	Soil erosion	Construction	Direct		Reversible				L			M		X	As per management plan (Table 10.2)
6	Noise pollution	Construction & operation	Direct		Reversible				L			M		X	As per management plan (Table 10.2)
7	Disposal of construction wastes	Construction & operation	Direct		Reversible			M				M		XX	As per management plan (Table 10.2)
11	Disturbance to the public (noise, workers)	Construction	Direct		Reversible			M					S	XX	As per management plan (Table 10.2)
12	Uptake of land areas	Construction	Direct		Reversible			M				M		X	As per management plan (Table 10.2)
13	Air quality issues from construction works	Construction	Direct		Reversible				L			M		X	As per management plan (Table 10.2)
14	Loss of aesthetic value	Construction	Direct		Reversible			M				M		X	As per management plan (Table 10.2)
16	Surface runoff and waste water	Construction & operation	Direct		Reversible		H				L			XX	Embankment, re-vegetation, Proper drainage systems
18	Dust emission	Construction	Direct		Reversible			M					S	X	Watering of area Dust masks to workers,
Environmental and social impacts due to design
19	Increase in project running costs	Construction & operation	Direct		Reversible			M				M		X	Budget for contingencies
Environmental and social impacts due to mobilization of personnel and equipment
21	Traffic flow impairment	Construction & operation	Direct		Reversible			M				M		X	Traffic management plan

Construction and management 353

22	Waste management concerns especially solid and effluent in the camp site	Construction	Direct	Reversible	M	L	XXX	Embedded in works Contract
23	Occupational safety and health concerns for the workforce	Construction & operation	Indirect	Irreversible	H	L	XXX	Health and safety management plan
24	HIV/AIDS	Construction	Indirect	Irreversible	M	L	XXX	Education awareness
25	Gender mainstreaming		Indirect	Reversible	L	S	X	Checking for compliance with legislations
27	Hazardous waste during operation	Operation phase	Direct	Reversible	H	L	XXX	Develop appropriate disposal plan

Environmental and social impacts due to demolition of existing buildings

28	Disruption of public utility services		Direct	Reversible	H	L	XXX	Obtain utility maps from authorities
29	Air pollution		Direct	Irreversible		L	X	As per management plan (Table 10.2)
30	Improper handling and disposal of solid wastes		Direct	Reversible	M	L	X	As per management plan (Table 10.2)
31	Physical hazards and accidents		Direct	Reversible	M	L	X	Health, safety, security and environment (HSSE) plan

Environmental and social impacts due to project construction

32	Loss of flora and fauna		Direct	Irreversible	M	L	X	As per management plan (Table 10.2)
35	Traffic flow impairment		Direct	Reversible	H	L	XXX	Traffic management plan
36	Inadequate occupational safety and health for workers		Direct	Reversible	H	L	XXX	Training & awareness
37	Accidents and injury		Direct	Reversible	H	L	XXX	(Health, safety, security and environment) HSSE Plan

(continued)

354 Integrated Drainage Systems Planning and Design for Municipal Engineers

Table 10.1 Cont.

#	Project activities	Project phase	Nature of impact		Permanency of impact		Magnitude of the impact				Duration of impact				Overall Impact	Mitigation approach
			Direct	Indirect	Reversible	Irreversible	H	M	L	N	L	M	S			
38	Sourcing and transport of materials		Direct		Reversible			M					S	X		As per management plan (Table 10.2)
39	Noise and vibration nuisance		Direct		Reversible			M					S	X		As per management plan (Table 10.2)
40	Air pollution		Direct			Irreversible										
41	Improper handling and disposal of solid wastes		Direct		Reversible			M				L			XXX	Management Plan
42	Disruption of public utilities		Direct		Reversible		H								XXX	Obtain utility maps from authorities
43	Noncompliance with the site instructions and recommendations		Direct		Reversible			M				L			XXX	Site meetings and compliance

Environmental and social impacts due to project operation

44	Increase in volume of traffic		Direct		Reversible			M					S	X		As per management plan (Table 10.2)
45	Improper handling and disposal of solid wastes		Direct		Reversible			M				L			XXX	Disposal management plan
46	Improper occupational health and safety provisions		Direct		Reversible			M				L			XXX	
47	Risk of fire		Direct		Reversible			M					S	X		HSSE Plan
48	Damage to the buildings and/or loss of lives		Direct		Reversible			M					S	X		HSSE Plan

Construction and management 355

construction, and operation and maintenance in reference to drainage and stormwater works. It provides the list of those impacts that may be reversible and those that may be irreversible.

The overall impact of an activity of the project is established based on a combination of considerations such as magnitude of its impact, duration, and their permanency. These are related on a continuous scale between extremes of very large positive impacts and very large negative impacts. Due to drainage works, areas likely to be affected by little impacts are always defined as overall impact assessment of minimal/no impact or small negative impact depending on the specific characteristics as summarized:

Scale		Narrative
++++	:	Very large positive
+++	:	Large positive
++	:	Medium positive
+	:	Small positive
0	:	Minimal/no impact
x	:	Small negative
Xx	:	Medium negative
Xxx	:	Large negative
xxxx	:	Very large negative

10.6.2 Environmental and social management plan

The necessary objectives, activities, mitigation measures, and allocation of costs and responsibilities pertaining to prevention, minimization, and monitoring of significant negative impacts and maximization of positive impacts associated with the construction and operational phases of the drainage development projects are outlined in Table 10.2.

10.6.2.1 Construction phase

Table 10.2 gives the potential environmental management plan for the construction phase.

10.6.2.2 Operation, repair and maintenance phase

Expected negative impacts	Recommended mitigation measures	Responsible party	Time frame
Effects on water resources	a) Limiting vegetation clearance to only the areas where it is absolutely necessary b) Putting erosion controls mechanisms in place c) Avoiding disturbing areas close to wetlands as much as possible d) Replanting areas cleared of vegetation and not paved after the project's construction phase with species indeginous to the site.	Operation and maintenance team/client	During repairs, and maintenance works, (periodic and routine)

Table 10.2 Detailed environment management plan for the construction phase of a drainage system

Expected negative impacts	Recommended mitigation measures	Responsible party	Time frame
High demand of construction materials	a. Ensure accurate estimation of actual construction material requirements. b. Ensure reinstatement of all material sources as applicable or conversion to positive use before project close. c. Conduct regular consultative meetings with stakeholders.	Contractor	Throughout construction period
Vegetation and animal disturbance	Avoid disturbance of areas not be used for project development; design and implement an appropriate landscaping program to help in revegetating affected areas the project area after construction.	Contractor, environment officer, local council leaders, and the community members	Throughout construction period
Increased stormwater, runoff, and soil erosion	a. A stormwater management plan that minimizes impervious area infiltration by use of recharge areas and use of detention and/or retention with graduated outlet control structure is recommended. b. Site excavation works to be planned such that a section is completed and rehabilitated before another section begins. c. The contractor should start constructing the drainage system from the outfalls so that when it rains, the runoff can safely route off without flooding the working area and excavated trenches.	Contractor/public authority such as a municipal council.	Throughout construction period
Dust emission	a. Ensure strict enforcement of on-site speed limit regulations, sensitize truck drivers to avoid unnecessary racing of vehicles. b. Sprinkle water on sites and access routes when necessary to reduce dust generation by construction vehicles and activities. c. Provide personal protective equipment, e.g., dust masks for those working in dusty areas.	Contractor	Throughout construction period
Noise and vibration	a. Sensitize construction vehicle drivers and machine operators to switch off engines when idle. b. Sensitize construction drivers to avoid unnecessary hooting especially when passing through sensitive areas such as mosques, schools, residential areas, markets and hospitals.	Contractor	Throughout Construction period

Occupational health and safety risks	a. Ensure that all plans and equipment to be used are approved by the relevant authority and the local Occupational Health and Safety Office. b. Ensure that provisions for reporting incidents, accidents, and dangerous occurrences during construction using prescribed forms obtainable from the local Occupational Health and Safety Office are in place.	Contractor; project health safety officer, and municipal environment officer	Throughout construction period
Incidents, accidents and dangerous occurrences	a. Arrangements must be in place to train and supervise inexperienced workers regarding construction machinery use and other procedures/operations. b. Equipment such as fire extinguishers must be examined by a government authorized person. The equipment should only be used if a certificate of examination has been issued.	Contractor; project health safety officer.	Throughout construction period
Possible exposure of workers to diseases including infectious diseases such as HIV/AIDs	a. Collaborate with other players in community training and sensitization on disease control during construction. b. Provide counseling and testing for HIV/AIDS to incoming construction personnel. c. Provide easy access to condoms.	Contractor; project health safety officer, and client.	Continuous
Insecurity	a. Coordinate with local authorities and the district administration to appoint security personnel operating 24 hours where needed. b. Ensure only authorized personnel get to the sites.	Contractor; community leadership, and contracted security body.	Continuous
Air emissions	a) Complying with the air quality regulations and emission standards. b) Using cleaner energy sources and promoting their use. c) Limiting land conversion to only necessary areas. d) Managing wastes according to regulations in addition to employing the 3Rs (reduce, reuse, and recycle) to ensure they are managed sustainably. e) Developing complementary waste management facilities including a sanitary landfill and a hazardous waste disposal facility. f) Integrating lifecycle assessments into the project level environmental risk management.	Contractor	Continuous

(continued)

Table 10.2 Cont.

Expected negative impacts	Recommended mitigation measures	Responsible party	Time frame
Effects on water resources	a) Limiting vegetation clearance to only the areas where it is absolutely necessary. b) Putting erosion control mechanisms in place. c) Avoiding areas close to wetlands as much as possible. d) Replanting areas cleared of vegetation and not paved after the project's construction phase with species indeginous to the site.	Contractor	Throughout construction period
Habitat alteration and biodiversity impacts (flora and fauna)	a) Forested/tree areas should be avoided. Where they cannot be avoided, then vegetation clearance would have to be limited to the area that is absolutely necessary. b) Forested/tree areas that are the major habitat to wildlife (mammals, reptiles, amphibians) should as much as possible be avoided. Where they cannot be avoided, the following should be done. c) Include pedestrian underpasses into the project designs. d) Aligning new drainage infrastructure with existing right-of-ways (RoWs) or defined corridors. e) Limiting the size of construction as per design and RoWs where possible. f) Complying with existing land use and PA management plans.	Contractor	Throughout construction period
Loss of forest and agricultural land	a) If the location is already a built-up area, try as much as possible to avoid traversing ecosystems such as grass areas. Where they cannot be avoided, then pedestrian crossings should be incorporated into project designs. b) Avoid cutting across shrubs/grass areas that are a major habitat for wildlife.	Client and designer/consultant.	Throughout construction period
Landscape modification	a) Locating drainage infrastructure with existing land use plans. b) Complying with ecosystem management plans of protected areas. c) Limiting the size of construction RoWs. d) Avoiding protected areas and settlements.	Contractor	Throughout construction period

Soil erosion	a) Implementing runoff and water management measures. b) Road drainage to be put in place. c) Strategies to reduce soil erosion prioritized. d) Limiting excavations to only necessary areas. e) Implementing soil conservation strategies in areas with high soil erosion potential. f) Complying with waste management regulations. g) Complying with regulations and guidelines on soil conservation such as those provided by land use plan, ecosystem management plans, and those gazetted by the Agriculture and Food Authority.	Contractor	Continuous
Impacts due to waste	a) Strategies to waste management should be prioritized.	Contractor	Throughout construction period
Vibrations and noise	a) Keeping noise levels to the minimum possible. b) Regular servicing of machinery and vehicles likely to generate noise if ill-serviced.	Contractor	Throughout construction period
Disposal of excavated loose material	Managing wastes (collection, transport, and disposal) in accordance to the provisions of the waste management statutes providing adequate equipment and facilities to do so,	Contractor	Throughout construction period
Dust and other emissions	a) Wetting of dusty surfaces as much as possible. b) Regular servicing of vehicles and machinery.	Contractor	Throughout construction period
Occupational health, safety, gender, and other social issues	a) Develop and implement safety, traffic management, HIV/AIDS, and gender plans.	Contractor	Continuous
Natural resource demand	a) Integrating lifecycle assessment into the project level environmental risk management, complying with regulations governing resource extraction. b) Ensuring building materials are sourced from sustainable sources. c) Implementing demand management and resource efficiency measures for water, electricity, and materials.	Contractor	Continuous

(*continued*)

Construction and management 359

Table 10.2 Cont.

Expected negative impacts	Recommended mitigation measures	Responsible party	Time frame
Spills	a) Developing project level emergency/disaster preparedness and response plans as well as spill contingency plans and providing resources to respond to spills. b) Complying with the National Oil Spill Response Plan by integrating its provisions to any project level contingency plans. c) Implementing and supporting monitoring programs at the project level. d) Providing training to project staff on spill prevention and management. e) Adhering to industry guidelines on the design and maintenance of any fluid storage, loading and conveyance equipment, and infrastructure.	Contractor	Continuous
Accidents	a) Complying with industry guidelines and regulations in the design of infrastructure. b) Implementing and supporting programs to ensure vehicles are maintained to regulatory approved standards. c) Developing project level emergency/disaster preparedness and response plans.	Contractor and client	Continuous
Natural hazard	a) Integrating considerations for seismicity in the engineering design of Drainage infrastructure in seismically active areas. b) Integrating the provisions of the National Disaster Response Plan into project level disaster/emergency preparedness and response plans and coordinating with the pertinent authorities.	Client and contractor	Continuous

Construction and management 361

Loss of cultural heritage	a) The presence of cultural heritage assets would need to be confirmed in detailed studies associated with each potential project; this could influence the design and location of drainage infrastructure. b) Incorporation of heritage sites into drainage master plans as a way of preserving such sites. c) Projects associated with the construction and operation of drainage infrastructure should be subject to environmental and social impact assessment (ESIA) commensurate with the scale of the project and impacts which includes consideration of cultural heritage and the development of appropriate mitigation and management plans. d) In terms of locally important cultural heritage sites, any loss or alteration to such sites should be consulted on, and agreed, with the local communities and the custodians of the site. If necessary, appropriate rituals should be undertaken to move the cultural asset, or to otherwise expiate disturbance or loss of the site. e) A framework 'chance finds procedure' should be developed to support the drainage system which involves and references all relevant ministries, other agencies, and major cultural heritage stakeholders in the country. The framework procedure should be deployed within the construction and management planning for all developments implemented under the auspices of the plan.	Client and contractor	Continuous
Livelihood	Institutional strengthening and capacity building for agencies across the project who are responsible for promoting and coordinating commercial developments to ensure that social risks are adequately understood and addressed through mitigation.	Client and contractor	Continuous
Rural–urban migration	a) Drainage infrastructure projects should be subject to ESIA undertaken in line with international standards such as those of the World Bank or the International Finance Corporation (IFC). The scope of the ESIA should always include consideration of rural–urban migration. b) Urban development plans should also cater for rural–urban migration.	Client	Continuous

(*continued*)

Table 10.2 Cont.

Expected negative impacts	Recommended mitigation measures	Responsible party	Time frame
Public health	a) Drainage infrastructure projects should be subject to ESIA undertaken in line with international standards such as those of the World Bank or the IFC. The scope of the ESIA should always include consideration of health-related impacts. For large projects this may require that appropriately qualified international experts are appointed to address impacts on health. b) The development and implementation of HIV/AIDS/malaria policies and information documents for all workers directly related to drainage projects. The information document addresses factual health issues as well as behavior change issues around the transmission and infection of HIV/AIDS as well as malaria. c) All projects should have a *worker code of conduct* for all project personnel that include guidelines on worker–worker interactions, worker–community interactions, and development of personal relationships with members of the local communities. As part of the worker code of conduct, all project personnel should be prohibited from engaging in illegal activities including the use of commercial sex workers and transactional sex. Anyone caught engaging in illegal activities should be subject to disciplinary proceedings. If workers are found to be in contravention of the code of conduct, which they are required to sign at the commencement of their contract, they would face disciplinary procedures that could result in dismissal. d) Working in conjunction with relevant partners (e.g., health authorities, NGOs, development agencies), information, education, and communication campaigns around diseases and health practices should be developed as part of the drainage infrastructure implementation.	Client and contractor	Continuous
Land use and settlement pattern	Institutions responsible for land use plans should ensure that they contain measures relating to infrastructure provision that are robust and fit for purpose, with a focus on the poorest and most vulnerable communities.	Client and contractor	Continuous

Expected negative impacts	Recommended mitigation measures	Responsible party	Time frame
Dust and other emissions	a) Wetting of dusty surfaces as much as possible b) Regular servicing of vehicles and machinery	Operation and maintenance team/client	Throughout construction routine/periodic maintenance time frame

10.6.2.3 Decommissioning phase

In addition to the mitigation measures provided in Table 10.3, it is necessary to outline some basic mitigation measures that may be required to be undertaken once all operational activities of the project have ceased. The necessary objectives, mitigation measures, allocation of responsibilities, time frames pertaining to prevention, minimization, and monitoring of all potential impacts associated with the decommissioning and closure phase of the drainage project are outlined in Table 10.3.

Table 10.3 Environmental management/monitoring plan for the decommissioning phase

Recommended mitigation measures	Responsible party		Time frame
1. Demolition of waste management			
All machinery, equipment, structures, and partitions that shall not be used for other purposes must be removed and recycled/reused as much as possible.	Community, environment/ resource office	Client/public authority	After decommissioning phase & continuous
Donate reusable demolition waste/equipment to charitable organizations, individuals, and institutions.	Contractor		After decommissioning phase & continuous
2. Rehabilitation of project site			
Implement an appropriate revegetation program to restore the site to its original status.	Community, environment/ resource office	Client/public authority	After decommissioning phase & continuous
Consider use of indigenous plant species in revegetation.	Community, environment/ resource office	Client/public authority	After decommissioning phase & continuous
Indigenous tree species should be planted at suitable locations so as to interrupt driveways and the development plan.	Community, environment/ resource office	Client/public authority	After decommissioning phase & continuous

10.6.3 Stakeholder's responsibilities

10.6.3.1 Contractor's project manager

a. The project manager has the ultimate responsibility for implementation of ESMP.
b. The project manager is responsible to ensure staff are adequately inducted and trained at site regarding environmental and social management including emergency procedures. The same applies to subcontractors.
c. Overall overseer on the contractor's side for the implementation of ESMP.

10.6.3.2 Contractor's environmental and social manager

a. Develops, implements, and reviews environmental management systems and plans.
b. Provides leadership to ensure all contractor's staff comply with ESMP.
c. Works with site engineer to develop site-specific environmental plans.
d. Notifies the engineers' environmental specialist of any noncompliance.
e. Responsible for reporting major defects and noncompliances and arranging for appropriate corrective actions.
f. Initiates and coordinates monitoring and auditing.
g. Monitors the effectiveness of environmental management plan.
h. Trains contractor's staff in environmental objectives and procedures.
i. Designs site-specific environmental and social plans in collaboration with contractor's environmental and social manager, site engineer, and other subcontractors.
j. Conducts monitoring and auditing and maintains relevant records.
k. Conducts daily/weekly site inspections of measuring devices.
l. Monitors and carries out routine maintenance of measuring facilities and to ensure that they are in good working conditions.
m. Ensures staff on-site adhere to laid down environmental requirements at all times.

10.6.3.3 Contractor's site supervisor

a. Ensures environmental works are implemented and maintained.
b. Leads the emergency response crew with advice from the environmental manager.

10.6.3.4 Contractor's staff

a. Responsible for reporting incidents, defects, and other problem areas to senior site staff as they arise on-site. Special forms are used for all incident reporting.
b. Carry out routine maintenance and emergency work when directed.
c. Care for all environmental works.
d. Ensure the site is kept tidy and litter is placed in bins.
e. Act in an environmentally responsible manner at all times to reflect the contractor's commitment to environmentally responsible environmental practices.

10.6.4 ESIA conclusion and recommendations

10.6.4.1 Conclusion

The drainage development project presents the impacts of implementing the drainage system and also includes an ESMP to be implemented by the developer and stakeholders and proposes social interventions and mitigation measures. The following are the conclusions as far as sustainability is concerned going forward:

a. That the development does not trigger any policies on involuntary resettlement.
b. Erosion and sediment control during construction is key.
c. No cultural sites should be traversed by the drainage system.
d. Periodic monitoring is recommended as per environmental management and monitoring plan by the engineer and the environment officer to ensure that environmental issues are adequately addressed during the construction phase.

10.6.4.2 General recommendations

Environmental and socioeconomic management issues at various stages in the life of the drainage project from detailed design through decommissioning are governed or guided by a number of 'standards', including those contained in national legislation. The environmental impact study must provide a detailed presentation of the likely impacts that could be associated with the drainage project in the short and long term. The following are generally recommended:

a. **Management systems:** The subsidiary plans that should be developed to support the environmental management plan include ambient air quality and emissions monitoring plan; construction site management plan; conservation management plan; spill prevention control, containment and emergency response plan; operational discharge management plan; integrated waste management plan; vehicle and traffic management plan; and employee and subcontractor training plans.
b. **Resource allocation:** The relevant human and financial resources should be effectively planned during the project lifecycle.
c. **Effective site management** is crucial in the mitigation measures proposed in the environment impact statement (EIS) and must be implemented. In this regard, construction plan incorporating erosion best management practices, health and safety, traffic, air, and water quality strategies should be developed and implemented. All effluent and sewerage, as the case may be, should be appropriately handled and in accordance with prescribed regulations.
d. **Operational measures:** Public involvement and stakeholder coordination are crucial for project planning and implementation. This helps capture new environmental concerns during the implementation and in decision making that can be used to update the current plans and designs. Regular meetings with stakeholders regarding land use planning, environmental management, civil works, and labor help improve decision making and avoid delays in the construction period.

10.7 SAFETY IN DRAINAGE CONSTRUCTION

10.7.1 Introduction

Safety in drainage construction primarily deals with excavations and confined spaces; the hazards and precautions required to ensure safe work commences and progresses at or in excavations and confined spaces. This section identifies hazards and risks associated with those hazards in drainage systems planning, design, and construction. It also provides methods to control and manage the risks.

The section further provides the hierarchy of preventive and protective measures in drainage systems planning, design, and construction emphasizing the fact that safety and health are integral parts in the design and development of solutions to solve a flooding or drainage problem but not an add-on.

Excavations that lead to enormous amount of earthworks present a highly likely dangerous environment for the construction workforce. The fact that drainage construction primarily comes with longitudinal trench excavations, which can significantly vary in width and depth depending on the size and complexity of the scheme, it presents a potentially dangerous environment for both construction workers and nearby communities. Several hazards are usually associated with drainage sewer excavations which include the following:

a. Movement of earthmoving and other plant, particularly at speed and reversing.
b. Blasting operations.
c. Collapse of unsupported excavations.
d. Soft compressible soils.
e. Asphyxiating gases in confined spaces.
f. Slips, trips, and falls on uneven ground.
g. Falling objects from overhead plant and demolition/construction activities.
h. Groundwater ingress into excavations.
i. Damage to (and from) public utilities such as electricity, gas, and water mains.
j. Contaminants that may be found on brownfield sites.
k. Chemical additives for earthworks conditioning, for example, quicklime and cement.
l. Plant falling into excavations.

In the UK, through the Construction (Design and Management) Regulations 2007 (CDM regulations (2007)) and Health and Safety Executive's (HSE's) Approved Code of Practice (ACoP), L144, Managing Health and Safety in Construction gives practical advice on how the law is to be complied with.[4] It states that if you follow the advice given, you will be doing enough to comply with the law as far as the CDM regulations are concerned. The contractor, client, and contractor's workers all have a role to play to ensure that the project commences and progresses safely. Subject to some transitional provisions, CDM 2015 replaced CDM 2007 from April 6, 2015. From this date, the ACoP), L144, which provided supporting guidance on CDM 2007 was withdrawn.

10.7.2 Hierarchy of preventive and protective measures in drainage system planning, design, and construction

As part of the risk assessment process, a drainage systems planner or designer needs to follow the general hierarchy of preventive and protective measures. The hierarchy of controls as presented in standard ISO 45001 can be seen as a typical system: eliminate the hazard; substitute with less hazardous materials, processes, operations, or equipment; use engineering controls; use administrative controls; and provide and ensure use of adequate personal protective equipment (PPE). ISO 45001 is an ISO standard for management systems of occupational health and safety (OHS), published in March 2018. The goal of ISO 45001 is the reduction of occupational injuries and diseases, including promoting and protecting physical and mental health.[5] The ISO 45001 general outline of hierarchy of controls is outlined below.

10.7.2.1 Elimination

This refers to complete removal of the hazard that creates a risk. For example, in drainage planning and design, the decision to take a given route may be completely dropped to adopt an alternative route in case there exists a **variance of underground utility networks.** During construction, cast-in-place concrete production may present a greater risk as result of manual handling of materials. Therefore, prefabricated units may be used to construct drainage schemes because they present a lower risk. These two examples clearly provide an insight that safety precautions should be an integral part of design and development of engineering solutions but not an add-on.

The risk may come from transportation of materials, weak soils risking collapsing, or sloping ground that needs a robust retaining wall. **Reinforced concrete structures** may have to be developed substituting for **flexible structures** in order to retain the soils before drainage structures are built to completely eliminate the risk of soils collapsing. In some instances, the sustainability goals may be well-supported with local materials such as cobblestones because of being economical, readily available, and lower transport costs associated with less emissions. However, the component of safety and reliability might favor concrete structures. This kind of engineering judgment is conducted in the evaluation process, Chapter 2, taking safety as an integral part but not an add-on.

10.7.2.2 Substitution

This refers to substitution of a hazard with another that creates a less risk. Construction materials may be substituted in case they pose a great risk. Some SuDS such as bioretention gardens and retention systems may become hazards when developed in the tropics because they can be breeding sites for misquotes presenting a risk of spreading malaria to communities. A proper SuDS may be selected based on the level of risk it presents in allowing mosquitoes to breed in Africa's tropics.

10.7.2.3 Engineering controls

These are engineering solutions that avoid exposure to the hazard. Drainage systems can be hazards with a significant level of risk during construction and after in case they are not properly developed. Engineering controls are designed so that the drainage systems perform reliably and safely. The main engineering controls employed in integrated drainage systems are **isolation or total enclosure** and **separation or segregation.**

- **Isolation or total enclosure**

The primary goal is to isolate the hazard so that people do not get exposed to it. For example:

a. Fencing wider channels.
b. Covering open channels with reinforced concrete slabs. Enhancing safety postconstruction activities needs channels to be covered with reinforced concrete slabs or other suitable material.
c. Installing kerb lines on road sides to safeguard people from falling into the open drains.
d. Use of grates to safeguard bicycle users and pedestrians while allowing stormwater to enter the sewer or drain.
e. Providing warning and prohibitory safety signs.
- **Separation or segregation**

This refers to placing the hazard in an inaccessible location. For example, relocation of services such as gas, telecom, electricity, and portable water pipelines to allow for a drainage route to pass. Also, footbridges are designed and developed across drainage systems that cross roads to segregate pedestrians from moving vehicles. See Figure 10.3 for a footbridge crossing a box culvert.

- **Partial enclosure**—this refers to partially enclosing the hazard.
- **Safety devices**—this refers to using devices that are considered safe with almost no risk.

Figure 10.3 Footbridge segregates pedestrians from moving vehicles.

10.7.2.4 Administrative controls

These are based on behaviors and procedures:

- **Safe systems of work.** These are important to prevent accidents and incidents. Safe systems of work define appropriate methods aimed at eliminating the hazard or minimizing the risk associated with the hazard. The method statements for constructing drainage systems must be sound enough providing safe systems of work to ensure safety. In some cases, complex drainage systems may require specialist contractors with permit-to-work systems as part of their safe systems of work. Micro-tunneling, cofferdam development, and huge excavations may call for specialized contractors. Therefore, any activity involving a significant level of risk needs a particular procedure which must be provided for in the safe system of work. Where hazards cannot be completely eliminated, and a significant level of risk remains, safe systems of work must be well defined.
- **Reduce exposure.** This refers to duration and frequency of exposure to a hazard with a significant level of risk. For example, a person who is exposed to a hazard once a week is at low risk compared to another exposed to the hazard six times a week. Also, a person exposed to a hazard 3 hours a day 6 days a week is at a lower risk compared to a person who is exposed to a hazard 6 hours a day 2 days a week.

10.7.2.5 Personal protective equipment

These are used primarily during construction, operation, and maintenance of drainage schemes, for example, hard hats, hand gloves, reflectors, safety boots, and sunglasses.

10.7.3 Excavations

Excavation present several hazards such as striking buried services, falls and falling objects, collapse of adjacent structures, collapse of sides, flooding, contaminated ground, toxic and asphyxiating atmospheres, and mechanical hazards. Consideration should be given to overhead hazards including power lines. Risk assessment factors to consider are depth, type of soil, type of work, use of mechanical equipment, proximity of roadways and structures, the presence of the public, and weather conditions. Risk assessment must be conducted before conducting excavations. Risk assessments should be carried out in accordance with the requirements of national regulations, with due consideration to the excavation as a construction activity.

10.7.3.1 Hazards and risk assessment for excavation work

The hazards associated with excavation work include the following:

a. **Mechanical hazards.** These primarily arise from the use of plant and equipment around or in the excavation. Hazards include vibration from plant operating may cause collapse. The excavating machinery such as excavators may create hazards, such as striking persons in or around excavations.

b. **Water ingress.** This includes surface water during heavy rain or snow. Water ingress can also result from groundwater (a high-area water table), nearby rivers, streams and watercourses (especially if breached), or a burst water main caused by the excavation activities.
c. **Contaminated ground.** On sites that previously housed chemical works or storage areas. Containing methane or hydrogen sulphide gas (both from microbial decay). Contaminants can be varied in range and include, for example, areas well-known for petrochemicals may be possibly contaminated by hydrocarbons, benzene, phenol, acids, alkalis. Areas well-known for power stations or gasworks may be possibly contaminated by coal, sulphur, phenol, asbestos, or cyanides.
d. **Buried services.** These hazards include striking services such as high-voltage electricity cables, gas pipes, water mains, or other buried services (e.g., telephone and cable TV lines). This can lead to electric shock, arcing, burns and fire, or gas explosion or rapid flooding of the excavation, as well as major business disruption to service users in the locale.
e. **Collapse of sides.** When the unsupported sides of an excavation slip and cave in (often due to poor support systems of the excavation sides). Severe crush injuries can result from even relatively small collapses because soil is very heavy, especially when wet. Workers buried or trapped in soil can asphyxiate in minutes and do not have to be completely buried for this to occur, being buried up to the chest can lock the rib cage and have the same effect.
f. **Objects and materials falling in excavations.** Tools or materials (bricks, timber, etc.) falling into an excavation or onto persons from an unprotected edge. This can also mean vehicles driving too close to the side of an excavation, collapsing the sides or tipping in. Spoil (loose soil) or stacked loose sand piled too close to the sides of an excavation may collapse. Adjacent structures (e.g., wall, bridge, or roads) undermined by an excavation and collapsing in.
g. **Collapse of adjacent structures.** This refers to digging too close to, or under, the foundations which support nearby buildings or structures which may undermine their support and cause collapse of the building or structure into the excavation. This would be even more dangerous if the excavation itself were also to collapse.
h. **People falling in trenches** because of an unfenced edge or while climbing in or out from ladders or other access equipment.
i. **Cofferdams and caissons.** These are structures (usually watertight enclosures) which are pumped dry to allow work to be carried out inside them below the waterline on bridge building, etc. They must be suitably designed and constructed to prevent the ingress of water (or other materials), or appropriately equipped to pump out water and provide shelter and escape should water or materials enter it.
j. **Toxic and asphyxiating atmospheres.** This arises from contaminated grounds and from gases used on-site. Gases heavier than air, gas such as liquid petroleum gas (LPG) and carbon dioxide can infiltrate an excavation. The combustion gases from nearby construction equipment such as diesel generators and motor vehicles can seep into excavations with the same effect.
k. **Overhead hazards including power lines.** Before excavation work starts, all overhead services should be identified, and any diversions or disconnections ensured before

excavation work begins. Service providers should be contacted to obtain accurate plans of supplies. As mentioned in the evaluation process (Chapter 2), Section 2.4.2, utility maps are essential components of planning. Three situations arise in construction work at overhead power lines:
- Excavation plant and equipment will pass beneath the power lines.
- No scheduled excavation work or passage of plant to take place under the lines.
- Excavation work will take place beneath the power lines.

10.7.3.2 Control measures for excavation work

Precautions must be taken to prevent people falling into, or being injured while working in, excavations. In common with other construction activities, control of the risks involved in excavation is based on effective site management. National and local regulations apply in this respect, and excavations must be carried out under the supervision of a competent workforce. Competent people conversant with national legislation must be deployed. For example, OSHA standards require that before any worker entry, that employers have a competent person inspect trenches daily and as conditions change to ensure elimination of excavation hazards. A competent person is an individual who is capable of identifying existing and predictable hazards or working conditions that are hazardous, unsanitary, or dangerous to workers, soil types, and protective systems required, and who is authorized to take prompt corrective measures to eliminate these hazards and conditions. An outline of control measures for excavation work is provided below:

10.7.3.2.1 Deploy safe digging elements and methods

Safe digging methods and precautions include the following:

a. Plan ahead.
b. Excavate alongside the service rather than directly above it. Final exposure of the service by horizontal digging is recommended, as the force applied to hand tools can be controlled more effectively.[6]
c. Hand digging should be employed when nearing the assumed line of the pipe or cable.
d. Insulated tools should be used when hand digging near electric cables.
e. Exposed cables and pipes should be supported and protected against damage by backfiling. They should never be used as hand-and-foot-holds.
f. Using locators to determine the position and route of pipes or cables (frequently using them during the course of the work).
g. If contact is made with any unidentified service pipe or cable, stop work until it is safe to proceed.
h. Regard all buried cables as live until disconnected and proven-pot-ended cables cannot be asumed to be dead or disused.
i. Excavators and power tools should not be used within 0.5 m of the indicated line of a cable/pipe.
j. Using spades and shovel (preferbly with curved adges) rather than other tools, for example, forks or picks.

k. Arrange a meeting with the facility locators and the excavator.
l. Keep a careful watch for evidence of pipes or cables. Remember that plastic pipes cannot be detected by normal locating equipment.
m. Report any damage to the appropriate agency and keeping personnel clear until it is reparied.
n. Having an emergency plan to deal with any damage to pipes or cables.

10.7.3.2.2 Supporting excavations

See Section 10.5. Excavation supports will prevent the collpse of the side walls of the excavation and allow work to continue uninturrupted. The type of support structure used will vary depending on the type of ground being excavated, length of time the excavation will be open and in use, type of work being caried out, groundwater conditions and potential for flooding, depth of the excavation, and number of people in the exavation. Supporting the sides of excavations to prevent collapse, and providing workers with suitable access and egress, and crossing points to pass over excavations is essential. Excavation supports must be inspected before each shift, and after any event that could affect the integrity of the excavation, and reports made and kept. Spoils (the ground dug from an excavation) to be removed to and stored at a safe distance from the excavation to prevent it collapsing back in.

10.7.3.2.3 Safe means of access

Excavations should only be crossed at designated points (Figure 10.4). The crossing point should be of sound construction and suitable to support all the types of vehicle and equipment likely to use it. Gangways across excavations should have gauradrails and toeboards.

Ladders are used as the main method for access to and egress from an excavation. They must be suitably secured to prevent undue movement and extend above the excavation to give the necesssary height required for a safe handhold (at least 1.05 m). Ladders must be kept in good condition, be fit for the purpose and of adequate strength. There must be adequate means of escape in an emergency: one ladder every 15 m is an average to work to (more may be required depending on the number of workers and the potential risk, e.g., where there might be a possibility of flooding from a rising water table.) Climbing in and out of the excavation using other means should be prohibited.

10.7.3.2.4 Installing barriers at the site

Barriers should be used to demarcate danger areas under excavations. These should be planned to consist of guardrails (as for a scaffold work platfom) to prevnt people falling in, and toeboards to prevent objects being kicked down into the excavation. In additon to prevent vehicles from falling in, logs or concrete blocks are laid some distance from the edge to act as a buffer. Excavations may need to be covered, especially at night. Such covers need to be capable of bearing a person's weight and be held securely in place. Fencing or hoarding may be required to protect members of the public as well as constuction workers and staffs.

Figure 10.4 Designated crossing point.

10.7.3.2.5 Installing lighting and warning and prohibitory signs

Lights and signs should be used to demarcate danger areas. Signs should be used to warn people of the excavation hazards, and any special precautions required. They should be placed in clearly visible spots at all potential access points. Appropriate lighting should be installed to ensure that there is an adequate level of illumination, without distracting shadows, to ensure safety of work activities both within the excavation and on the surface. High-powered electric lights or those which operate from LPG will be required for general workplace illumination, and consideration may need to be given to smaller, personal lights for individual workers. These should be battery-operated to avoid the risks associated with trailing electrical leads. When working on a roadway, the police or the local authority need to be consulted about traffic lights or stop/go signs. There are rules for the placing of cones to warn motorists of the hazard.

10.7.3.2.6 Controlling, positioning, and routing the movement of vehicles, plants, and equipment

Vehicles and materials on-site should be kept away from excavations to prevent them from falling into or collapsing the excavation.

10.7.3.2.7 Providing personal protective gear for workers

The workers in excavations should wear items of PPE such as safety helmets, safety footwear, respirators or breathing apparatus, and hearing protection.

10.7.3.2.8 Contaminated ground

Testing for contaminated ground and providing extra welfare facilities to accommodate workers (separate from normal site facilities). Health surveillance may be appropriate for contaminants such as asbestos, lead, or radioactive materials.
Water from excavations.

10.7.3.2.9 Dewatering excavations

Dewatering and freezing techniques used to remove water from excavations should be applied. See Section 10.4.

10.7.3.2.10 Detection, identification, and marking of buried services

Before work starts, the location and configuration of underground services should be identified. Buried services must be identified and safe digging methods adopted to avoid contact with them. It may well be possible to avoid cable routes at the planning stage of work. Before work commences, the following action should be taken:

   a. Check any available plans.
   b. Contact local services providers and owners for services such as gas, water, electricity, and telecommunications.
   c. The positions of known services should be marked on plans and also on the ground itself. All workers and staffs must receive adequate information and instruction about the nature of the risks.
   d. Survey the site and surrounding areas to identify indicators of the existence of cables, for example, streetlights or junction boxes.
   e. Use cable locators with trained operators. Plastic and nonmetallic underground services cannot be identified by conventional locators but could be identified by the use of metallic tracer wire laid with the pipe or by using a signal transmitter inserted and pushed along the pipe itself.
   f. Where appropriate, arrangements must be made with the services providers to isolate the cables/pipes and ensure that it is safe to work in the vicinity of them.

10.8 QUALITY CONTROL AND MANAGEMENT IN DRAINAGE CONSTRUCTION

10.8.1 Introduction

Drainage systems quality management is approached benchmarking on standards such as ISO 9001:2015. The Quality Management System (ISO 9001:2015) specifies requirements for a quality management system when an organization needs to demonstrate its ability to consistently provide products and services that meet customer and applicable statutory and regulatory requirements.[7] The development of guidelines and reporting procedures for the quality control of drainage work must be done in compliance with ISO 9001:2015 and several other applicable specifications, national standards and codes of practice. The quality control and quality assurance system for the drainage system refers to the organizational structure, responsibilities, procedures, processes, and resources for implementing quality management in designing and developing drainage systems. Quality checking mechanisms, handling and installation of storm sewer pipes requires utmost care. It is required to employ suitable material quality and appropriate construction practices to construct satisfactory drainage systems. The evaluation of storm sewer pipe quality along with its handling and installation practices[8] are discussed in the following sections.

10.8.2 Quality of storm sewer pipe

Storm sewer pipe producer and independent inspection and testing laboratories are responsible to carry out proper sewer sanitary pipe inspection and testing. Storm sewer pipe inspection includes visual inspection of pipe surface finish, workmanship and markings, sanitary sewer pipe thickness, length, diameter, and joint tolerances. Representative specimen tests and confirmation of pipe stiffness design material tests in the case of flexible pipes or crushing strength design material tests in the case of rigid pipe are conducted. It is recommended to inspect sanitary sewer pipe at production plant because delivery of unacceptable pipe would increase the total cost of the construction. As far as sanitary sewer precast concrete pipe is concerned, the pipe strength can be estimated using either three edge bearing test, core test, or standard cylinder test.

It should be known that standard cylinder test is not suitable with mixes utilized number of production techniques. That is the reason core test should be adopted if three edge bearing test is not used. This test would also examine installation tolerances of reinforcing cages. Certificate of compliance with applicable standards should be provided by manufacturer. This certificate is normally checked immediately after the loads of sanitary sewer pipe reach the project site. As sanitary sewer pipes reach construction site, visual inspection should be conducted to check whether the pipes have suffered damages during transportation and similar inspection is needed while it is placed in storage and handled.

10.8.3 Handling of storm sewer pipes

It is required to practice great caution while storm sewer pipes are handled during loading and transportation to the project site or unloading and storage or installing in the trenches. This is because damages and deterioration of storm sewer pipes during these activities due to mishandling are likely to impose high impacts or point loading is highly likely. The entire phases of construction process should be carried out with adequate accuracy to obtain conditions specified according to the design. This will ensure that the loading conditions accounted for in the design will not be exceeded. It is required to specifically pay attention to the joint elements to avoid damages.

10.8.4 Installation of storm sewer pipes

Storm sewer pipe should be installed on bedding of adequate strength but show flexibility to a certain extent under pressure. The bedding should have proper grading and consistent bearing under the whole length of storm sewer pipe barrel. The disturbance of placed subgrade should be prevented while pipe joints are fastened. This can be achieved using supports for the pipes and keep it away from subgrade. Proper excavation should be created to accommodate sanitary sewer pipe collar and bells, therefore supports and bottom reactions are merely assigned to storm sewer pipe barrel.

Necessary configurations should be done by removing excessive bedding or adding sufficiently consolidated foundation materials beneath sanitary sewer pipe. The use of blocks, wedging, and hitting sewer pipes to realize required adjustments should be strongly prohibited. As far as joining operation is concerned, cleaning of spigots and bells should be ensured, and any adhered materials must be removed and eliminated. Joint materials used for various types of sewer pipes are not the same, so each type needs certain care and considerations. Therefore, the specified material types and determined jointing procedure should be used for each type of sewer pipe joint. All types of storm sewer pipe joints are required to be sufficiently secured and pass exfiltration of infiltration tests. It would be necessary to apply considerable force to insert the spigot of compression type joint of large-diameter pipe into the bell. So, it is possible to use a crane, which is employed to handle large sewer sanitary pipe to obtain the required force.

Activities of machineries over small-diameter pipes should be prevented after the pipe jointing. This is because such operations and other similar activities will impose undesired influence on the installed pipes. Sanitary sewer pipes, which have not been installed and lift at construction site at the end of working day, should be sealed properly to keep the pipes clean. And the construction works should not start the next day unless the elevation of last installed pipe is verified. Good workmanship, material testing, specification and standards, legal and policy framework are aspects to take seriously.

10.9 EQUIPMENT AND MACHINERY USED IN CONSTRUCTION AND SUSTAINABLE DRAINAGE STRUCTURES MAINTENANCE

The equipment used in drainage construction include excavators, plate compactors, microtunneling and pipe jacking systems, etc.

10.9.1 Excavators

Excavators are essential construction equipment widely used in the civil construction industry. Their general purpose is excavation, but other than that, they are also used for many purposes like heavy lifting, demolition, river dredging, cutting of trees, etc. Excavators contain a long arm, with a digging bucket attached at the end.

10.9.2 Backhoes

A backhoe is designed primarily for excavation below ground, and it is especially employed for trench excavation works. It digs by forcing the bucket into soils and pulling it toward the machine and possesses the positive digging action and accurate lateral control.

10.9.3 Draglines

A dragline excavator is a piece of heavy equipment used in drainage construction and surface mining. A dragline bucket system consists of a large bucket which is suspended from a boom (a large truss-like structure) with wire ropes. The bucket is maneuvered by means of a number of ropes and chains. The hoist rope, powered by large diesel or electric motors, supports the bucket and hoist-coupler assembly from the boom.

10.9.4 Compactors

Compaction is an important site job required for almost all construction projects. It involves compressing the soil on-site to improve its load-bearing capacity. The load-bearing capacity is essentially improved when the air voids in the soil are removed because of compaction. There is a variety of equipment available in the market for compaction that ranges from small handheld compactors to large compacting units. A Roller compactor or sometimes also termed as a road roller or just a roller is one such type of equipment used for compaction. It is two- or three-wheeled vehicles having a heavy mass commonly used to compact soil, gravel, concrete, or other road materials like asphalt.

10.9.5 Tunnel boring machine

A tunnel boring machine (TBM) is a machine used to excavate tunnels with a circular cross section through a variety of soil and rock strata. They may also be used for micro-tunneling. They can be designed to bore through anything from hard rock to sand. Tunnel diameters can range from 1 m (done with micro-TBMs) to 17.6 m.

10.9.6 Sewer cleaning truck

Specialized vehicles used to clean sewers are known as sewer cleaning trucks. A typical sewer cleaning truck should be able to remove sand, stones, bottles, trash, grease, sludge, roots, and other debris from sanitary sewers and storm drain pipes, manholes, inlets, and vaults. Professionals use two systems of cleaning—jetting and vacuuming. Jetting is a process in which a high-pressure water hose with a sewer-jetting nozzle is pushed into the dirty pipeline. High-pressure water is released from the sewer jet truck to dislodge all of the dirt and flush it away into a gully hole. As the dirt is removed, the nozzle continues to move deeper into the sewer. Sewer cleaners also use vacuuming through which all of the dirt is extracted with a vacuuming system. All the solids and water removed this way are then disposed into a sludge tank.

10.9.7 Tree spades

As it turns out, there are some specialized machines called tree spades that make the process relatively simple, easy, and very quick. Most of these machines are mounted on some form of truck or trailer that can carry the tree, once extracted, from one location to another and replant them. These are applicable in SuDS, especially tree trenches, for relocating trees.

10.10 CONCLUSION

The construction and management of a drainage system primarily focuses on quality management, safety and health, and environmental management. The quality is achieved through sound project management where reliability is built in the drainage system at the earliest start of the project. Also, sound project management can support the economization of resources so that they are used sparingly and significantly reducing waste and innovating appropriate methods and/or new materials. For the drainage system to function efficiently and effectively throughout the design life, reliability metrics must be incorporated from conception, planning, design, construction, operation, and maintenance phases.

The collection of sound data and information, particularly the survey and hydrologic data during the planning and design, is essential in building a reliable drainage system, among others. Also, employing experienced, skilled, and competent personnel during design and construction will help build reliability in the drainage system. The use of quality materials to the desired standards and specifications builds reliability in the project. Also, the use of modern machinery and equipment improves the quality of work, the efficiency, and effectiveness of the drainage system.

Environmental management is a key pillar of sustainability of the drainage system. The designer owes a duty to integrate safety and environmental issues in the design and development of drainage solutions by specifying safe and reliable construction materials and minimizing the use of carbon-intensive materials. The choice to specific drainage routes is primarily influenced by the environmental measures which might include safeguards for environmentally sensitive areas, such as cultural heritage sites, sacred sites, and wetlands. These might have unique approaches that necessitate the designer and planner to think solutions that contribute to the sustainable management of the environment. During construction, the contractor owes a duty through the environment and social management plan to follow mitigation measures to safeguard the environment from negative impacts due to drainage systems development.

In the safety perspective, which forms a great part of the social component, the drainage designer and contractor owe a duty to communities, workers, and staffs throughout the design and development of drainage solutions. In case of negligence on the part of the designer or contractor, drainage-related cases can attract civil suits that can be costly. Construction (Design and Management) Regulations (2015) of the UK outline the role of designers and contractors and level of care that should be exercised by each party from design through the construction phase of the drainage scheme.

References

1. What is timbering | Importance of timbering | Timbering methods (civilengineeringweb.com). www.civilengineeringweb.com/2020/07/what-is-timbering.html#
2. ISO 14001 Environmental management system. Accessed May 14, 2022. www.iso.org/iso-14001-environmental-management.html
3. ISO—ISO 19011:2018—Guidelines for auditing management systems. Accessed May 15, 2022. www.iso.org/standard/70017.html
4. The Construction (Design and Management) Regulations 2015. UK Statutory Instruments, 2015 No. 51. Accessed May 14, 2022. www.legislation.gov.uk/uksi/2015/51/contents/made
5. ISO 45001:2018 Occupational health and safety management systems—Requirements with guidance for use. Accessed May 14, 2022. www.iso.org/standard/63787.html

6. Health and Safety Executive. Excavation and underground services. Accessed May 14, 2022. www.hse.gov.uk/electricity/information/excavations.htm
7. ISO - ISO 9001:2015 - Quality management systems — Requirements. Accessed May 14, 2022. www.iso.org/standard/62085.html
8. Quality, Handling and Installation of Sewer Sanitary Pipes (theconstructor.org). Accessed May 15, 2022. https://theconstructor.org/construction/quality-handling-installation-sewer-pipes/18234/

Chapter 11

Operation and maintenance

11.1 INTRODUCTION

The drainage systems need routine maintenance and repairs to reliably perform the intended purpose. Routine inspection of sewers, for example, is essential to avoid severe blockages. Once constructed, the drainage systems do not last forever. Without proper maintenance, drainage systems clog and crack. Sustainable drainage systems (SuDS), on the other hand, require to constantly identify common problems that require constant overseeing and maintenance to aid inspectors in the field. For example, periodic maintenance of SuDS vegetation is required to ensure that it remains healthy and well-established.

Drainage system maintenance refers to work done to maintain the system to function as required. It is conducted to delay, prevent, or correct deterioration and to practically keep the facilities performing their intended function throughout the design life of the drainage structure. Maintenance also includes emergency repairs due to accidents, inclement weather, and other unexpected damages caused to drainage structures. Routine oversight and monitoring of sustainable drainage schemes is vital. It is essential to take operation and maintenance as a critical requirement for the sustainability of drainage facilities. In many cases, communities fail to keep maintenance standards, thus lowering the functionality of drainage facilities. Maintenance practices vary widely, and the attitude toward maintenance is sometimes not so good in some areas.

The problem is primarily due to inadequate planning focusing on construction and waiting to rehabilitate or resurrect drainage schemes gone into hibernation after total breakdown. This is a poor maintenance culture and SuDS cannot function in such situations as they require routine oversight and monitoring. Maintenance is critical to ensure the longevity and continued effectiveness of SuDS. In several countries worst hit by climate change, routine oversight and monitoring of drainage assets is not prioritized leading to increased lifecycle costs hence affecting drainage systems' functionality. This affects the sustainability goals going forward.

The current modern practice is viewing drainage schemes through an asset management approach viewing the water cycle as an asset and planning for upholding drainage effectiveness through the lifecycle phases. This is possible through proper routine and periodic maintenance planning for hardware and software activities in annual budgets and work plans for cities, municipals, and towns. Municipals maintain records of inspections and conditions, remedial actions taken and presented in annual reports, if required. Routine and periodical drainage asset evaluation could form a valuable part of asset management strategies in order to address maintenance challenges to uphold the functionality of drainage systems. The efficiency and effectiveness of drainage systems is constantly kept in check through routine oversight, monitoring, and proper maintenance practices.

Continuously updating the asset register and management information systems for drainage systems for municipals, cities, and towns is vital. This constantly tracks the inefficiencies that develop on drainage systems from time to time and remedial actions taken immediately. Corrective actions and preventive maintenance would support the drainage systems reliability and sustainability in real time. This is because it supports upholding maintainability and operability of the drainage systems and consequently improves the functionality and reliability going forward. For drainage schemes, both gray and SuDS infrastructure to function efficiently and effectively, a maintenance plan must be in place to schedule maintenance activities. These activities would guide the maintenance crews on how to monitor drainage systems. A maintenance plan outlines the necessary routine and periodic inspections, aimed at either corrective or preventive maintenance. Whenever inspections are made, the maintenance crew identifies hotspots that need remedial actions. These can be corrective or preventive actions. Remedial actions such as repairing a broken culvert are considered corrective while removal of sediment or trash from blocking the sewer entrance/exit is a preventive measure aimed at upholding the hydraulic efficiency of the sewer line.

11.2 STORM SEWER APPURTENANCES

Sewer appurtenances are those structures and devices of a storm sewer system which are constructed at suitable intervals along a storm sewer line to assist in the efficient operation and maintenance of the system. This chapter addresses the technical details of selected sewer appurtenances briefly. Detailed technical information about storm sewer appurtenances is out of the scope of this book. The following are the most important sewer appurtenances:

11.2.1 Inlets

An inlet is a small box-like chamber made of brickwork or concrete, with an opening at the top in vertical or horizontal direction for the entry of stormwater and sometimes surface wash. Typically, stormwater enters the sewers through inlets. The water from this inlet chamber leaves through an outlet provided at its bottom and carried by a storm sewer pipeline to a nearby manhole. Inlets are the devices meant to admit stormwater and sometimes surface wash typically flowing along the roads and streets and convey the same to stormwater sewers or combined sewers. Flow in streets is calculated by using Manning's equation, modified for a triangular gutter cross section Equation (11.1):

$$Q = K \frac{z}{n} S^{1/2} y^{8/3} \qquad (11.1)$$

Where:

Q : Gutter flow
Z : Reciprocal of the cross slope of the gutter
n : Roughness coefficient

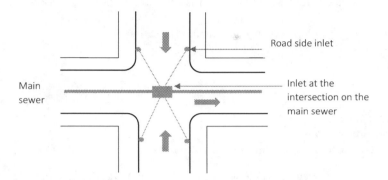

Figure 11.1 Manhole inlet located at the junction receiving flows from minor sewers that feed the main sewer.

S : Gutter slope
y : Depth at the curb
K : Constant depending on units, for metric system $K = 0.38$ (m³/s)

The design and location of inlet chambers take into consideration how far water will be permitted to extend into the street under various conditions. The permissible depth of water in the gutter in most cities is limited to 150 mm on residential streets and to that depth which will leave two lanes clear of standing water on arterials and one lane on major streets.

On curved streets, the gutter depth must be decreased to prevent the flow from jumping the kerb at driveways or other openings. These criteria are applied to the design storm which is used to size storm sewers. During high-intensity storms with recurrence intervals greater than that of the design storm, the streets are expected to be flooded.

In case Manning's equation that expresses flow velocity with the channel geometry is used to describe the flow condition in the street. In that case, using the flow, the street cross section, and slope, we can calculate the depth at the kerb. Therefore, the width to which the water will spread is equal to zy which can be compared to the criteria above.

In business areas where there is heavy pedestrian traffic, inlets may be located as shown in Figure 11.1 to keep the crosswalks relatively dry. Inlets may be classified according to location as inlets in sumps, inlets on grade, and inlets on grade with gutter depression. They may further be classified according to design as curb opening, grate, and combination inlets. Inlets in sumps are those which are located at low points in the street system where water which is not removed by the inlet will pond rather than pass by. Kerb inlets in sumps have a capacity equal to:

$$Q = K y^{1/2} L \tag{11.2}$$

Where:

y : Depth at the gutter
L : Length of kerb opening
K : Constant dependent on units and equal to 1.66 (metric system)

The flow calculated from Equation (11.2) is often reduced by 10% as an allowance for clogging.

Grate inlets in sumps have a capacity equal to (Equation (11.3)):

$$Q = KAy^{1/2} \tag{11.3}$$

Where:

- y : Depth at kerb
- A : Open area of grate
- K : Constant dependent on units and equal to 2.96 (metric system)

The capacity of grate inlets on grade is given by Equation (11.3). If the gutter is depressed, the value of y should be replaced by $(y+a)$. The calculated capacity may be reduced by up to 25% to allow for clogging. A combined inlet on grade has a capacity equal to the sum of the flows given by Equations (11.3) and (11.4). The flow intercepted per unit length by a kerb inlet on grade is given by:

$$\frac{Q}{L} = \frac{K}{y}\left[(a+y)^{5/2} - a^{5/2}\right] \tag{11.4}$$

Where:

- $\dfrac{Q}{L}$: Flow intercepted per unit length
- y : Depth at kerb above normal gutter grade
- a : Depression of gutter at inlet below its normal level elsewhere
- K : Constant dependent on units and equal to 0.39 (metric system)

The flow given by Equation (11.4) is constantly reduced by 25% or more to account for clogging. Inlets in sumps must be sized to accommodate the entire flow which will reach them under design conditions, since no flow will pass by to other inlets. Inlets on grade are usually designed to permit between 5% and 15% of the upstream flow to pass the structure. This results in an affordable and uncostly design that would result if the streets were dewatered completely at each inlet.

11.2.2 Types of inlets

There are three types of inlets: kerb inlets, gutter inlets, and combination inlets. These are outlined below.

11.2.2.1 Kerb inlets

Kerb inlets (also called vertical inlets) have vertical openings in the road kerbs through which stormwater enters and flows. The openings are provided with gratings of closely placed bars. These inlets are preferred where heavy traffic is anticipated. The kerb inlets are termed as deflector inlets when equipped with diagonal notches cast into the gutter along the kerb opening to form a series of ridges or deflectors. The deflector inlet also does not interfere with the flow or traffic as the top level of the deflectors lies in the plane of the pavement.

11.2.2.2 Gutter inlets

Gutter inlets (also called horizontal inlets) have horizontal openings in the gutter which are covered by gratings through which stormwater flows. The clear opening between the bars of the gratings should not be more than 25 mm and the gratings should be capable of sustaining heavy traffic loads.

11.2.2.3 Combination inlets

Combination inlets are composed of a kerb and gutter inlet acting as a single unit. Normally, the gutter inlet is placed right in front of the kerb inlet but it may be displaced in an overlapping or end-to-end position. Each of the above noted three types of inlets may be either undepressed (or flush) or depressed depending on their elevation with reference to the pavement surface. The depressed inlets may, however, result in some interference with traffic. The inlets are usually located by the side of the roads. Maximum spacing of inlets depends on conditions of road surface, size and type of inlet, and rainfall. A maximum spacing of 30 m may be adopted for locating the inlets. Further inlets are also located at intersections of roads and are so placed that crosswalks are not flooded. However, road corners are avoided for this purpose.

11.2.3 Catch basins or catch pits

A catch basin or catch pit is a device that serves to retain heavy debris in stormwater that would be carried into the storm sewer. The catch basin or pit is an inlet whose outlet is well above its bottom. Because its outlet is above the invert level of the inlet, heavy debris flowing along with stormwater is allowed to settle down and thus prevented from entering the storm sewer. Because catch basins are predominantly developed on combined systems, the outlet from the basin is provided with a hood or it is trapped to prevent escape of foul gases from the sewer and to retain floating matter. At the bottom of the basin enough space is provided so that the accumulated matter settles.

Catch basins or pits are periodically cleaned because the settled organic matter may decompose, producing foul odors, and may also become a breeding place for mosquitoes. Catch basins are mostly essential in areas where large amounts of sediment load is expected. However, where streets are well paved and not much sediment or garbage is expected, catch basins or pits are not considered very essential. In many cases, self-cleansing velocities can be achieved, which makes it possible to leave out catch basins or pits in the drainage system. Previously, catch basins were considered indispensable parts of a combined sewerage system. However, now that uncombined systems are largely developed, their uses has declined over the years.

11.2.4 Clean-outs

Clean-outs are the devices meant for cleaning the sewers. They are used to permit cleaning of small sewers, particularly at the upper end of laterals. These are generally provided at the upper ends of lateral sewers in place of manholes. A light cast iron cover provides access to a line of pipe leading to sewer. A clean-out consists of an inclined pipe, one end of which is connected to the underground sewer and the other end brought up to ground level. A cover is provided at the top end of the clean-out pipe at the ground level. For cleaning the storm sewer, the cover of the clean-out pipe is removed and water is forced through clean-out pipe to the sewer to remove obstacles in the sewer. If obstructions are large enough, flexible rod may be inserted through the clean-out pipe and pushed backward and forward to remove such obstacles.

The connection to the sewer is made through a special wye with its side outlet making an angle of 270° with the main rather than the standard 60°. When the sewer is fairly deep, a 1/16 bend is installed above the wye to bring the clean-out to the surface at a 45° angle. The sewer may be flushed with water through such structures and small television cameras can be inserted to permit inspection of the condition of the pipe and joints.

11.2.5 Manholes

Manholes provide access to storm sewers for inspection and cleaning. Manholes are provided at every change in alignment of sewers (changes in direction), at every change in gradient of sewers, at every junction of two or more sewers, at head of all sewers or branches, and wherever there is a change in size of sewer and at intervals of 90 to 150 m in straight lines. Sewers larger than 1.5 m in diameter can be entered readily and thus may need fewer manholes. On straight reaches of sewers, manholes are provided at regular spacing which depends on the size of the sewers. The larger the diameter of the sewer, the greater may be the spacing between two manholes. Further, the spacing between manholes also depends on the type of equipment to be used for cleaning sewers.

Manholes can be constructed out of masonry or reinforced cement concrete (RCC) chambers and constructed at suitable intervals along the alignment of sewers to provide access to the sewers for the purpose of inspection, testing, cleaning, and removal of obstructions from the sewer lines. They also help in joining sewer lines and in changing the direction or alignment as well as gradient of sewer lines. Manholes may receive sewage from sewers coming from various directions and also from sewers of various sizes. Manholes are usually constructed directly over the centerline of the sewer. They are circular, rectangular, or square in shape. The design of manholes is standardized in most municipalities and cities. Typically brick and concrete manholes are the commonest. The interior of brick manholes is often plastered with Portland cement or mortar. The three primary types or manholes include brick manholes, cast-in-place concrete manholes, and precast concrete manholes.

11.2.5.1 Precast concrete manholes

Figure 11.2 shows a typical precast manhole. Precast concrete manholes can be constructed with fewer man hours of skilled labor than brick manholes. Their use is much more suited to the tempo of modern construction using machine excavation for trenches and flexibly jointed pipe. Engineers once used brick, clay, and cast-in-place concrete to build access points to sewers. These materials, while durable, require underground, on-site construction, a challenging and dangerous proposal. Precast concrete sets the standard for utility holes. It is the material of choice for both private and public construction projects. From a city street to a sewer drain, precast manholes are the best option in many cities globally. It is possible that brick manholes will become obsolete except for special purposes, such as overflow chambers. The chamber and shaft are formed from precast tubes with ogee joints. The chamber roof is

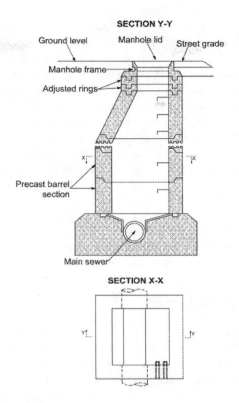

Figure 11.2 Typical precast manhole.

either a precast slab with a circular roof, is either a precast slab with a circular hole for the shaft or, if more depth is available, a tapering section. The shaft is capped with a precast block on which the manhole frame is mounted.

With the form of construction shown, the lower chamber rings are specially ordered with pipe opening in the required positions. It is usual to prepare a schedule of these as one of the contract drawings. Surveying and setting out have to be done carefully and changes due to the presence of water mains, etc., found during excavation sometimes cause difficulties. The complications of having purpose-made chamber rings have been avoided by one authority at least by constructing the part of the chamber below the pipe crown in brickwork. Though the brickwork is circular on plan, there is no need to use special bricks as they are covered by the benching concrete.

Precast concrete manholes are an integral component of any modern storm sewer system. Precast concrete manhole sections are produced in a controlled environment, thus exhibiting high quality and uniformity. A properly designed and installed precast concrete manhole system provides superior watertight performance and will provide the long-term solution required. Precast reinforced manholes have an unmatched reputation for its superior strength, durability, and hydraulic efficiency.

Manhole covers and frames are manufactured in several standard weights for different traffic conditions. The heaviest covers and frames weigh about 340 kg. Covers are generally cast with a raised pattern on their surface to make them less slippery when wet. Openings through the cover should not be allowed, since these contribute to infiltration during rainfall events. The roof slab is of reinforced concrete, designed to carry traffic and earth overburden. For the typical slab, a single heavy-wheel load is taken as being uniformly distributed over the clear span and the earth load added. As the slab is supported on all four sides, the usual method for two-way slabs should be used in calculating bending moments.

Concrete is an inherently durable material. And unlike other structure types, concrete has increased strength over time. Studies show that precast concrete manholes can last for more than a century. We feel they are the best option for a long-term solution. Concrete manholes are not only durable but nonflammable and watertight under many different installation conditions. Structural performance benefits of precast concrete utility holes also include the following:

a. Concrete is inherently strong.
b. Compression tests will verify the strength.
c. Strength will be supported from the structure itself.
d. Customized higher strengths are available.

11.2.5.1.1 Advantages of precast manholes

Perfect base	Precast channels installed at the base of the manhole make for easier transitions from sewer pipe to manhole to sewer pipe. This decreases the number of blockages throughout the line and makes for a more efficient line. These bases are built using exact specifications for easy installation and can be made in many configurations.
High durability	The consistency and durability of precast materials give manholes the probability of a long life. Precast concrete does not lose strength the way other materials can. Estimates are that it has a service life of 100 years.
Manufactured in controlled conditions	The quality of precast concrete is much more reliable than cast-in-place or brick manholes. Controlled conditions in the factory are stringently regulated. Temperature, humidity, and other conditions that affect the curing of the concrete are constantly monitored. That uniformity means that each precast concrete product can be manufactured to exact specifications.
Do not leak	Manholes must be able to endure rough conditions. They live underground, in dark, wet conditions, beneath roads and parking lots, allowing workers access to sewerage sewers and storm sewer systems. To fulfill their function and protect the environment, they need to be watertight. Over the years, engineers have perfected designs, sealants, and gaskets to ensure precast concrete manholes won't leak. Water tightness can be tested in the factory to ensure the quality of products.

Easy to install   Precast concrete utility holes are, by comparison, much easier to install. Contractors use lifting devices and machinery. A minimal crew handles the installation. Once the manhole is placed in the trench, workers can backfill immediately, completing the project. There is no need to wait for the concrete or mortar to cure. All the difficulties of building a brick or poured-in-place concrete utility manhole are eliminated by installing the precast manholes. In cast-in-place or brick manholes, the workers must labor for extended periods inside the trench. Soil compaction may complicate construction. Cold and wet weather can delay the work.

11.2.5.2 Brick manholes

A typical manhole constructed in brickwork is shown in Figure 11.3. The base concrete completely fills the bottom of the excavation. The bottom of the manhole is normally concrete, sloping toward an open channel, which is an extension of the lowest sewer. Conventionally it is shown as extending 75 mm outside the wall of the chamber, though, in fact, it will often extend further. Measurements of excavation and concrete for payment are based on the plan dimensions shown on the drawing, the contractor allowing for any excess over these dimensions in fixing his rate. A typical brick manhole, illustrated in Figure 11.3, has a cast iron frame and cover with a 500- to 600-mm opening. The frame rests on brickwork which is corbeled as shown to form a cylinder from 1 to 1.25 m in diameter which extends downward to the lowest sewer. The walls are typically 200-mm thick for depth up to 4 m and increase by 100 mm for each additional 2 m of depth. The interior of brick manholes is often plastered with Portland cement or mortar.

Manhole brickwork is always laid in English bond, courses of headers and stretchers alternating. The fair face is on the inside and 1:3 cement sand mortar is used with second-class engineering bricks. To ensure watertightness, the brickwork must be laid to engineering rather than building standards, all joints being well flushed up with mortar, no internal voids being left. Where pipes pass through the walls, relieving arches are constructed to two rings for larger ones. The overall dimensions of bricks are to be used as far as possible. The top of the benching concrete is at pipe soft level and is sloped toward the channel. Often the specification for benching concrete differs from that for base concrete. Thus, the maximum size of aggregate might be 18 or 25 mm for the former and 40 mm for the latter. The mix would be 4:2:1 for base concrete but might be richer for the benching. The channel is most conveniently formed by using half pipes, but some engineers specify first-class engineering bricks. These are used to form both the invert and the vertical walls of the channel which are surmounted by bull-nosed bricks on edge.

11.2.5.3 Cast-in-place concrete manholes

Cast-in-place concrete is a construction technique that utilizes a temporary formwork to shape the concrete slurry until it hardens. It has many applications including housing construction, manhole and drainage works, gutters, traditional open-trench pipeline construction, and the

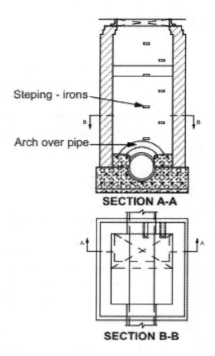

Figure 11.3 Brick manhole.

manufacture of concrete pipes used in the trenchless construction industry. Newer methods of cast-in-place concrete installation include the use of machines that use centrifugal force to blast concrete onto the walls of existing metal pipes. Cast-in-place concrete is also known as poured-in-place concrete. The present invention relates to a method and apparatus for casting concrete liners in existing, deteriorated manholes so that they will have many years of useful, substantially maintenance-free life.

The roof slab is of reinforced concrete, designed to carry traffic and earth overburden. For the typical slab a single heavy-wheel load is taken as being uniformly distributed over the clear span and the earth load added. As the slab is supported on all four sides, the usual method for two-way slabs should be used in calculating being moments. Shear is seldom critical. The square hole left in the corner for the shaft presents a difficult structural problem fundamentally, but this is often overcome simply by concentrating the steel which would have crossed the hole had it not been there under the shaft walls. An alternative is to design two 1-ft-wide strips under the shaft walls, crossing at right angles as though they were beams, putting normal slab reinforcement elsewhere. The use of mesh reinforcement has many practical advantages over individual bars, but bars are needed to supplement it under the shaft walls.

11.2.6 Mini-manholes

These are small-diameter manholes. Usually in HDPE, plastic, and steel, manholes installed on storm sewers with relatively low flows. Available to suit installations from 600- to

1800-mm invert. In some cases, mini-manholes are lightweight and used when the job leads to a manhole structure especially when the working trench needs to become wider before you can continue with the pipe installation. These special trench boxes provide complete protection for the workers.

11.2.7 Deep manholes

A manhole having a depth of up to 0.9 m is termed a shallow manhole. A deep manhole has a depth greater than 1.50 m. Figure 11.4 shows a deep manhole. Steel ladders are used instead of step irons in very deep manholes and rest platforms are provided at intervals. In very deep manholes, it is advisable to stagger successive lengths of ladder so that there is no possibility of a man falling in the depth. Where intermediate platforms are provided, they should be arranged so that it is possible to raise bucket or small skip directly from the bottom. This will involve having two small holes in each platform: one for man access and one for skips. The former will be vertically below each other and should have guard-chains or rails. Shafts in such cases need to be quite large and may need multiple covers. Since deep sewers are usually constructed in heading, the manholes are built in the working shafts and are often constructed in heading; the manholes are built in the working shafts and are often constructed of bolted segments of precast concrete, circular on plan, fixed as the excavation proceeds. All manholes deeper than 3 m are usually provided with a bolted stainless steel grid beneath the manhole cover unless the manhole cover is of the hinged type fitted with an approved lockable device. A warning sign is usually provided on the stainless steel grid or the back of the manhole cover, as appropriate. The manhole bases are usually benched and haunched with concrete, formed and vibrated to a smooth finish to accommodate the inlet and outlet culverts. The concrete is usually about 17.5 MPa or 1:2:4 mix (cement: sand: 20-mm aggregate).

Plastering of benching to achieve a smooth finish is usually not permitted. Severely honeycombed benching is usually rejected and replaced fully and any minor defects are made good using epoxy mortar. In some cases, the height of the manhole throat (access shaft) is not greater than 350 mm. The throat and any subsequent extensions to the throat are cast in situ using **17.5-MPa** compacted watertight concrete to a smooth finish. Precast throat is not permitted. Watertight bonding is provided between the throat and the lid, and between the existing part of the throat and subsequent extension. If a further extension (or lift) is required, such that the throat is greater than 350 mm, then the original precast concrete lid is removed and an additional manhole riser installed and sealed as for normal construction. For large-diameter culverts, recessed steps without rungs may be permitted below pipe benching level, provided the lowest rung can be easily reached by a person standing at invert level.

Cast in situ manhole bases are usually constructed using ordinary grade concrete (17.5 MPa) vibrated to give maximum density and watertight construction. The method of joining the precast sections is strictly in accordance with the recommendations of the manufacturer, and using a proprietary jointing compound or adhesive, in conformity with the manufacturer's instructions. The joint provide a watertight structure to the satisfaction of the public authority. Large manholes (with a minimum diameter > 1.8 m) on stormwater lines of 1.05-m diameter and above may be constructed using offset intakes which may also be used in conjunction with bends, formed using epoxy mortar adhesive. Where culvert diameters increase at the manhole, the soffit of the outlet culvert can be at the same level, or below that of the inlet culvert. Energy losses associated with angular deviations must be allowed for.

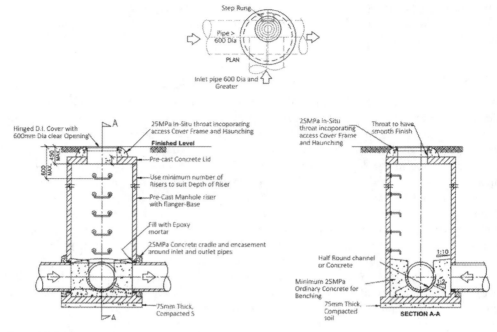

Figure 11.4 Standard deep manhole up to 5-m depth to invert.

11.2.7.1 Drop manholes

A drop manhole is constructed on a storm sewer line where a sewer at a high level is to be connected to another sewer at a lower level (see Figure 2.3). A drop manhole is usually provided when the difference in elevation between the high and low sewer exceeds 0.6 m. Situations that call for drop manholes may arise as indicated below:

a. Branch sewers are generally situated at lower depths below the ground level whereas main sewers are laid at greater depths below the ground level. When a branch sewer located at a higher level is to be connected to a main sewer located at a lower level, then if ordinary manhole is provided, the stormwater runoff from the branch storm sewer will fall from above into the manhole which is not desirable and is to be avoided. Thus, at the junction of branch storm sewer and main storm sewer, when the difference between the invert level of branch sewer and peak flow level of main sewer is more than 600 mm, a drop manhole is provided. The construction of drop manhole permits the runoff from branch storm sewer to be discharged at the bottom of the manhole without necessitating steep gradient for branch storm sewer and thus reduces the quantity of earth work.
b. In case of ground having steep slope it is not possible to lay the sewer line at a uniform gradient that will not produce scouring velocity. In such cases, when a drop of more than 600 mm is required to be given in the same sewer line, a drop manhole is provided

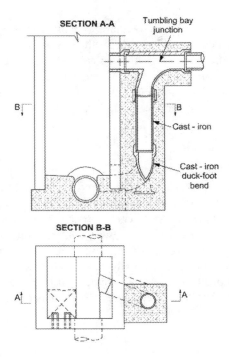

Figure 11.5 Drop-junction manhole.

at suitable place so as to keep the sewer line below the ground level and to lay it at a limiting gradient.

The most common application of drop manholes is to limit velocities. It is necessary to lay sewers at much flatter gradients than those of the steep land. Figure 11.5 shows a typical 'side-drop' manhole on a storm sewer. The bottom of the drop is a cast iron duck-foot bend. The duck-foot makes building easer but is not absolutely necessary. In the figure, a 'tumbling-bay' junction is used for the top. This differs from a standard junction in that it has spigots in place of the latter's sockets and vice versa. When large sewer has to be dropped, a ramp is used. Sewers of intermediate size (1.0-m diameter) are rather large for vertical backdrops and too small for a specially constructed ramp to be warranted. When large sewerage flows must fall long distances in order to reach a lower sewer, the fall is generally interrupted by staggered horizontal plates within the shaft or by a step-like arrangement. These devices prevent excessive kinetic energy from damaging the bottom of the hydraulic structure. When large sewer has to be dropped, a ramp is used. An alternative is a cascade of steps. The advantage claimed for cascades is that the energy is dissipated a little at a time, but this is probably true only at low flows when the problem is not serious. At high flows the flow is much the same as whether the steps are there or not and a good deal of energy remains at the foot of the cascades. Steps are certainly subject to more wear and if they are constructed of brickwork, very frequent inspection is desirable. Once a brick is loosened, deterioration is rapid.

11.2.7.2 Junction boxes

Junction boxes are nodes where different hydraulic structures such as storm sewers and drains interconnect. They are built to simply conserve the flood waves so that they are routed safely from one hydraulic structure to another. In most cases, the junction box is a belowground round or square structure made of precast concrete. The purpose of these structures is to interconnect storm sewer or other piping together to provide for change-in direction, joining piping of different sizes, or for sewer access and inspection.

11.2.8 Outfalls and outlets

The outfall lines are constructed of either iron or reinforced concrete and may be placed from barges or joined by divers. Iron is generally preferred for outfalls 610 mm in diameter or less. In bodies of water, which are sufficiently large to permit heavy wave action, the outfall may be protected by being placed in a dredged trench or by being supported on pile bents. Outlets to small streams are similar to the outlets of highway culverts, consisting of a simple concrete headwall and apron to prevent erosion. Gravity discharge lines in such circumstances must be protected by flap gates or other automatically closed valves which prevent the stream flow from backing up into the plant. Sewers discharging into large bodies of water are usually extended beyond the banks into fairly deep water where dispersion and diffusion will aid in mixing the discharge with the surrounding water.

11.2.9 Lamp-holes

A lamp-hole consists of a vertical stone ware or concrete or cast iron pipe of 225 to 300-mm diameter, connected to the sewer line through a 'T' junction as shown in Figure 11.6. Lamp-holes are small openings provided on sewer lines mainly to permit the insertion of a lamp into the storm sewer for the purpose of inspection of sewer lines and detecting the presence of any obstructions inside the storm sewers. The pipe is surrounded by concrete to make it stable. At the ground level the lamp-hole is provided with a manhole cover with frame strong enough to take up the load of traffic. The manhole cover with frame is carried on a concrete-surround which is separated from the lamp-hole by 15-cm vertical clay joint and thus permits the concrete-surround to bear the traffic load passing on the cover without causing the pipe-column to be subjected to heavy loads resulting in its possible collapse.

A lamp-hole may serve the following purposes:

a. **Flushing.** Under certain circumstances when no other flushing devices are available, lamp-holes may be used for flushing the sewers.
b. **Inspection.** For inspection of sewer line an electric lamp is inserted in the lamp-hole and the light of the lamp is observed from the two manholes, one upstream and the other downstream of the lamp-hole. If the sewer length is unobstructed, the light of the lamp will be seen.
c. **Ventilation.** If the cover at the top of the lamp-hole is perforated, the ventilation of the sewer is affected. Such a lamp-hole is also known as fresh-air inlet.

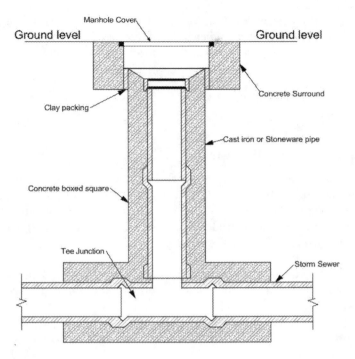

Figure 11.6 Lamp-hole.

Lamp-holes are found suitable for use under the following situations:

a. A lamp-hole may be provided when there is change in the alignment or gradient of the sewer line between two manholes that are a short distance apart.
b. When the sewer is straight for a considerable distance beyond the usual spacing of manholes, a lamp-hole may be provided.
c. When the construction of a manhole is difficult, a lamp-hole may be provided in the place of manhole.

Note: lamp-holes have become more or less obsolete and hence at present times these are rarely provided.

11.2.10 Flushing devices

These are mostly applicable in partially or combined systems. When sewers are to be laid in a flat country, it is not possible to obtain a self-cleansing velocity even once a day due to flatness of gradient especially at the top ends of branch sewers which receive very little flow. Similarly near the dead ends of the sewer lines, the self-cleansing velocities cannot be achieved because the discharge coming at the starting point happens to be small. In both these cases, the discharge is required to be increased and this is done by adding a certain quantity of water by means of flushing devices. Thus flushing devices help to prevent clogging of sewers and permit the adoption of flatter gradients than those required to maintain self-cleansing velocity. The various flushing devices may be broadly classified under the following two categories:

a. Hand operated flushing devices.
b. Flushing tanks.

11.2.11 Grease and oil traps

Grease and oil traps are the chambers provided on the storm sewer line to exclude grease and oil from stormwater runoff before it enters the storm sewer line. These are located near the sources contributing grease and oil to sewage, such as automobile repair workshops, garages, kitchens of hotels, grease- and oil-producing industries, etc. It is essential to exclude grease and oil from sewage due to following reasons[1]:

a. If grease and oil are allowed to enter the storm sewer, they stick to the inner surface of the sewer and become hard, thus cause obstruction to flow and reduce the sewer capacity.
b. The suspended matter, which would have otherwise flown along with stormwater runoff, sticks to the inner surface of the sewer pipe due to sticky nature of grease and oil, thus further reducing the sewer capacity.
c. The presence of grease and oil in storm sewage makes the sewage treatment difficult as they adversely affect the biochemical reactions, especially in partially or combined systems.
d. The presence of a layer of grease and oil on the surface of storm sewage does not allow oxygen to penetrate due to which aerobic bacteria will not survive and hence organic matter will not be decomposed. This will give rise to bad odors.
e. The presence of grease and oil in storm sewage increases the possibility of explosion in the storm sewer line.

Oil water separators can efficiently aid in the removal of gasoline, diesel fuel, crude, vegetable, and almost any type of oil that is lighter than water. The effluent from oil/water separators is typically discharged to either a sanitary sewer system or a storm sewer. Grease and oil traps are the chambers provided on the storm sewer line to exclude grease and oil from sewage before it enters the sewer line. These would be located near the sources contributing grease and oil to sewage, such as automobile repair workshops, garages, kitchens of hotels, grease- and oil-producing industries, etc.

The principle on which grease and oil traps work is simple. The grease and oil being light in weight float on the surface of sewage. Hence, if outlet draws the sewage from lower level, grease and oil are excluded. Thus grease and oil trap is a chamber with outlet provided at a lower level near the bottom of the chamber and inlet provided at a higher level near the top of the chamber. However, in addition to grease and oil if it is desired to exclude sand, space should be kept at the bottom of the chamber for sand to be deposited. Figure 11.7 shows a typical grease and oil trap. It consists of two chambers interconnected through a pipe. The inlet

Operation and maintenance 397

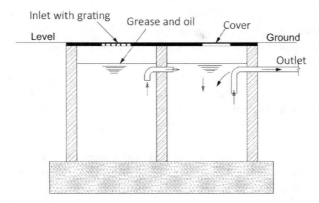

Figure 11.7 Grease and oil trap.

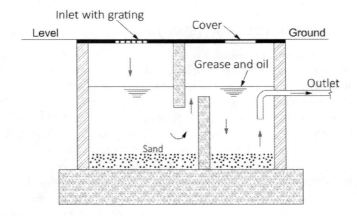

Figure 11.8 Combined sand, grease, and oil trap.

with grating is provided near the top of one of the chambers while the outlet is provided in the other chamber. The end of the outlet is located at a height of about 0.6 m above the bottom of the chamber and it is held submerged. The wastewater obtained from garages, particularly from floor drains and wash racks, contains grease, oil, sand, and mud. To trap all these, a combined sand, grease, and oil trap is provided which is shown in Figure 11.8.

11.2.12 Stormwater regulators

In the case of combined system of sewerage a large quantity of stormwater flows through the storm sewer along with domestic sewage. Usually, it is neither advisable nor practicable to pump and/or treat such a large quantity of sewage. Moreover, the percentage of domestic sewage in a combined system is very much less, and hence the domestic sewage gets diluted by stormwater to a considerable extent.

It is, therefore, possible to divert a portion of this combined sewage and discharge it into a watercourse such as stream, lake, river, etc., without any trouble. This is achieved by providing stormwater overflow devices or stormwater regulators in an intercepting sewer so that the outfall sewer will carry only a portion of the combined sewage to the sewage treatment plant.

For example, if the combined sewage is six times dry weather flow, three times dry weather flow will be taken for treatment as in the case of a separate system of sewerage and the remaining three times dry weather flow will be discharged directly into a watercourse.

Thus stormwater overflow devices or stormwater regulators are the devices which are provided in a combined system of sewerage to permit the diversion of excess sewage through relief sewer to a watercourse and thus prevent overloading of sewers, pumping stations, treatment plants, or of disposal arrangements. Following are the three types of stormwater overflow devices or stormwater regulators:

- Side flow weirs or overflow weirs.
- Leaping weirs or jumping weirs.
- Siphon spillways.

11.3 LEGAL AND POLICY FRAMEWORKS

Communities globally conduct drainage asset planning, design, and management following national legal frameworks from which policies are derived to promote efficiency, effectiveness, and sustainability. Sometimes, policies lead to new laws. For instance, environmental protection statutes and policies govern physical and urban planning practices to promote sustainability worldwide. In the interest of sustainability, policies are framed for achieving certain goals. Therefore, to guide ongoing maintenance of drainage assets, policies are formulated and strictly complied with, for example, solid waste management policies and byelaws, building permit requirements, environmental protection, and physical and urban planning. Policies on operation and maintenance of drainage assets must be properly followed, for example, continuously updating the asset register, routine oversight and monitoring, preparing monitoring plans and monitoring reports, checking the functionality of drainage systems, and effective planning for software activities. The legal and policy framework guides the operation and maintenance crews for drainage systems.

11.4 STORMWATER DRAINAGE NETWORK POLICY FRAMEWORK

The drainage system is an essential part of living in a city or urban area, as it reduces flood damage by carrying stormwater runoff away from properties. Therefore, a drainage master plan is a live document, which is constantly reviewed and updated. It is always good practice to update the asset register for municipal drainage systems including green and gray infrastructure. An inventory of hydraulic structures has to be done from time to time clearly defining the lined and unlined drains in kilometers, sewer appurtenances, and associated infrastructure. Structural and nonstructural SuDS should be included in asset registers to aid proper maintenance scheduling.

Increased urbanization triggers huge runoffs from impervious surfaces, which is dealt with through SuDS and standard pipework sewers in modern drainage master plans, which adopt integrated solutions to meet sustainability goals. The flow discharged in receiving waters is of great significance and routine inspection of drainage practices has to be planned to ensure that the quality of waters discharged in receiving waters is kept to the desired standards.

11.4.1 How does a standard pipework drainage system work?

The municipal council has a big role to play in removing excess water from urban areas through constructing safe and adequate drainage networks. Part of rainwater naturally seeps into the ground when it rains and SuDS also play an important role to mimic the natural environment thereby absorbing part of the rainwater. The rest of the rainwater finds its way through drainage systems into receiving water bodies, that is, rivers and streams and finally into lakes, swamps, and ponds. In some areas, huge amounts of water can generate so quickly when it rains heavily and without adequate drainage systems the runoff flows in low-lying areas leading to flooding that causes property damage and safety risks. The municipal authority is tasked to provide a safe drainage system to manage stormwater and stormwater runoffs from the built-up areas for reuse or discharging into rivers and streams. Figure 11.9 illustrates how a drainage system network operates from the source up to the lake.

a. Stormwater runoff from residential properties and their roofs through house gutters and downpipes and into residential drains.
b. Residential drains connect to municipal council drains which could be roadside drains or small council drains.
c. Municipal council drains connect to big or master drains.
d. Division/master drains direct water into the nearest river, river tributaries, and streams
e. Rivers and streams eventually empty the waters into lakes or swamps.

11.4.2 Who's responsible for what?

Municipal authorities and individuals or private developers have a responsibility for the drainage systems as follows:

a. SuDS can be developed in both private and public spaces. The responsibility of managing these SuDS is rested on the respective individual owner or the public authority.
b. Individual property owners are responsible for drainage systems on their property such as roof gutters, downpipes, and pipelines which generally connect to council drainage systems.

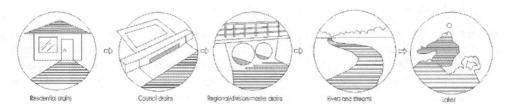

Figure 11.9 Ideal stormwater movement pattern for residential areas.

c. Large drains and stormwater infrastructure connecting with rivers, streams, lakes, and ponds are managed by the municipal authorities. The municipal is responsible for installing and maintaining the municipal drainage network. It also manages the local drainage network, including street gutters and drains.

11.5 HOW DO SUDS WORK?

Sustainable drainage systems aim to serve four primary goals: stormwater runoff quantity reduction, peak flow reduction, runoff quality control, and sinking carbon. This is achieved through physical, chemical, and biological processes that SuDS employ. Stormwater is collected, stored, conveyed, and treated in a way that mimics the natural environment. SuDS employ the following processes to that which mimic predevelopment conditions: evaporation/transpiration using native vegetation, infiltration (allowing water to slowly sink into the soil), and rainwater capture and reuse (storing runoff to water plants, flush toilets, feeding animals, etc.).

11.6 A GUIDE TO DRAINAGE DESIGN

Several design manuals specify annual exceedance probability (AEP) or average recurrence interval (ARI) for design purposes targeting individual and residential sewers and drains, council and master drains, and rivers and streams as follows:

a. Individual property developers' drains or residential drains and conduits are usually designed to cope with an ARI between 5- and 25 years. In some countries/regions ARI of 25 years is considered for residential drains. However, this depends on the magnitude and frequency of storms for a given area to determine the most appropriate AEP to adopt for residential or individual properties' drainage.
b. Municipal divisions and council drains or pubic sewers are usually designed to cope with an ARI of 25 years, in most cases.
c. Streams and river restoration mechanisms are usually handled considering an ARI of 100 years.

Individual or residential drainage systems are generally designed to cope with more frequent storms—those with a 20% (5-year ARI) chance of occurring in a year. Any excess water travels along planned overland flow paths that carry water away from residential properties. This prevents them from flooding in the majority of storms—up to those with a 1% (100-year ARI) chance of occurring in a year.

11.7 LEGAL POINT OF STORMWATER DISCHARGE

The legal point of stormwater discharge (LPSD) is a point specified by the municipal where stormwater from a property must be discharged. In order not to strain the municipal council drains such planning is vital. Therefore, every new development would be required to produce an LPSD report. Here, the municipal is tasked to keep closely monitoring development in the municipal as far as stormwater management is concerned. The report would provide an account about the likely stormwater quality, and where the quality is in doubt, the proposed remedial options to improve the quality of water discharged to the public sewer would be evaluated.

The report would also provide the actual point where the stormwater running from an estate or property will discharge to the public sewer and how underground utilities are dealt with to avoid interruption of services. In some cases, stormwater management includes sewers and SuDS situated in the road reserve and may require approval from the municipal. Just like other applications submitted to the council to approve construction drawings, issuing building occupation permits and planning permits, the legal point of stormwater discharge permit would require the developer to apply to the council and to pay required fees.

As the applicant applies to make a connection to the public sewers, designs are checked for compliance with the region's codes and standards of practice including AEP requirements. The LPSD report would be comprehensive providing the size, depth, and offset of the municipal drain or sewer on-site ensuring that council drain or sewer will not be damaged by the proposed development or works.

The legal point of stormwater discharge is usually to a municipal managed drain or the street kerb and channel. A legal point of discharge report from council filled closely with the developer shall be required when applying for a building permit. Not only individual property developers shall be assessed but also new project infrastructure developments such as roads need to be assessed for final stormwater discharge points.

11.8 DRAINAGE INFORMATION, REPORTING, AND RECORD KEEPING

Proper documentation and record keeping ensures that the municipal stormwater department is adequately performing drainage asset inspection and maintenance responsibilities as required. Checklists or monitoring tools (inspection and maintenance checklists) are filled whenever inspections are made for a drainage asset. This information is entered into the municipal stormwater management information system (SMIS) or stormwater asset mapping database (SAMD) as provided by the municipal's stormwater management department.

The current practice allows the inspector to fill the inspection and maintenance checklist at the time of inspection using digital platforms. These are usually customized mobile applications used to enter collected data into the SMIS. The checklists or monitoring tools are used to conduct field observations, collect enough drainage asset data, spot maintenance triggers for synthesizing into further remedial actions as appropriate, for example, repairing, cleaning, and replacement of drainage assets. When these tools or checklists are filled, they are sent to the maintenance office for further review and incorporation into the future work plan and budget. This helps the municipal to prioritize flooding hotspots, identify pollutant hotspots, and meet the annual performance reporting requirements for stormwater management.

Publicizing the information on noticeboards and on municipal websites including those for lower local governments within a municipal is vital. It is good practice for municipals to keep proper records on stormwater management such as record drawings of building projects, drainage studies, etc. The prepared standard drawings and specifications for catch basins, drop inlets, junctions, manholes, culverts, etc., may be uploaded on municipal website as appropriate. The role of publicizing and keeping drainage information in a discoverable fashion is to help keep track of drainage assets, schedule proper maintenance, and guide future drainage studies.

In case drainage information is made readily available, planners and designers will quickly access it while appraising new developments and retrofitting existing areas. This can be vitally important especially where SuDS are required to be retrofitted in areas where they never existed.

Standard drawings, specifications, and manuals may be uploaded on municipal websites as appropriate for ready access and this supports the work of consultants, engineers, and planners. Keeping stormwater management records improves drainage reliability and contributes to the achievement of sustainability goals going forward, and it is highly recommended for municipals for drainage performance improvement planning.

11.9 ENFORCING LAWS, POLICIES, RULES, AND REGULATIONS FOR EFFECTIVE DRAINAGE ASSET MANAGEMENT

Physical planning acts and land management laws are relevant for drainage system networks to effectively perform. The relevant laws about physical planning, land management and acquisition, and environmental management should be properly enforced. Physical planning can be described as follows: physical planning in its broadest sense refers to a set of actions aimed at improving the physical, social, and economic welfare of a place and its dwellers. It entails the organization of land uses so that people enjoy the highest achievable degree of efficiency in resource utilization, functionality of places, and aesthetic quality. The main concerns of urban planning therefore include spatial orderliness, aesthetics of the urban places, efficiency of operations in the social, economic, and other arena, and most importantly, human's well-being.

In Uganda, a country that largely adopts UK standards, the Physical Planning Act of 2010 and Land Act of 2010 summarize all the above into a policy and legal framework to guide the municipals and local governments. Such kind of Acts must be properly enforced to avoid irregular and haphazard construction practices appearing, for example, in several low-income countries like Uganda. Iregular and haphazard construction has a significant impact on stormwater drainage status of the municipal. Therefore, the approved physical plan has to be strictly enforced.

The detailed physical plans are recommended to be produced for the entire municipal. A more realistic structure development plan takes a further step to produce 3-D model master plans. The future promises to shift to 3-D planning worldwide because when 3-D model master plans (MMPs) are produced, they produce a virtual reality of the appearance of the municipal in a pictorial format and SuDS development can be clearly illustrated. Such models simplify the work of stormwater management teams and maintnence crews. The former could be council teams or consultants enforcing policies on stormwater or conducting studies, etc., while the later could be maintenance crews, officials, and consultants locating sewer lines and portable water pipelines for maintenance, upgrading, or expansion.

The other advantages of 3-D models include giving a good insight into the cost of implementing the projects and making nontechnical people appreciate the proposed development. With 3-D MMPs, the technical teams don't find difficulty thinking together with the nontechnical teams. 3-D MMPs also provide the architectural impression of the type of buildings, etc., required in a given section of the municipal. 3-D models can also aid the

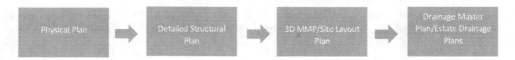

Figure 11.10 Moving from 2-D to 3-D planning.

planning and design of underground drainage structures so that the council moves away from a dominance of open drains with associated open-cut technologies to closed underground hydrualic structures which require trenchless technology, see Figure 11.10.

11.10 MAINTENANCE OF SUDS

SuDS require constant maintenance to perform effectively. Similar to standard pipework storm sewers, SuDS require maintenance to ensure optimal performance. Many SuDS require regular maintenance, whether related to vegetation (pruning, weeding, mulching) or operational maintenance/repair such as cleaning permeable pavers. The SuDS lifecycle of the technology or vegetation used in the GI technique must be taken into account when preparing a maintenance plan.

The common practices usually employed for SuDS routine maintenance are similar to general landscape maintenance. These include cleaning out accumulated sediment and pollutants, maintaining plants healthy, and removing trash, leaf litter, and debris. Routine oversight and monitoring would support upholding the functionality of SuDS. Preparing a SuDS maintenance plan would take on a holistic approach spanning from available budgets, personnel, and equipment required. Below are some of the ways to ensure that SuDS are effectively maintained over the long term.

11.10.1 Procure equipment

The equipment needed for maintaining SuDS should be planned ensuring that it is accessible whenever required. Several customized equipment are usually considered depending on the landscape and type of SuDS. In some instances, relatively heavy equipment is not recommended for routine maintenance due to the fact that it is more likely to cause soil compaction reducing the effectiveness of the SuDS.

The equipment includes tree watering bags, hose, and irrigation systems used for watering plants during drought periods and plant establishment. The removal of debris, leaf litter, trash, and sediment can be executed with rakes, trash grabbers, and shovels.

Green roofs are inspected with the help of a ladder. This includes cleaning the roof drains that connect to rainwater harvesting systems.

On golf courses where infiltration systems are largely employed, the turf grass is kept at appropriate height by the aid of mowers. In some cases, vegetation in bioretention systems, which largely function as infiltration systems, is kept aesthetically appealing and healthy employing pruning shears and weed pullers. The same equipment is employed to maintain healthy vegetation in vegetated swales and kerbs.

Infiltration systems such as porous pavement is maintained employing a vacuum-powered street sweeper, and replacement pavers are usually needed for repairs.

In several instances, heavy equipment such as front-end loaders and backhoes are usually employed where huge amounts of sediment loads accumulate and where SuDS need replacement. Also, flat-blade shovels are used to scrap accumulated sediment off inlets and curbs.

11.10.2 Maintenance budgets

Effective maintenance requires maintenance funds to be readily available. In many cases, the conditional and unconditional grants and loans from the state or central government may not be used for routine maintenance of SuDS. In that case, local funding through collecting tax revenues and utility fees provides a relatively firm and stable maintenance fund for SuDS.

11.10.3 Working with partners

In some cases, routine maintenance practices such as eliminating trash and weeds from bioretention areas and other SuDS may be conducted by partnering with local organizations, garden clubs, and greenway groups to leverage their funds.

11.10.4 Inspection and maintenance personnel

It is vitally important to know the number of maintenance personnel available to conduct routine oversight and monitoring for a particular SuDS and be able to determine if additional staff is required. This evaluation would help predetermine whether hiring a specialist and experienced contractor is essential or training available to personnel is more cost-effective to maintain the SuDS. In some cases, public authorities such as municipals usually consider departments that have the skill set and equipment to constantly inspect SuDS. In addition, periodic trainings for their staffs and the sourced contractors is conducted developing SuDS reliability, functionality, and sustainability through imparting knowledge and skills in the involved personnel.

11.10.5 Identify SuDS maintenance triggers

Scheduling maintenance requires identifying maintenance triggers that would elicit a response from inspectors conducting routine oversight and monitoring. Nonroutine maintenance practices that trigger the onset of maintenance include overgrown vegetation, structural damage, standing water present for more than three days, trash and debris, excess sediment accumulation, and signs of erosion. These are very important aspects written down in a monitoring plan to aid inspectors. These triggers aid the inspectors who would get down to the field checking for their presence and likely future occurrence and prepare monitoring reports for further synthesis coming up with remedial actions. In that case, routine and period maintenance is well-organized and planned going forward.

11.10.6 Update the infrastructure asset register and standard operating procedures

Public authorities continuously update the infrastructure asset register. It is vitally important that SuDS infrastructure assets be updated together with other infrastructure assets for effective routine and periodic maintenance activities. In case the authorities have streamlined standard operating procedures for routine and periodic infrastructure maintenance, it is great to consider updating the asset registers to incorporate SuDS, remedial actions, and SuDS maintenance triggers. Maintenance plans are always provided in explicit language in contract documents, and in some cases, specialized contractors are employed requiring constant training of maintenance teams. Scheduling maintenance follows specific SuDS and a tracking system should be in place to ensure that maintenance is conducted as required.

11.11 SPECIFIC SUDS MAINTENANCE PRACTICES

This section provides specific maintenance practices for a variety of SuDS. It provides the range of tools, equipment, and machinery required for routine and nonroutine maintenance practices, the required personnel, scheduling of monitoring plans, maintenance schedules, and monitoring reports that should be rich in sustainability, social, and environmental aspects.

11.11.1 Postconstruction

11.11.1.1 Tree trenches

The designer should prepare a site-specific operation and maintenance plan prior to putting the stormwater practice into operation. The plan should provide any operating procedures related to the practices. The plan should also provide clear maintenance expectations, activities, and schedules. If possible, photos should be included. Be clear who is responsible for maintenance and the type of expertise needed for distinct operation and maintenance activities.

The routine maintenance practices include watering, mulching, treating diseased trees, and removing litter as needed. Annual inspection for erosion, sediment buildup, and vegetative conditions is required for healthy tree trenches. Also, biannual inspection of clean-outs, inlets, and outlets is essential. The maintenance cost for a prefabricated tree pit is about US$100 to 500 per year.

Maintenance of tree trenches and tree boxes does not require large or heavy equipment, but routine maintenance should be expected once or twice a year. Designers can incorporate solutions to facilitate the following maintenance activities. Incorporating multiple and easy access points is essential. Placement of tree trenches near supportive companion plants can help to prevent diseases. A site-specific Operations and Maintenance Plan should be prepared by the designer prior to putting the stormwater practice into operation. This plan should provide any operating procedures related to the practices (Minnesota Stormwater Manual, 2021)

11.11.1.2 Bioretention gardens

Maintenance costs will vary depending on the level of attention the bioretention cell warrants. The maintenance cost is similar to traditional landscaping. As with any garden, the more the care is given, the more able the plants will be to survive. As in any natural setting, all vegetation will eventually die, so the cells will have to be replanted over time. This is not expected to be a yearly occurrence, however. It is reasonable to assume that vegetation will require replacement approximately every 10 years. The maintenance practices include the following:

a. Watering: One time for two to three days for first one to two months, then as needed. During drought times it may be important to water the plants in the bioretention garden.
b. Spot weeding, pruning, erosion repair, trash removal, and mulch raking: Twice during growing season.
c. As needed, add reinforcement planting to maintain desired density (remove dead plants), remove invasive plants, and stabilize contributing drainage area.
d. At a minimum, it is necessary to yearly inspect the bioretention garden's underdrains to ensure they are not clogged.
e. Annual: Spring inspection and cleanup, supplement mulch to maintain **a 75-mm** layer, and prune trees and shrubs.
f. Mulch will need to be added to the bioretention cell periodically. This keeps up the appearance of the cell (minimizing weeds) and continues to provide a key water quality function. This maintenance will need to be performed one to two times per year.
g. At least once in every three years: Remove sediment in pretreatment cells/inflow points and replace the mulch layer.

11.11.1.3 Infiltration systems

There are a few general maintenance practices that should be followed for infiltration SuDS. These include:

a. All catch basins and inlets should be inspected and cleaned at least twice per year.
b. The overlying vegetation of subsurface infiltration feature should be maintained in good condition and any bare spots revegetated as soon as possible.
c. Vehicular access on subsurface infiltration areas should be prohibited (unless designed to allow vehicles) and care should be taken to avoid excessive compaction by mowers.

11.11.1.4 Porous pavers

There are a few general maintenance practices for porous pavers. These include:

a. Routine cleaning of inlets.
b. Permeable pavement is best maintained using a vacuum-powered street sweeper.
c. Vacuuming annually. The primary goal of porous pavement maintenance is to prevent the pavement surface and/or the underlying infiltration bed from being clogged with fine sediments. To keep the system clean throughout the year and prolong its life span, the pavement surface should be vacuumed biannually with a commercial cleaning unit. Approximately US$400 to 500 per year for vacuum sweeping of a half-acre parking lot is needed for maintenance.
d. Maintain adjacent landscaping/planting beds.
e. Periodic replacement of paving blocks is essential.
f. Periodic replacement of paver blocks. Replacement pavers are sometimes needed for repairs.

11.11.1.5 Vegetated swales

There are a few general maintenance practices for vegetated swales listed below:

a. Inspect soil and repair eroded areas, remove litter and debris, and clear leaves and debris from overflow.
b. Inspect trees and shrubs to assess their health.
c. Re-mulch void areas as much as applicable, treat or replace diseased trees and shrubs, and keep overflow free and clear of leaves as needed.
d. Add additional mulch, inspect for sediment buildup, erosion, vegetative conditions, etc. annually.
e. Maintenance cost: Approximately $200 per year for an 80 m^2 vegetated swale.

11.11.1.6 Open/green spaces

Routine maintenance practices on open/green spaces is largely similar to general landscape maintenance, that is, removing trash, leaf litter, and debris; keeping plants healthy; and cleaning out accumulated sediment and pollutants. Parks and open space operations and maintenance activities commonly involve the operation of equipment such as periodic mowers and tractors, disposal of waste from mowing, planting, weeding, raking, pruning, and trash collection, application of pesticides, herbicides, and fertilizers, cleaning and maintenance of park amenities such as play equipment, restrooms, and structures; and snow removal. Regular inspections will indicate if the practices are not functioning as intended.

a. Watering during the plant establishment period and in extended droughts should be done with a hose, irrigation system, or tree watering bags.
b. Irrigation. Repair broken sprinkler heads as soon as possible and only irrigate at a rate that can infiltrate into the soil to limit runoff.
c. Avoid irrigating close to impervious surfaces such as parking lots and sidewalks.
d. Repair damage to landscaped or mulch or vegetated bare areas as soon as possible to prevent erosion. If there are areas of erosion or poor vegetation, repair them as soon as possible, especially if they are within 15 m of a surface water (e.g., pond, lake, or river).
e. Remove (sweep or shovel) materials such as soil, mulch, and grass clippings from parking lots, streets, curbs, gutters, sidewalks, and drainageways.
f. Do not clean up any unidentified or possibly hazardous materials found during maintenance; notify a supervisor immediately.
g. Heavy equipment, such as backhoes and front-end loaders, may be needed occasionally if the facilities need to be replaced or if large amounts of sediment have accumulated.
h. Leaf litter, trash, debris, and sediment should be removed with rakes, shovels, and trash grabbers.
i. Flat-blade shovels are especially useful for scraping accumulated sediment from inlets and along curbs/gutters.
j. Vegetation can be kept healthy and attractive using pruning shears and weed pullers, and mowers can be used to maintain turf grass at an appropriate height.

11.11.1.7 Downspout disconnection

The operation and maintenance practices include the following:

a. Inspect pipes, bends, and connections for leaks and defects.
b. Remove accumulated debris, especially from gutters.
c. Check materials for leaks and defects.
d. Remove accumulated debris, especially from gutters.

11.11.1.8 Vegetated kerb extension

Maintenance practices include the following:

a. Removal of accumulated debris.
b. Clean inlets.
c. Relatively inexpensive to retrofit ($300/m^2 for new construction).

11.11.1.9 Rainwater harvesting systems

a. To do before the rainy season:
   - Continuously clean the tank, roof, gutters, the mesh filters, and the downpipe.
   - It is advisable to discharge before next storm event.
b. Clean annually and check for loose valves, etc.
c. May require flow bypass valves during the winter season.
d. At the end of the rainy season, make sure the hatch is closed properly or when the tank is full, make sure no animals, mosquitoes, or light can enter the tank (this can contaminate the water).
e. A ladder may be needed for inspecting roof drains that connect to rainwater harvesting systems.

11.11.1.10 Green roofs (vegetated rooftops)

General maintenance practices for green roofs include the following:

a. Green roofs require some support during establishment and yearly maintenance thereafter. Plants or sprigs should be irrigated until established, and additional plants or sprigs added to ensure good plant coverage, if necessary. With drought-resistant vegetation, irrigation of an extensive green roof is rarely necessary after the two-year establishment period.
b. Weeding and mulching may be needed during the establishment period and periodically thereafter over the life of the roof. Any woody plant which becomes established on the roof need to be removed regularly.
c. If necessary (many roofs can survive on deposition of airborne nitrogen and biomass breakdown), application of a slow-release fertilizer once a year will ensure continued vigorous growth of the vegetation. Soluble nitrogen fertilizers and compost should not be used due to the potential for nutrient and bacteria export.

Operation and maintenance 409

d. Once vegetation is established, little to no maintenance needed for the extensive system.
e. Maintenance cost is similar to traditional landscaping, $2.5 to 12.5/m².

11.11.1.11 Detention and retention ponds

Retention pond or detention pond maintenance checklist (Community Association Management).[3]

a. Keep the earth and dam around your retention pond in good order. The vegetation around your retention pond will reduce the pollutants in the stormwater; however, the vegetation should be well maintained, and any overgrowth should be reduced. It is also a good idea to remove any new trees that may cause future problems.
b. On a periodic basis remove any debris and any silt buildup from your retention pond.
c. As the detention ponds are located close to residential areas, it would pose an added threat to the main ponds on algae bloom by such common activities as fertilizing and watering lawns, washing cars, and painting houses. The simple act of water runoff from fertilization will increase the buildup of nutrients in the main ponds. The ponds with shallow water under the warm weather can create an algae bloom that will cause the main ponds to attain a green scum or large clumps of algae floating on the surface. This bloom can remove the oxygen from the water and kill off any fish or other aquatic inhabitants that are present, creating an unsightly and smelly mess. The main ponds must, therefore, be regularly inspected and cleaned up to avoid the problem.
d. Communicate with the homeowners in your neighborhood or the tenants in your commercial space and make sure everyone understands the importance of reducing the chemicals, pollutants, and waste products that make their way down the storm drains in the neighborhood or office park.
e. Once constructed, the ponds should be inspected after several storm events to confirm drainage system functions, bank stability, and vegetation growth. The outlet structure should be inspected for evidence of clogging or outflow release velocities that are greater than design flow.
f. Inspect the headwall, the weir, the exhaust, and other key components of the retention pond on a regular basis to ensure the pond is operating as intended.
g. During the rainy season, the main ponds retain about 1-m depth of water. Warning signs and safety barriers must be provided to prevent any children from playing close to the ponds. More frequent inspections must be carried out to avoid any accidents. Alternatively, CCTV monitoring should be provided.
h. Remove any silt or sediment that may have accumulated at the basin forebay on a regular basis.
i. Inspect the stormwater drains that are delivering water to the retention or detention pond and make sure they are free of debris and in good working order.
j. Sediment removal from detention basins. Dry basins accumulate sediment with time that will gradually reduce the storage capacity available and can, in some cases, also reduce sediment trapping efficiency. Also, sediment may tend to accumulate around the control device, which increases the risk that either the orifice may become clogged

or that sediment may become re-entrained into the outflow. Where basins are amenity features, sediment accumulation is likely to be unsightly and reduce the amenity value of the component. Sediment accumulation should be monitored as part of the inspection regime for the surface water management system and appropriate frequencies determined for removal and disposal. Small volumes of sediment can usually be removed by landscape contractors using hand tools. Sediment excavation using front-end loaders or backhoes is simple, if appropriate access is available for the equipment. Sediment removal will usually damage the vegetation, and reestablishment may be required.

k. At least twice during the rainy season, accumulated trash and debris should be removed from the side slopes, embankment, and spillway. All pond outlet devices should be protected from clogging. Sediment should be removed from the main ponds as necessary and at least once in every two years (usually sooner rather than later).

11.11.1.12 Soakaways

A soakaway system normally includes screens, a catch pit, a septic tank, soakaway pit and trenches, and the associated dung channels. Farms which can achieve very high levels of dry removal of solid waste, and lowlevels of water use (i.e., chicken and small duck farms), will not normally require septic tanks. Maintenance requirement: Periodic sludge removal, at least once in every six months but once every three months is normally expected. The soakaway pit/trench purpose: Percolation of wastewater to surrounding subsoil.

11.11.1.13 Planter boxes

Some maintenance practices of a planter box resemble those of a bioretention garden. See rain garden maintenance. Bypass valve during winter. Maintenance cost: $400 to 500 per year for a 150-m² planter, varies based on type, size, plant selection, etc.

11.12 MAINTENANCE OF STORM SEWERS AND DRAINS

There is a need to prepare routine oversight and monitoring plans for storm sewers and drains. Sewers with flat slopes may be inspected more frequently than those with mild or steep slopes. Inspections are visually done from manhole to manhole by placing a bright light in the manhole toward which the inspector is looking and a mirror is lowered on a pole into the manhole so that the inspector can view the inside of the sewer while standing at the street level. Rodding from manholes is used to locate damaged sewer sections. Some of the items that cause damage to storm sewers and would need routine or periodic inspection to delay, prevent, or correct the defect on the storm sewer include the following.

11.12.1 Excessive surface loads

Excessive surface loads risk pipes to fail by developing longitudinal fracture at the springing level, followed by cracks at the invert and crown. The excessive loads can result from heavy trucks passing above the pie beyond the limit specified for the sewer pipe, for example, pipes laid under roads and parking areas.

11.12.2 Corrosion of sewers

Corrosion is the primary factor affecting the longevity and reliability of offshore and onshore buried pipelines throughout the world. As a maintenance practice, epoxy coatings offer excellent resistance to high temperatures, chemicals, and corrosion. In concrete pipes, this attack can happen both above and below the water line. The corrosion is predominantly caused by acid attack on the cement paste in the concrete. Biogenic acid attack has been identified as the prominent concrete ailment in sewer, which is accounted for almost 40% of the damage in concrete and brickwork sewers (Shima et al., 2018).[4]

Biogenic sulphuric acid corrosion is often a problem in sewer environment; it can lead to a fast degradation of the concrete structures. Since the involvement of bacteria in the corrosion process was discovered, considerable microbiological research has been devoted to the understanding of the corrosive process (Monteny et al., 2000).[5] The resistance of several concrete mixes against biogenic sulfuric acid corrosion varies. Corrosion can make pipe walls thinner, making pipes more vulnerable to breaks. Therefore, removing the corrosion can restore water to flow well.

11.12.3 Root intrusion

Root intrusion is another issue that blocks storm sewers. Openings in sewers may cause soil to enter and root invasion affecting the performance of the sewer. When pipes develop cracks, holes and joints weaken, they become susceptible to root invasion. Sometimes, roots find way through poorly jointed sewers. And when the roots find their way inside the pipe, they significantly reduce the hydraulic efficiency of the pipe as they keep multiplying and spreading working their way through the system at a rapid pace. Slowly, clogs start to develop due to debris, sediment, grit, and grease failing to pass through the sewer and becoming stuck among the roots, causing a blockage.

11.12.4 Sediment and grit

Sediment and grit deposits may come with the runoff. The construction slopes are a determinant factor for how much sediment gets deposited in sewers. Some areas are prone to erosion and it becomes inevitable for the flow to carry sediment loads. Sediment traps are particularly installed in sewers to localize sediment loads. Inspectors always single out specific points along sewers to check for accumulated sediment and grit deposits. Also, construction sites, disturbed areas, and gravel roads tend to produce huge loads of sediment and immediate drainage lines get clogged with sand, silt, and other materials.

11.12.5 Trash and debris

These include garbage, plastics, bottles, paper, etc., usually dropped in sewers. Trash and litter pickup is vital. Trash should be removed on a routine basis as part of maintenance activities to reduce the potential for clogging during storm events. Trash can be concentrated on outlet orifices, trash racks, basin and swale floors and side slopes, and other components, as well as from the area surrounding the SuDS. Trash and other debris can pollute surface waters and damage or constrict stormwater control devices.

11.12.6 Hazardous materials

Grease and oil are not allowed to go down the sewer or drain unless the concentration is less than 500 mg/L. However, in some cases, parking areas and garages leaking/excess oils and grease might mix with rainwater producing a relatively high concentration as automotive oil, gear oil, and machinery grease runoff. These are hazardous chemical wastes that need grease traps installed in sewers to remove the pollutant loads. Grease and oil block sewers.

Other materials that cause environmental concern when they find access to storm sewers include dead animals, garbage, gases and vapors, and flammable and explosive liquids.

Vegetable oils and lard mainly from pork, if not placed in regular trashcans, could block sewers if it mixes with stormwater runoff.

Industrial and commercial areas, hotels and restaurants may constantly produce wastes that could pass through a partially combined system and needs special attention treating the runoff-foul water mix. Such areas may include bakeries, restaurants, creameries, and hotels. Sewers in these places are most susceptible to blockages due to grease on pipe walls. Although uncombined systems are advocated for globally, once the risk of runoff-foul water mix is detected, grease traps or grit separation facilities must be installed in sewers and drains. Ordinances and byelaws on grease removal are particularly designed in many municipals globally aiming at environmental protection.

11.13 STORM SEWER REPAIR AND MAINTENANCE PRACTICES

Specialized equipment and machinery are used for storm sewer maintenance, especially clearing plugged or partially plugged pipes. Routine sewer inspections are usually conducted following a monitoring plan to avoid blockages. Inspections are intended to prevent any potential discharges of pollutants into waterways and to plan for remedial actions following the existing condition of the inspected drainage assets.

The damage to sewers resulting from surface loads, corrosion or differential settlement should quickly be repaired without delay. A broken section of sewer pipe can be removed and the flow temporarily diverted by pumping from one manhole to the next. Drainage sewers are usually connected with a network of manholes, junction boxes, inlets, outlets, outfalls, kerbs, etc.; these are inspected to evaluate the existing condition and functional and physical state of the asset. In many cases, inspections are conducted from manhole to manhole with bright light placed in the manhole toward which the inspector is looking. Then, a mirror on a pole is lowered into the manhole and the inspector can ably view the inside of the sewer while standing on the street level. Sometimes ground sensor technologies are used to keep track of the performance of sewers. Also, television cameras are passed through sewers locating the blockages, breakage, and lose joints.

Broken concrete sewers are repaired by removing the damaged section of the pipe and replace it with a new one and pumping flow from one manhole to the next. It is quick to replace a plain-end concrete pipe compared with the socket and spigot type because the upper section of the socket of the section downstream has to be chipped off.

Making new connections to existing sewers should be inspected by municipal crews to ensure that the work is certified to conform to required standards. New connections should be carefully made by redesigning nodes such as manholes and junction boxes. In cases where uPVC or HDPE pipes are used, wyes and tees should be inserted or wye or tee saddles should be used. To uphold the operability and functionality of drainage assets, the above hazards are dealt with as follows:

11.13.1 Roots invasion in sewers

These are removed using a water jetting machine or a hydro-jetter. The hydro-jetter is a giant pressure washer used for the inside of piping. It has nozzles that rotate 360 degrees as they clean, ensuring that every part of the inner piping is thoroughly cleaned. The high-pressured water can cut the roots out of the pipes without damaging the pipe.

Avoiding roots from getting into the drainage system should be the first step to eliminate the risk of damaging the system. It is always better to repair broken pipes as soon as possible avoiding seepage and leaks since roots follow the water into the pipe. Through sewer inspections, spots such as cracks, holes, and weakening pipe joints can be identified on sewers and immediate action taken to correct the defect. Large storm sewers may require a cutting drag pulled through by and cable and winch.

11.13.2 Corrosion control measures

Due to variations in concrete composition, one treatment for corrosion might not be suitable for every sewer pipe and the efficacy of a treatment should be examined prior to application in order to determine the most effective form of treatment for that particular situation (Shima et al., 2018).[4] However, building reliability in the drainage system should be done at inception and design stages to ensure longevity of the pipes. The right materials are specified during design conforming to site and environmental conditions. The following measures should be considered during design, operation, and maintenance of storm sewers for corrosion control in sewers. The design is expected to provide for self-cleansing velocity, good ventilation, low turbulence, flushing facilities, minimal periods of flow, and minimum stagnation.

1. The design of sewer section with a depth of flow of about 0.8 D will minimize the chance of corrosion.
2. Pipes made of inert materials are preferable.
3. In case of large-diameter pipes, RCC with sacrificial lining of 25- to 50-mm thick is the suitable pipe material.
4. Lining the inside of the RCC pipe with sulphate-resistant or high alumina cement as sacrificial layer may increase the life expectancy of the pipe by three to five times.
5. RCC pipes are manufactured with sulphate-resistant cement when the soil contains sulphur and other corrosive substances.
6. For metallic pipes (DI or MS) the acceptable linings are cement mortar lining either with sulphate-resistant or high alumina cement.
7. Good ventilation usually removes condensation in the immediate vicinity of the air inlet.
8. Periodic flushing of sewers is necessary to remove solid accumulation and control their subsequent anaerobic decomposition and hydrogen sulphide (H_2S) formation.

11.13.3 Grease and oils

Grease traps are used to block grease and oils to flow with runoff to receiving waters. The removal of grease from the sewers is done by employing hand-driven or motor-driven rotating tools mounted on the end of a flexible tape. Grease is removed from relatively small sewers by cutting tools followed by brushing with a wire brush.

11.13.4 Sediment and grit

Buckets and scoops perform well removing sand and grit. These can be pulled through the sewer by a cable and winch. If the deposits are extensive, vacuum trucks are used to clean the sewers. In relatively small deposits, turbine cleaners with water-powered rotating cutter is drawn through the pipe by a cable as it flushes the deposits from the line. Sometimes, flushing with water from fire hydrants is enough to remove grit but cannot remove grease and root invasion. Enough care must be exercised while flushing with fire hydrant water to prevent pipe surcharging causing water to back up in basement drains or household drains.

In some cases, inflated soft rubber balls adjusted very close to the size of the sewer are used to remove grit deposits. The ball self-adjusts conforming to irregularities in the sewer while the water held behind it exits around its edges at high velocities, flushing the deposits. The ball can also be relatively effective in removing grease deposits and breaking root intrusion of up to about 6 mm especially when the flow head behind the ball is slightly above 0.3 to 1.2 m, although sometimes this head may cause the flow to back up into adjoining assets and lower heads up to 0.05 m would be considered safe.

11.13.5 Trash and debris

Trash and litter can be picked regularly by hand or specialized trucks. Flow-through devices remove trash, debris, and coarse sediment in the water by capturing it in a tubular entrapment device installed in drainage sewer system. Debris can also be removed at a manhole by inserting an ell section in the outlet with the open end turned upward. In that case, the manhole serves as a trap and retains the solids. To avoid material removed during sewer cleaning, it should be immediately removed in the next manhole hence preventing it from forming a blockage due to cleaning operations in a downstream line.

11.14 COMMONLY RECOMMENDED OPERATIONAL AND ROUTINE MAINTENANCE MANAGEMENT PRACTICES FOR STANDARD PIPEWORK DRAINAGE SOLUTIONS

The tasks associated with routine maintenance include the following:

1. **Employee training**—public authorities need to consistently teach their staff about stormwater management, potential pollution sources, and best management practices.
2. **Improvements in housing standards.** An organized housing plan leaves space for drainage structures, that is, drainage easements.
3. **Community sensitization.** The population must feel they are a part of the drainage project so that they can maintain it and not look at it as an intrusion.

4. **Identify and prohibit illegal or illicit discharges to storm drains**—inventory of outfalls, field investigations, and follow-up visits for suspected problems within the storm sewers and drains are some of the steps cities/municipals should undertake to pinpoint sources and implement corrective actions.
5. **Clean and maintain storm inlets, catch basins, and drain channels**—inlets, catch basins, and manholes should be periodically inspected and cleaned out using a vacuum truck.
6. **Street sweeping**—Remove sediment on roads and reduces the amount of pollutants entering local waters.
7. **Permanent/temporary seeding**—An inexpensive, yet effective, method to stabilize flat areas and slopes. Cities/municipals adopt this method and promote its use to the community.
8. **Vegetative buffers at watercourse**—A specified width of buffer between a construction site and an adjacent watercourse. The buffer acts as a filter to reduce soil erosion and sedimentation from entering the watercourse.
9. **Vegetated buffer strips**—A method to reduce sheet flow velocities which may create rilling and gullying. Also useful to establish permanent vegetative cover and prevent sloughing and loss of seed.
10. **Sediment traps and garbage traps**–A device used to intercept concentrated sediment or garbage disposed in channels and prevents it from being transported off-site or into a waterway or wetland. Screens should be put at recommended intervals and regularly cleaned during the operational phase of the project. This is to avoid the solid waste materials that are normally carried by stormwater runoff from entering wetlands. During the operational phase, the drainage channels should be regularly be desilted to reduce the amount of silt entering the wetland.
11. **Use of screens.** Screens should be put at recommended intervals and regularly cleaned during the operational phase of the project. This is to avoid the solid waste materials that are normally carried by storm runoff from entering wetlands.
12. **Regular maintenance of channels and culverts.** This involves desilting, cutting of vegetation around drains, repairing broken edges, etc. During the operational phase, the drainage channels should regularly be desilted to reduce the amount of silt entering the wetland.
13. **Proper disposal of garbage.** This is the most significant challenge of drainage systems especially in developing countries. Improper disposal of garbage destroys the hydraulic efficiency of drainage systems. Garbage should be disposed of properly, and this should be a routine activity to keep drainage systems functional.
14. **Upholding the desired wetland hydrology.** When stormwater discharges into the wetland, it can alter wetland hydrology, topography, and vegetative community. It is therefore advisable that the developer should ensure that soil, polythene materials, and dangerous wastes are removed or filtered from the stormwater before it is discharged into the wetland.
15. **The construction phase** of this project should ensure that wetland vegetation remains intact.
16. **Covering storm sewers.** Nutrients and sediment increase the wetland's productivity, and if poorly managed, it will lead to eutrophication of the recipient water body.

The drainage channels should therefore be covered to reduce the amount of sediment entering wetland system through stormwater.

17. **Safe discharge of effluents.** It should be a requirement for all industries to have an effluent discharge permit and should conduct a preliminary treatment of effluent before connecting their effluent to the drainage channel. This is to reduce the prevalence of dangerous chemicals being transported within the environment especially to sensitive areas like the wetlands where bioaccumulation can occur.

11.15 ROUTINE OVERSIGHT, INSPECTION, MONITORING, AND MAINTENANCE PLAN

Routine oversight and monitoring of drainage schemes is vitally important for both gray infrastructure and SuDS. It avoids blockage of sewers and improves the functionality of SuDS. The frequency of monitoring is determined by the age, type, slope of the drainage asset, pollutant potential, location and condition of the drainage asset, among other factors.

In some cases, drainage asset life spans can be up to 20 years with inspections recommended every five years. Drainage assets are highly recommended for routine inspections to uphold the functionality through routine maintenance as appropriate. If an asset is not functioning as required, the cause must be determined after inspection and the site be restored to working order as soon as practicable. Inspections are usually conducted through visual observations to evaluate how the asset is functioning relative to its intended design purpose.

Nonroutine inspections may be necessary following a heavy storm or an accident. In such cases of heavy storms, inspectors may look out for areas susceptible to flooding, erosion, and huge sediment deposits after it rains. Accidents may destroy the drainage asset calling for immediate inspections and corrective actions. Relatively flat sewers may require quarterly inspections while those with good slopes may be inspected once or twice a year.

Routine oversight and monitoring is conducted by city crew teams using inspection and maintenance checklists (monitoring tools). After the data are collected, it is entered in drainage management information systems, as appropriate.

11.15.1 What constitutes an inspection and maintenance checklist for a drainage asset?

The following are typical items considered during an inspection of a stormwater or drainage asset:

11.15.1.1 Site conditions

- Check for presence of sediment accumulation, obstructions of the inlet or outlet devices by trash, debris, sediment, and vegetative growth.
- Check for root invasion into sewers.
- Check for trash and debris disposed of in sewers, drains, and/or SuDS.
- Check for algae, water stagnation, and odors.
- Check for animal burrows.
- Check whether drainage system is vandalized.

11.15.1.2 Vegetation

- Check for overgrown vegetation cover that requires slashing.
- Check for any unwanted vegetative growth that affects functionality.
- Check for poor and distressed stands of grass.
- Check for overgrown trees.
- Evaluate the condition of the bare ground.

11.15.1.3 Structural conditions

- Check for corrosion of sewers.
- Check for cracks and deterioration of inlets, outlets, pipes, and catch basins.
- Check for water seepage or ponding.
- Check for slow draining infiltration devices that need to be corrected.
- Check for malfunctioning valves, sluice gates, locks, and access hatches.
- Check for inadequate outlet and inlet protection.

11.15.1.4 Earthworks

- Check for deterioration of downstream channels.
- Check for collapsing embankments.
- Check the condition of retaining structures.
- Check presence of excessive erosion or sedimentation, primarily in emergency spillways, filter strips, or forebays.
- Check cracks or settling in the embankment or berms.

11.15.1.5 Spills/releases

- Check for hazardous spill materials.
- Check for illegal dumping, solid waste, dead animals, garbage, glasses, etc.
- Check for illicit discharge.

11.16 CONCLUSION

Maintenance is a key component for the sustainability of drainage systems. Without proper maintenance, the reliability of the drainage system gets compromised with time and the hydraulic efficiency significantly drops. It is typically recommended, through best management practices, to continuously keep maintaining drainage structures to uphold their functionality.

When designing and developing drainage systems, it is essential to ensure that maintenance budgets are also not forgotten. This is because to uphold the desired functionality of drainage systems, the drainage structures must be properly maintained. Following the operation and maintenance plans, maintenance crews are tasked to schedule maintenance and to quickly intervene through best maintenance practices to repair broken sections on a drainage system before the entire system is damaged.

References

1. List of sewer appurtenances: Waste management (engineeringenotes.com). Accessed August 5, 2022. www.engineeringenotes.com/waste-management/sewer-appurtenances/list-of-sewer-appurtenances-waste-management/40080
2. Operation and maintenance (O&M) of tree trenches and tree boxes. Minnesota Stormwater Manual. (2021). Accessed July 29, 2022. https://stormwater.pca.state.mn.us/index.php?title=Operation_and_maintenance_of_tree_trenches_and_tree_boxes
3. Community Association Management. Maintaining Your Retention Pond. Accessed July 19, 2022. https://communityassociationmanagement.com/c29-facilities-a-maintenance/c31-amenities/maintaining-your-retention-pond/#print
4. S. Taheri, M. Ams, H. Bustamante, L. Vorreiter, M. Withford, & S. Martin Clark, A practical methodology to assess corrosion in concrete sewer pipes. MATEC Web of Conferences 199, 06010 (2018), *ICCRRR 2018*. Accessed July 29, 2022 https://doi.org/10.1051/matecconf/201819906010
5. J. Monteny, E. Vincke, A. Beeldens, N. De Belie, L Taerwe, D. Van Gemert, & W. Verstraete (2000). Chemical, microbiological, and in situ test methods for biogenic sulfuric acid corrosion of concrete. *Elsevier/Cement and Concrete Research*, *30*(4), pp 623–634. https://doi.org/10.1016/S0008-8846(00)00219-2

Further reading

1. SuDS Manual C753 Chapter List (ciria.org). www.ciria.org//Memberships/The_SuDs_Manual_C753_Chapters.aspx

Chapter 12

The future of urban drainage

12.1 LIKELY FUTURE DEVELOPMENTS IN THE DRAINAGE FIELD

Floods constantly hit several parts of the world due to changing climate as a result of human activities on the environment. The effects of floods are devastating and call for innovative and creative ways of dealing with stormwater to reduce runoff quantities or delay the movement of floods following a storm while improving flow qualities. Holistic drainage planning interventions will be at the climax in the near future requiring public authorities to enforce stormwater best management practices through all-around legal and policy frameworks as much as possible. Sustainable drainage systems (SuDS) will become a primary design consideration for most civil engineering projects, globally, necessitating an extensive shift from overreliance on standard pipework drainage solutions to SuDS, thus controlling stormwater at the source. In that case, developers will consider stormwater management a priority right from the inception project stages.

Public authorities will increasingly be tasked to design policy frameworks that require all property developers in an urban setting to incorporate SuDS in their Greenfield developments or upgrade old facilities. This requirement will need to be enforced by city crews that would inspect the SuDS developed at specific sites to ascertain whether they conform to the required standards. There will be a need to issue planning and development permits for SuDS, and property development plans will only be approved when SuDS are a part of the master plan.

In line with the legal point of stormwater discharge (LPSD) permits, SuDS approval would be a success when design sheets, together with calculations, are appended to reports. This would provide the basis to understand whether sustainability metrics are sound enough to guarantee Greenfield development or upgrading old facilities. The 'polluter must pay' sustainable development principle would gain more enforcement in future owing to the demand to meet sustainability goals.

With the world getting increasingly anxious about the sustainability of life on earth we know it, digital planning is gaining momentum. The future is destined to promote digital technologies as much as possible so that drainage asset management is conducted on digital platforms. In several communities, drainage systems inspectors fill the collected data in mobile applications customized for drainage maintenance and repairs. This will be extensively promoted to reduce carbon associated with stationery.

As municipals are increasingly required to reduce the frequency and volume of combined sewer flow (CSO) events, emphasis is being placed on implementing alternative ways of managing urban stormwater runoff employing SuDS. Through legislation, ordinances, and byelaws, the future promises that public authorities are likely to get highly precipitated and issue stormwater runoff reduction policy measures such as installing rainwater harvesting (RWH) and infiltration systems on Greenfield and retrofitting projects. The RWH technique reduces water runoff as it impedes water from roofs and pavements from entering the drains. RWH systems can either be underground or above the ground. In future, public authorities are likely to enact policies requiring developers to utilize the RWH and infiltration techniques to reduce stormwater runoff. This will help control excessive pollution of the receiving waters.

12.1.1 The concept of a sponge city

The concept of sponge city is very similar to that of SuDS. The idea of sponge city originated from China. Prof. Yu Kongjian, who is considered the brain behind the idea of a sponge city, believes China's coastal cities, and other places around the world with a similar climate, have adopted an unsustainable model for building cities. He said, 'The technique that evolved in European countries cannot adapt to the monsoon climate. These cities fail because they have been colonized by Western culture and copy their infrastructure and urban model'.[1]

Prof Yu began advocating an urban design philosophy based on traditional Chinese concepts. He was convinced that the 'grey, lifeless infrastructure' was unsustainable and could not solve the problem of flooding, a problem he attributed to copying Western culture and their infrastructure and urban model. The sponge city concept as developed in China emphasizes the traditional practices of managing and conserving stormwater and stormwater runoff in a way that mimics the natural environment and that considers the climate of the area. Besides sponge cities, for instance, he calls for natural rustic landscaping or a 'big feet revolution', in opposition to overly manicured parks which he likens to the outdated Chinese practice of binding women's feet.

The idea of a sponge city has since faced numerous criticisms from several experts and many experts are not sure whether with heavier storms the sponge city can be effective. For example, flood management expert Faith Chan of the University of Nottingham Ningbo said, 'Sponge cities may only be good for mild or small rainstorms, but with the very extreme weather we are seeing now, we still need to combine it with infrastructure such as drains, pipes, and tanks'. He also pointed out that for many dense cities where space is a premium, it may be difficult to implement some of Yu's ideas such as providing land for floodplains.

The special attribute about a sponge city is that, as an alternative use of the urban subsoil, a sponge city has potential for reducing the effects of climate change. By adopting the sponge city principle, the subsoil of a city becomes a space for storing stormwater and green spaces become the city's natural 'refrigerators'. The promotion of the sponge city principle and the development of sustainable storage and irrigation systems are central tasks for the future for climate-adapted cities. These sponge city structures include storage tunnels, rain gardens, wetlands, and bio-swales that help the city to soak up extra water.

One of the main challenges China faces is the growing urban population. Out of 1 billion people in China, 896 million reside in cities which requires stronger infrastructure, which in turn leaves very little space for the rain to be absorbed. The Chinese government plans to build 30 sponge cities around the country with the aim of urban areas absorbing and reusing at least 70% of rainwater by 2030.

In 2015, the Chinese government announced a multimillion Yuan plan and an ambitious goal to achieve by 2030 where about 80% of China's municipal areas would have elements of a sponge city and to recycle at least 70% of rainfall. One of the China's biggest sponge city is in the Pudong district in Shanghai. In Linglang, for example, the sponge park central reservations work as rain gardens, allowing rain to be absorbed by soil and plants while water-absorbent bricks take the place of concrete sidewalks.

Around the world, more places are struggling to cope with more extreme rainfall, a phenomenon that scientists have linked to climate change. As temperatures rise with global warming, more moisture evaporates into the atmosphere, causing heavier rain. And they say it will only get worse; in the future, rainfall will be more intense and severe than previously expected. Therefore, immediately, the concept of a sponge city would be embraced by many cities.

12.1.2 Flood warning systems

Automatic warnings are usually sent to emergency stations to allow for enactment of emergency operation plans for flooding. An effective flood warning system enables public authorities to take appropriate action such as evacuation and control of dams in advance of a flood event, often leading to much less catastrophic outcomes when severe weather strikes. It is well-known that flood warning systems save people's lives and property and reduce the limit of damage to infrastructure through monitoring rainfall and water levels. This is done by alerting management officials when conditions approach dangerous levels.

Flood warning systems are early warning that are implemented as a chain of information communication systems and comprises sensors, event detection, and decision subsystems.

Following the threat of climate change, the future promises an increase in flood warning systems installed in many places around the world to monitor weather conditions for effective flood risk monitoring and flood emergency actions to safeguard communities from the devastating floods. Several communities increase their emergency preparedness by employing flood warning systems and delineating and regulating development in floodplains.

Emergency actions to safeguard the communities can range from evacuating neighborhoods to barricading roadways that are prone to flooding. The measurement tools are installed such as water pressure transducers, rain gauges, temperature and humidity sensors, and water level sensors that feed data to the central computer system. These data are fed in real time to a public website to create awareness to the public. Satellites are used to deliver early warning of dam failures using remote sensing and the other using quantum physics to provide better insight to ground conditions. Flood warning systems primarily rely on radar and rainfall and stream flow gauges which are connected by satellite transmitters to feed real-time data to central computers that can be accessed on the internet.

The computers constantly monitor water levels in rivers or lakes, rain depth, and intensities. Therefore, when preset thresholds are reached, the computer combines the real-time data, weather forecasts, and watershed models to predict when and where the water levels will crest.

There is urgent need to understand current flood risk to plan better for future resilience of cities. By studying past floods, several agencies can profile under which conditions certain streams and rivers overflow and actions can be implemented. Thus, an effective flood monitoring and risk mitigation is achieved through a network of flood warning stations installed at numerous points within a watershed. Monitoring stations log both rainfall and water levels and provide advanced warning of dangerous conditions.

The effects of floods can be devastating. It can generally lead to both pecuniary and non-pecuniary losses, an outbreak of diseases, depression, and trauma in affected peoples, loss of property, and loss of life, destroy structures and property, inundate agricultural fields, and spread disease. As climate change increases the frequency of extreme weather events, unprecedented floods and their consequences will become a more serious concern worldwide. In history, several remarkable flood incidences happened, which are outlined below.

12.1.2.1 Remarkable flood impacts

a. In July 2021, the Henan province of China experienced torrential rains that affected tens of millions of people forcing thousands to evacuate their homes—signaling the need to address sustainability globally. This is greatly attributed to climate change, whose consequences lead to floods.
b. 2020—Midland, Michigan: Two dam failures released a flood that damaged 2500 buildings and caused over $200,000,000 in damage.
c. 2017—Texas: Hurricane Harvey destroyed over 200,000 buildings and caused over $125 billion in damage in a torrential flood.
d. In July 2016, China's Hunan Province experienced traumatic floods that swept away property, left many people dead, and registered direct and indirect heavy pecuniary losses.
e. 1931—China: Record rainfall and floods killed hundreds of thousands of people through drowning, famine, and disease.
f. 1913—Dayton, Ohio: Torrential rains and flooding killed over 400 people, destroyed thousands of buildings, and caused billions of dollars' worth of damage (adjusted for inflation) throughout the Miami River valley.

12.1.3 Drainage easements

In order for a municipal drainage system to work effectively, the municipality must ensure that drainage easements are in place so that city workers are able to access private property in order to maintain and repair drainage systems. For example, if a culvert runs through a property, the drainage easement allows the city to replace it if it is damaged. The drainage easement benefits the property owner as much as it benefits anyone else.

The increasing need to control the effects of climate change such as flooding, the size of drainage easements may need to be increased in several areas around the world to cater for the drainage network expansions, providing land for floodplains, and the potential development of SuDS in some parts of drainage easements.

By definition, a drainage easement is an attachment to a property deed which states that access to part of the property is given to a third party, usually a municipality, for the purpose of maintaining drainage. The drainage easement may include a culvert or drain which feeds into a drainage system or the easement may simply state that runoff needs to be allowed to flow freely over an area of the property. The easement cannot be lifted from the deed unless there are special circumstances, and it will be associated with the deed even when it is transferred or sold.

When a drainage easement is in place, there are restrictions on how the easement area can be used. People usually cannot degrade the soil or build structures because this could impede the free flow of stormwater over the property. If there is a culvert or pipe, trees cannot be planted over it, because their roots could block it, and people are expected to keep gratings clear so that the water can drain properly. If people would like to change the terms of the easement, they must reach an agreement with the agency which holds the easement.

If the drainage system does not work properly, the property developer can be at risk of flooding and other problems. In exchange for allowing the municipality to access the property, the property owner gets to enjoy a property which drains freely and knows that neighbors are also obliged to maintain their drainage easements to ensure that water intrusion will not occur along the property line.

Drainage easements can vary in size and location, depending on the property. Easement areas are generally near the perimeter—an edge of backyards, for instance. They can contain drain grates and underground pipes or be simple constructions with a culvert and carefully sloped land. An easement might take up just a small area of a property, but it all depends on what the municipality determines that it needs. Minimum easement widths of about 9 m are common, but some contracts may allow temporary work easements on either side of a permanent easement area.

Easements, in general, are used to address cases in which someone other than the property owner needs the right to access the property, and the easement comes with the right to use, but not full ownership rights. Another common example of an easement is a road easement, in which part of a property is used for a shared or even public road and an easement protects the rights of others to use that road. Easements are drafted by lawyers who work with the property owner and the parties who need access to ensure that the document is accurate and fair.

12.1.4 Vertical forests and forest cities

Vertical forests could be the breath of fresh air that pollution-choked cities desperately need. Vertical forests are high-rise buildings covered with trees and plants which absorb carbon dioxide, filter dust from pollution, and produce oxygen. They are also an ingenious way of planting more trees and creating habitats for wildlife in cities that are squeezed for space.

It has recently become part of modern construction practices to incorporate vertical forests in estate development. Vertical forests are designed and planted on going upward on a building structure. They are very good at absorbing carbon dioxide. Vertical forests are high-rise buildings covered with trees and plants which absorb carbon dioxide, filter dust from pollution, and produce oxygen. They are also an ingenious way of planting more trees and creating habitats for wildlife in cities that are squeezed for space. Vertical forests could be the breath of fresh air that pollution-choked cities desperately need.

In February 2017, China issued ambitious plans to build vertical forests as towers that would produce 60-kgs of oxygen per day. The towers would be home to more than 1,100 trees and 2,500 cascading plants, and they were estimated to absorb about 25-tons of carbon dioxide every year. To put things in perspective, saving 25 tons of CO_2 would be equivalent to taking five cars off the road for a year. Chinese cities have some of the most polluted air in the world (Hutt, 2017).[2]

Countries like China, Italy, and Singapore are fast-tracking vertical forests and forest cities to help absorb carbon dioxide and produce oxygen for their cities. This is increasingly becoming beneficial in building resilient cities and can equally form an integral part of SuDS because of mimicking the natural environment, creating cool temperature ranges, and contributing to the conservation of the hydrology of the place. The future promises vertical SuDS, and vertical forests integrated to solve drainage problems.

12.1.5 Verticulture

The future promises to invest more in verticulture and stormwater management. As part of controlling stormwater at the source and harnessing the benefits that come along with it, verticulture shall gain value over the coming years. Similar to vertical forests, verticulture can be practiced on buildings. Stormwater can be harvested and utilized in vertical farming practices. The potential location of vertical systems means that the cost of transport is nearly nullified as consumers may access them within urban areas. Also, the enclosed feature of verticulture means that pests and parasites are easily controlled, reducing the use of pesticides to a minimum.

12.1.6 The future of trenchless technology

Trenchless tunneling technology will continue to develop alongside the growth of pipeline infrastructure. The worldwide population growth, ongoing urbanization, and climate change demand suitable underground infrastructure. The successful installation of a pipeline using pipe jacking or micro-tunneling relies on a combination of planning, investigation, technology, and experienced application. The omission of one of these factors, or the incorrect approach to any of them, can result in the failure of the project or, at least, difficult recovery operations leading to a significant increase in costs. The designer needs to gather as much geological information as possible and seriously consider early contractor involvement to maximize any construction opportunities available. It is essential to find a reputable trenchless contractor with a robust track record and reap the benefits of their experience. It will almost always save your time and money. Of recent, trenchless technology is proved to be environmentally more viable, and it is becoming more cost-effective to bore major roads and rails.

The future promises interchangeable machines capable of installing a variety of pipe sizes through a wide range of geological conditions. To achieve this, available torque in a more compact drive unit will be required, and this will be realized as advances in electronics and hydraulics become commercially available. At present, rock micro-tunneling is generally successful from 900-mm diameter. For example, using air hammer technology and smaller cutters will allow small-diameter bores to be driven from 300- to 600 mm in medium hard rock. However, in the future, we see improved hardened surfaces and smaller cutter design below 150 mm to drill a rock, improved dewatering methods for fine sand, silt soil units, and clay, and real-time navigation systems that can help guide long-distance micro-tunneling.

12.1.7 Polyhedral pipes

Polyhedral pipes are one way to solve the problem of floods in cities. The advantage of polyhedral pipes is that they can have any cross-sectional shape, including a rectangular cross-sectional shape, which is possible only for reinforced concrete pipes. During storm sewer repairs, maintenance or upgrades, replacing the old pipes with new ones, the diameter of which is two times larger, usually increases the throughput of the system four times, requiring three times the area of the road.

Transverse elements are connected at their ends and/or through the corner elements into the shape of a regular convex polygon, and laterally connected into a polyhedral composite pipe. This pipe can have from three to any desired number of faces (Georgievich, 2016).[3]

In congested cities, it is almost difficult to occupy such a space because the entire area of roads in the city environment is already occupied by several other underground service networks such as water pipelines, sewers, hot water and heating pipes, and power and communication cables. Therefore, to do this, we may need to transfer all these networks to some other location. However, polyhedral composite pipes present a great opportunity to solve the problem.

12.1.7.1 Advantages of polyhedral composite pipes

a. Polyhedral composite pipes of rectangular cross section, having a width equal to the diameter of a round pipe, are able to have a cross-sectional area several times larger than that of a round pipe; the increase in the cross-sectional area occurs due to the larger size of the polyhedral pipe in height.
b. Flexible polyhedral tube manufacturing technology is similar to 3-D printing, where a polyhedral pipe is continuously assembled and glued from hard pieces of hollow composite profiles, which are cut and processed on a Computer Numerical Control (CNC) line before assembly.
c. The use of hollow composite profiles gives the greatest possible strength to the polyhedral pipe. Polyhedral pipes have less weight and material consumption compared to round pipes of the same strength. The difference in weight and material consumption can be multiple. Accordingly, the cost of polyhedral composite pipes is several times less.
d. Polyhedral composite pipes have high strength and can lay on the same level with the road surface, while round storm sewer pipes are usually laid at a depth of 1 m from the top of the pipeline to protect them from damage.
e. Polyhedral composite pipes can have any cross-sectional shape, including a rectangular cross-sectional shape, which is always possible only for reinforced concrete pipes.
f. A polyhedral pipe can produce with a gradually increasing cross section, not only in height but also in width. Such a drainage pipeline with an ever-increasing cross section will keep increasing its productivity, which will provide the necessary throughput along the entire length.
g. The technology for the production of polyhedral pipes allows you to manufacture any parts for multifaceted pipes: tees, transitions, bends with any turning radius. This technology can be used to make wells and chambers, as well as tanks for storing rainwater reserves underground.
h. A continually increasing cross section of a drain pipeline from polyhedral pipes will ensure its steadily increasing throughput and the intake of all rainwater from the surface in heavy rain.

12.1.8 Digital monitoring of sewers

Several countries have developed mobile applications used during the periodic inspection and monitoring storm sewers. During the operation and maintenance of storm sewers, the data collected about the condition of the drainage asset are fed into the smart phone application. These data are processed and information used to improve or upgrade storm sewers. Thus, helping to uphold the drainage asset functionality over the design period. Now that the world is fast-tracking sustainable drainage systems, an inventory of drainage assets and SuDS would form the basis of operation and maintenance planning.

12.1.9 Work planning and budgeting

It is primarily because drainage projects do not come with revenue streams; they are often not prioritized by the public authorities. Therefore, they are majorly funded on a cost-effectiveness approach, evaluating the lifecycle costs of the different materials or drainage schemes. In that case, the evaluation of lifecycle costs is based on initial capital costs, operation and maintenance costs, replacement costs, rehabilitation costs, and residual costs, which usually factor in the component of inflation and interest rates over the design period.

Therefore, annual work planning and budgeting for stormwater management would encompass software and hardware activities. This may include capacity building programs, among other institutional strengthening programs. The future promises advanced integrated SuDS to become interesting projects whereby they would be managed in the same way other municipal works, such as portable water, are managed. This would be achieved through fixing utility fees such as legal point of stormwater discharge fees, among others.

12.1.10 Nonconventional drainage design methods

For many years, the conventional design method has been used for the design of drainage systems. The problem with conventional approach is that it does not just translate to a conventional rainfall-runoff modeling for designing drainage system components, but also implements conventional management methods which ignore the interaction between drainage systems, the environment, and the society. Regardless of all improvement in drainage systems design and planning methods, growing problem implied a deficiency in the conventional approach. Various case studies investigated the causes and effects of urban drainage failures in different locations and reported that the climate change and urbanization were the most critical factors impacting on the performance of conventional drainage systems (Nie et al., 2009).[4] Global climate change could affect the drainage systems in many ways. It places a strong barrier to the provision of reliable hydrological input data for the design process (Yazdanfar & Sharma, 2015).[5]

12.2 FUNDING OPPORTUNITIES FOR DRAINAGE SCHEMES

Funding opportunities for drainage schemes include loans and grants, donations, and own source revenue collections. In some areas, own source revenue streams are the commonest, decentralized approaches which present opportunities to finance and maintain drainage projects. Drainage assets are community facilities that can be funded using direct loans from banks.

The applicants may include individuals, public bodies/entities, nonprofit organizations/corporations, and water and sewerage corporations. The objective of funding drainage schemes includes providing financing for essential community facilities for rural communities and wastewater and stormwater financing in rural areas. Several financing types are available across the world which include direct loans, grants and loan guarantees (up to about 90%), and own source revenue. The purpose and uses include build, repair, and improve stormwater systems and waste collection and treatment systems. Other related soft costs that need funding include engineering, land, legal, etc.

12.3 COST IMPLICATIONS OF DRAINAGE SCHEMES

The cost of drainage schemes varies wildly from place to place, globally. The cost is always derived summing up the direct and indirect costs. The cost of materials, transportation of materials, labor costs, and professional's fees in design and supervision are summed up to obtain the cost of drainage schemes. These costs vary from one place to another.

Due to climate change, storm sewers are now designed to cope with the ever-changing climate. The risk of flooding is lowered when storms of appropriate recurrence intervals are selected as design storms. When the risk is lowered and climate change safeguards and adaptability measures such as SuDS are integrated in the drainage schemes, the total project cost is derived. Nowadays, modern drainage systems planners and designers should think green and employ at least one of the many SuDS practices. For example, RWH systems and underground detention systems are the commonest in residential areas, especially in emerging economies.

The higher the flows accommodated by the drainage storm sewers, the higher the investment cost because the hydraulic capacity of the storm sewers is decided following the volumes and peak flows. The terrain on which the drainage scheme is to be built also influences the cost as controls may be required in the drainage channels and storm sewers to properly conserve the flood waves within the drainage system. The availability of materials, skilled personnel, equipment, and machinery influence the cost of the project. Both SuDS and gray infrastructure call for technical competence in design and construction. Therefore, enhancing reliability in drainage projects requires employing skillful people right from the start. This helps to avoid reliability costs that accrue when additional works become inevitable half-way the development of drainage schemes. That means maintenance crews must form a significant part of the planning process, especially for the SuDS, and the budget for operation and maintenance of a drainage system must be sought right from the start. Lifecycle cost analysis must be properly conducted to weight the initial cost against operation and maintenance costs for the integrated drainage project to be deemed feasible and sustainable.

The reasons for planning and developing drainage schemes vary wildly and this influences the cost of the schemes going forward and the overall sustainability metrics. In cases where sustainability is considered the overarching goal, SuDS must be an integral part of the drainage systems, but, of course, site-specific conditions influence the choice of SuDS. Standard storm sewers are now built and supplemented with SuDS with the overarching goal of mimicking the natural environment as far as possible. The total cost, therefore, weighs the initial cost investment in both SuDS and gray infrastructure drainage assets. Although the initial cost for an integrated drainage scheme may be high than that of only standard pipework solutions, the advantages that accrue in the long run are worth it, considering the fact that the world is fast-tracking the need to promote the green campaign. Therefore, initial cost implications for an integrated drainage system development is always higher than overreliance on gray infrastructure drainage assets alone and that means sound government policy is the tool to unlock SuDS worldwide.

12.4 FLASH FLOODS

A flash flood is a rapid flooding of low-lying areas, which are typical of washes, rivers, dry lakes, and depressions. It may be caused by heavy rain associated with a severe thunderstorm, hurricane, tropical storm, or melt water from ice or snow flowing over ice sheets or snowfields. Flash floods may also occur after the collapse of a natural ice or debris dam, or a human structure such as a man-made dam, as occurred before the Johnstown Flood of 1889. Flash floods are distinguished from regular floods by having a timescale of fewer than six hours between rainfall and the onset of flooding. Flash floods can also be human-induced due to the changes in land use patterns upstream. Most flash flooding is caused by slow-moving thunderstorms, thunderstorms repeatedly moving over the same area, or heavy rains from hurricanes and tropical storms.

Climate change is likely to cause changes in legislations globally to cope with the hazards that the development of areas on especially steep slopes cause to communities in the aspect of flash floods. There is an increasing occurrence of flash floods in several communities arising out of disturbing the environment while developing sites on steep slopes. Construction activities on hilly areas are likely to warranty standard practices to mitigate flash floods. Construction sites greatly distort the environment, and when heavy storms fall, the runoff carries sediment load blocking storm sewers leading to severe flash floods. The losses from flash floods have been devastating in the recent years, and both pecuniary and non-pecuniary losses have been registered in many areas around the globe. There is need to make changes in statutory regulations as well as how the workforce is managed and trained in order to curb the devastating incidences of flash floods, especially those that are human-induced.

Construction sites are the predominant areas most susceptible to human-induced floods. Human-induced floods can be devastating and can cause monetary and nonmonetary costs. Construction sites on relatively steep areas need to be more vigilant today than ever because of the changing climate patterns worldwide. It has always been a standard of practice to develop work plans that include inclement weather but nowadays this is fast becoming the top priority across the globe. The effects of urbanization on runoff are increasingly becoming terrible. The modifications of land use have varying effect on the runoff characteristics of a given area.

12.4.1 Case study 12.1: Flash floods that occurred at the construction of US$350 m Lubowa international specialized hospital, Kampala, Uganda (2019)

12.4.1.1 Introduction

This case provides an excellent example of how continuous flash floods through the months of August to December 2019 led to heavy pecuniary losses over US$300,000 and affected more than 20 homesteads. Figure 12.1 provides the extent of flooding on one of the fateful days. The flooding from the cleared site blocked roads for many hours and severely damaged the drainage system.

The clearing of vegetation cover to construct a US$350 million international specialized hospital in August 2019 at Lubowa, Uganda triggered generation of huge runoff volumes from

The future of urban drainage 429

Figure 12.1 Extent of flooding on one of the fateful days. The flooding from the cleared site blocked roads for many hours and severely damaged the drainage system.

the steep slope causing flash floods that destroyed homesteads and household properties. It is the responsibility of the contractor to adopt a methodology that is safe and unhazardous to the community. In this case, a flood defense control mechanism to protect nearby homesteads, properties, and people was supposed to be put in place prior to site clearance. Because of being a steep terrain, the risk of flash floods was high. However, the contractor did not provide safeguards to mitigate the flash floods. The contractor has to exercise reasonable care to the people involved with the works and those affected by the works. The UK Construction (Design and Management) (CDM) 2015 regulations place responsibility to the contractor in that regard.

Following the clearance of the site for construction of the specialized hospital on a steep terrain at Lubowa, several people's residences faced the consequences of flash floods because of the increased runoff volumes from the catchment, triggered by clearing the vegetation cover. The runoff waters destroyed houses and properties in the houses. Successive stormwater runoffs through the months of August to December 2019 weakened some of the residences' foundation calling for a remedial solution in the form of reconstruction.

The clearance of site for construction of the hospital increased the runoff volumes generated from the catchment. Because the site was steep, the stormwater runoff travel time to reach residences was reduced as a result of removing the vegetation cover, hence the short time of concentration yielded huge volumes of runoff that weakened the foundations of the residence and the immediate fences.

In an attempt to control the flooding, the contractor excavated a stormwater runoff collection point in the form of a detention pond (SuDS). This collection point was about 12.3 m from the affected residences. However, the collection point (detention pond) was not constructed to meet required standards of seepage and sediment control mechanisms. It did not meet the immediate purpose of safely holding the stormwater temporarily and conveying stormwater runoff to the designated discharge points. Therefore, standing water found way to seep through the stratified layers of the embarked soils and the soils deposited along the road down through the foundation of immediate residences, thereby severely damaging the houses. On the principle of underground water recharge, the standing waters on clayey soils at a higher altitude find a way to percolate through the soil layers down to the nearby streams or wetlands. At some point, the seeping waters fully saturated the soil strata on which the immediate residences were founded.

When stormwater runoff of considerable amount of acidic content finds way through a pore in concrete, it predisposes the concrete to chemical attacks hence weakening the steel reinforcement and overall concrete strength which reduces the building's life span. It is, therefore, upon this background that fresh reconstruction of the some buildings was inevitable because the flood damage was beyond repair, rendering some of the buildings unserviceable. A weakened foundation of a storied structure poses a high risk to the occupants.

12.4.1.2 Lessons from the case

A case for one of the flash floods victims provided lessons that are of interest to drainage planners and engineers on four legal topics: negligence, contributory negligence, strict liability, safe method of work. The victim founded the US$250,000 two-storey structure on waterlogged soils and provided no adequate foundation to safely control subsurface drainage. Lowering the water table with adequate rock fill material, a geotextile layer, and/or subsurface drains as appropriate was a must for a properly engineered foundation (considering a two-storey structure) in a waterlogged area.

The insurer agreed to compensate the victim but not wholly arguing the flash floods came but hit a poorly founded structure—attributing the blame partially to the victim. The insured also owed a duty while clearing vegetation from a steep site and his flood emergency/control mechanism was also faulted. As the insured made attempts to create a flood defense control channel to temporarily hold the flash floods and to divert the floods, he excavated a **detention pond** (SuDS) upstream of the victim's house, which was inadequate. The intent was to temporarily hold stormwater runoff. The contractor went ahead to place noncompacted stratified clayey soil layers with a few sandbags on the road between the pond and the house. Pond's standing waters seeped through the layers fully saturating some sections of the soils on which the house was founded. Note: Drainage conditions affect the groundwater table.

The insurer wished the insured to accept premium readjustment to pay more as she partially attributed part of the aftermath's outcome to the insured—claiming this was foreseeable as far as safe methods of work, good workmanship, and appropriate standards are concerned. So, the victim's claim went on a standstill while other victims were quickly compensated. The victim suffered from contributory negligence, because the house was constructed not to the required standards and codes of practice, and lacking approvals in several building annexures. Despite all circumstances brought by the contractor in failing to adequately plan to manage the risk of flooding before clearing the vegetation on the steep terrain, the victim had to have a well-built structure following national standards and codes duly approved. To be on a safe side, all civil engineering structures must be approved before erection including minor boundary walls. One of the primary lessons from this case is that the future is uncertain, and therefore, following national standards and codes of practice is always necessary as it protects the developer throughout the project lifecycle phases. It safeguards the developer from the risk of negligence/contributory negligible at present and in future when such human-induced flash flood incidences and similar things like natural catastrophes happen. Therefore, planning that fits in the national legal frameworks is essential and must be taken very seriously at all levels of project development. That means physical planning Acts and associated environment laws must be followed properly to enhance reliability in project developments. Lacking an approved plan and building in a flood-prone zone can rarely earn compensation when such flash floods happen to destroy the property in the area unless circumstances dictate otherwise. Therefore, the following were part of the lessons from the isolated victim, going forward:

a. Overall, the relevance of the contractor's all risk (CAR) insurance policy on the side of the contractor is of no doubt to be taken seriously. The CAR insurance policy offers coverage against any kind of loss or damage caused in the construction site to the worker, machine. This policy is designed to keep the safety of architects, construction engineers, workers, and financers. This policy contributes to reduce overall expenses of construction. It provides enough financial protection to the parties involved. Most of the plans under this insurance type cover projects of civil engineering nature where the civil work is more than 50% of the total value of the contract. The plan starts covering the project from the time of material storage, construction, and covers it until the project gets completed and is handed over to the client. The common benefits or coverage that most of the policies provide under this plan are as follows:
  i. **Material damage.** Under this loss, damage, or destruction of property incurred due to any cause other than those excluded in the policy are covered. Most of the policies pay for the losses or damage up to a certain amount that does not exceed the sum mentioned against every item and does not exceed the total sum insured.
  ii. **Third-party liability.** Under this, the coverage is toward legal liability related to accidental damage or loss caused to the property of the third person. Legal liability for nonfatal and fatal injuries caused to the third person (other than the own employees or workers of the insured) due to the construction of the property. Perils that are covered under CAR insurance policy are as follows:
    - Fire and allied perils.
    - Collapse.
    - Earthquake, shock, and fire.
    - Faults in construction.
    - Storm, flood, cyclone, and tempest.
    - Negligence and human errors.
    - Stormwater and water damage.
b. The need to follow safe systems of work policies in place (Section 10.7.2.4) is essential.
c. The need to employ technically competence workforce to build reliability in the project right from the start.
d. Negligence and contributory negligence repercussions can be devastating since a portion of the approved house plans was not available. Contributory negligence is the failure of an injured party to act prudently, considered to be a contributory factor in the injury/harm which they have suffered. In some common law jurisdictions, contributory negligence is a defense to a tort claim based on negligence. If it is available, the defense completely bars plaintiffs from any recovery if they contribute to their

432 Integrated Drainage Systems Planning and Design for Municipal Engineers

own injury through their own negligence. In this case, failure to provide an adequate foundation for the building structure and failure to provide approved plans for building annexure were contributory negligence and could not garner the victim full compensation for the damaged property.

12.4.2 Case study 12.2: The November 5, 2021, flash floods at seroma workshop premises: Seroma workshop flood damage evaluation and analysis, Kampala, Uganda

12.4.2.1 Background to the problem

US$61 million Kampala flyover construction and road upgrading project (Lot 1, Package 1: Clock Tower Flyover) brought about several activities in the area. The contractor carried out excavations, installed cofferdams, road diversions, used barricades as barriers to fence off intruders in the working area, and blocked existing storm sewer pipelines as construction of new drainage and other road works commenced and progressed. The November 5, 2021 storm occurred when several construction activities were underway. See Figure 12.2.

The November 5, 2021 flash floods caused excessive damage to the properties of Seroma Limited in Kampala central. Seroma workshop is seated on about 1.311 acres (5305.7 m^2). An investigation was launched to reveal the probable cause of the unprecedented flash floods that heavily damaged the workshop properties and caused Seroma Company a financial loss of US$244,177.105. An extensive evaluation was done to explore the probable cause of the excessive flooding that occurred on November 5, 2021 that had never been experienced before despite the fact that storms of a higher magnitude had constantly been received in the area. The findings indicated that the contractor faulted the evaluation process and the method of work prior to the start and during the drainage and other road construction works. The findings of interest to a drainge engineer and planner are summarized below:

Figure 12.2 Aerial view of Seroma workshop area.

1. There were several barriers at the construction site that diverted the stormwater runoff toward the Seroma workshop. For example, the wing walls on the newly constructed drainage corridor/channel on the western side of Seroma estate diverted the stormwater runoff toward Seroma estate. The contractor's working drawings indicated that Seroma workshop is surrounded by construction zone 1 and 2 on the southern and western sides, respectively, which is first-line evidence that a number of safeguards would be in place to mitigate consequences due to construction barriers and excavations—coming from a safety perspective and Seroma workshop being in a low-lying flood-prone zone. The contractor providing two drainage ditches/channels, that is, channel A and B on the eastern and western side of Seroma premises demonstrates appreciation that the area is a low-lying flood-prone zone that required adequate ditches to safely route the runoff to the Nakivubo channel. See Figure 12.3.
2. The contractor blocked the existing storm sewers, and when it rained, huge runoff discharges overflowed creating hydraulic jumps when rapidly moving runoff met with slowly moving runoff. This created extensive backwaters that found their way to Seroma's estate and accumulated therein. Therefore, the runoff got diverted to the lowest point which is in the Seroma estate. Through consultations, that is, with those that witnessed the November 5 floods and what transpired thereafter as contractor's remedial actions, it was reported that the constructed slopes of the newly constructed drainage channel on the western side of Seroma estate had to be reworked to avoid reoccurrence of backwaters. See Figure 12.3, point A. Therefore, on the fateful day, there was occurrence of backwaters in the immediate western drainage system, and barriers blocked the usual flow paths of the runoff—a combination of both provided sufficient energy enabling excess flash floods to destroy the Seroma southern fence, hence flooding the entire estate up to a depth of 1.25 m (Figure 12.4).

Figure 12.3 Kampala flyover construction and road upgrading project (the contractor built two drainage ditches in blue dots).

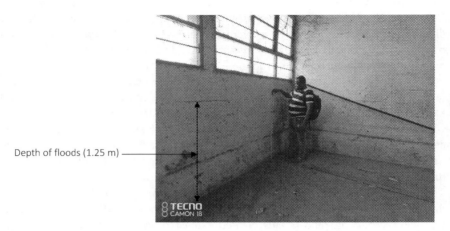

Figure 12.4 Flash floods reached a height of 1.25 m in one of the store rooms.

3. It was noted that on the fateful day the construction barricades used to keep away intruders from the site and enhance safety were kept very close to each other. Thus, the stormwater runoff would not pass in-between them but rather find its own path beneath the barricades and directed toward the Seroma workshop area. Also, the constructed drainage channel whose walls are up to **2.0 m** above the ground level, at the western side of Seroma premises (Pan African outfall relief channel), acted as a barrier preventing excess runoff from quickly entering Nakivubo channel. This, instead, diverted the runoff toward Seroma premises, especially the front. It is important to note that land-use changes can significantly modify the runoff process.
4. November is among the months well-known for high rainfalls in Kampala. Failure to plan and prepare for adequate runoff exit routes while constructing the drainage scheme and other works is negligence on the contractor's side. The area had experienced storms of a higher magnitude before. However, such floods had never been experienced. For example, the highest rainfall received in November 2020 was 72.5 mm compared to the 65.7-mm rainfall received on November 5, 2021 and which was the highest in the month. Therefore, excessive floods were not catastrophic; they were instead human-induced floods.
5. It was evidenced through photographs and videos taken on the fateful day that the contractor's personnel rushed to Seroma workshop to witness the aftermath and quickly rushed to open public storm sewer lines they had blocked and those that had since got excessively blocked due to poor maintenance. This indicated that their evaluation process was faulty because the maintenance of the existing drainage networks was supposed to be done before the start of drainage construction of the new scheme and construction of other road works. In addition, the functionality of the newly constructed drainage network posed technical questions about its hydraulic efficiency.
6. From topographical maps, the Seroma workshop is in a low-lying area nearing the Nakivubo channel, which is very susceptible to flooding. See Figures 12.2 and 12.8. It lies in the center of the drainage easement corridor and, therefore, the distortion in the drainage networks could inevitably cause excessive flooding at the Seroma workshop.

The average elevation at Seroma workshop is 1154 m asl compared to Kibuye roundabout at 1176 m asl and Katwe church at 1188 m asl and Lubiri ring road at 1176 m asl, and Nsambya hospital area at 1198 m. These surrounding areas discharge runoff into the Nakivubo channel but through the environs (easements) surrounding the Seroma workshop, which is a low-lying flood-prone zone.
7. From the field findings, oil marks left on walls as flash floods mixed with oils in the estate reached a height of about **1.25 m** indicating the extent of the flash flood volume that accumulated in the estate. Thus, many items were submerged into the floods, that is, generators, compressors, trucks' engines, computers, and spare parts, among others.
8. The catchment area whose stormwater runoff generated passes through the Seroma workshop nearing the Clock Tower corridor is about **333.017 hectare**, and a considerable fraction is impervious, implying that runoff that is intercepted and infiltrates into the ground is insignificant. Considering a return period of 25 years, and an intensity of 195 mm/h, an estimated volume of 975.78 m^3 for 80-hectare area, time = 30 seconds, gets routed through the Seroma/Clock Tower drainage easement corridor before discharging to the Nakivubo channel. This is a huge volume of runoff. This was an indicative estimate.
9. The contractor's activities caused Seroma company a financial loss of **US$244,177.105.**

The topographical survey shown in Figure 12.5 was used to delineate the effective catchment that contributes stormwater runoff toward Seroma premises. Figure 12.6 shows Clock Tower

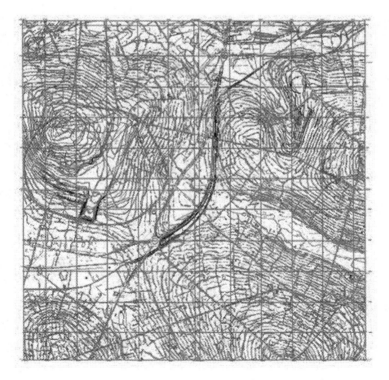

Figure 12.5 Topographical map.

area most affected by the floods from Kibuye roundabout (Figure 12.7). The runoff flow movement pattern was therefore evaluated from Figure 12.8.

12.4.2.2 Interpretation

Seroma premises is in a low-lying area that is prone to floods. Stormwater runoff from Lubiri ring road, Katwe, Kibuye, Makindye, and Nsambya finds its way to Nakivubo channel through Seroma area. The average catchment slope of 6% provides a good insight into the

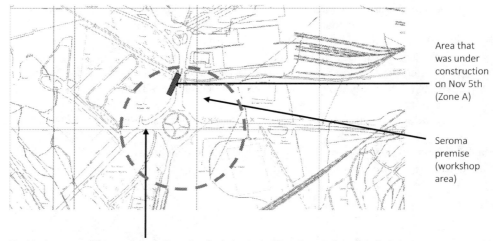

The blockage runoff flow path near the police fire brigade significantly contributed to the flooding of Seroma premises. All the area around clock tower discharges through on the Western side of Seroma premises which were blocked.

Figure 12.6 Clock Tower area.

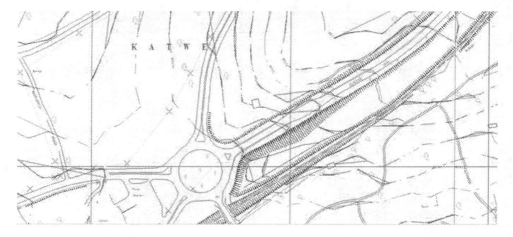

Figure 12.7 Kibuye roundabout.

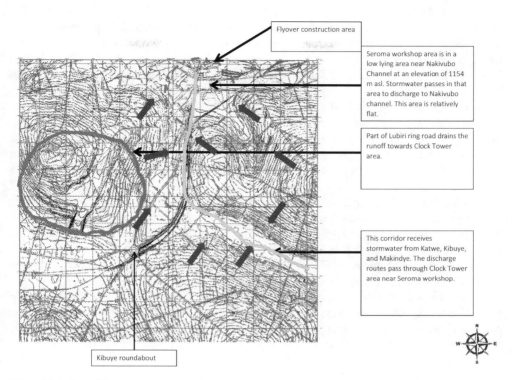

Figure 12.8 Runoff flow movement pattern.

available energy to route the stormwater runoff. This means the water propagates downstream without any impediments to stagnate or pond upstream, and the immediate discharge point before discharging to the Nakivubo channel is the Clock Tower area. However, the Clock Tower area was a construction zone with many barriers such as barricades, nets, and heaped soils. On November 5, 2021, these were placed in the drainage corridor (Figure 12.8); together with the blockage of existing sewer lines, the overflows were perpetuated from point to point leading to severe flooding at Seroma premises.

Nakivubo channel is the receiving stream for most of the runoff generated from Katwe, Kibuye, Makindye, and Nsambya areas as shown in Figure 12.7. The area around Clock Tower and Seroma workshop is in a low-lying zone, a corridor through which runoff passes to discharge in Nakivubo channel. Seroma workshop premises is one of the immediate flash flood victims if slight interruption of runoff upstream occurs, that is, in the areas of Katwe, Kibuye, Makindye, and Nsambya.

It is evident from the topographical map that a slight interruption in the drainage path would pose a significant flooding risk to the low-lying Seroma workshop area.

The rainfall received on November 5, 2021 was highly expected. It did not come as a surprise because the month of November is among the months that receive the highest rainfalls in the year. That means the contractor's work plan and corresponding methodology of work must have included safeguards to mitigate the worst scenarios.

For example, constructing temporary ditches to safely route off the flash floods during the construction period. Another safeguard would be ensuring the clearance and maintenance of drainage sewer lines to safely route the excess floods off the impervious surfaces would be a good management practice before any attempt was done to interrupt drainage paths. However, the contractor instead blocked the drainage sewer lines intercepting the flow of the runoff.

In the end, the runoff created hydraulic jumps as slowly flowing runoff mixed with speeding runoff. In addition, the runoff carried excessive sediment load from the construction site that perpetuated the blockage of drainage corridors. Consequently, in a couple of minutes, the entire area was flooded because of huge discharges accumulating in the estate as they overflowed at the blocked storm sewer entries causing excessive damage to Seroma workshop properties.

The contractor quickly opened storm sewers toward the end of the storm. The public storm sewer crossed through Seroma workshop estate. This is what should have been done prior to diverting and blocking the runoff as drainage construction commenced. Not doing so was a fault on the part of the contractor.

In one of the stores the volume accumulated to a depth of about **1.25 m**. Thus, the eastern fence which was under construction collapsed due to momentum change and hydrostatic pressures induced from the accumulated floods in the estate (see Figure 12.9). It fell outside onto the newly constructed eastern drainage channel. The southern wing fence (Figure 12.10) collapsed due to huge impact arising from fast change of momentum as slowly-moving waters mixed with speeding waters, amid barriers that the contractor had placed on the construction site.

Figure 12.9 Collapsed eastern wall fence.

The future of urban drainage 439

Figure 12.10 Fallen southern fence due to flash floods.

The distortion of the runoff flow path caused excessive flooding, which consequently damaged the Seroma workshop properties. The Seroma workshop premises and the entire area had experienced relatively higher storms before, but runoff volumes had never reached the depth of flow attained on that fateful day, that is, 1.25 m. This runoff volume is attributed to the distortion of drainage corridors. For example, the highest rainfall received in November 2020 was 72.5 mm too way above the 65.7 mm received on November 5, 2021, but there was no damage caused to Seroma properties.

When the contractor blocked the existing storm sewers diverting the runoff as drainage construction progressed, the runoff flow paths were distorted. As a result of drainage construction activities, the runoff carried excessive sediment load, trash, and garbage. This brought about unprecedented flash floods that damaged Seroma properties.

As a standard of practice, work plans and method statements are prepared to take into consideration inclement weather. From the topographical map, it is evident that the Seroma workshop exists in a low-lying area. It is a drainage corridor receiving runoff from the surrounding areas of Katwe, Kibuye, Nsambya, and Makindye. Therefore, it cannot be claimed as a surprise when the area flooded.

As the floods accumulated in the estate, the eastern boundary wall that the contractor had constructed collapsed. However, if the floods had been accompanied or preceded by winds, and in this case, they were unique and catastrophic, the impact resulting from the fast change of momentum would perhaps sweep away the entire estate and its properties. Since this never happened, it is evident that the flash floods accumulated slowly filing the estate. This evidences the fact that there was a blockage upstream.

As backwaters accumulated gaining sufficient energy, the immediate southern gate had to collapse because it was very close to the western drainage ditch 2.0-m high wing wall that acted as a barrier to divert the stormwater runoff toward the Seroma workshop. Therefore, the gate received the greatest impact. However, as the flow propagated downstream through Seroma premises, the available head dropped losing momentum, an explanation for being human-induced floods as a result of activities upstream but not catastrophic flooding. The eastern gate collapsed due to a combination of hydrostatic forces and runoff momentum change. There was sufficient evidence that the contractor's activities significantly contributed to the flooding of the Seroma premises on which the contractor based to compensate the flood victim.

12.4.2.3 Rainfall analysis

The analysis followed the data and information gathered to process and come up with meaningful insights into the probable cause of the flash floods that damaged Seroma workshop properties. Kampala city experiences bimodal rainfall. The highest rainfalls are received in the months of March, April, and June and again in October, November, and December as shown in Figure 12.11. For the last **ten years**, the highest rainfall received was **82.4 mm** in the month of May 2020. The highest rainfall received was **65.7 mm** on November 5, 2021. Interestingly, the preceding year had received much higher rainfall in the same month, that is, **72.5 mm**. See Table 12.1 for Kampala rainfall data.

The two-parameter Weibull probability distribution function indicated that the city would receive more rainfall in future. This technique provided sufficient evidence that the rainfall was highly expected since the likelihood is high. The shape parameter $\beta = -2.49 < 1$ (see case study 4.2) indicated a decreasing failure rate. In other words, the variability between the monthly maximum rainfall data for the month of November through the years 2012 to 2021 indicated that there is a more likely event for increased maximum rainfalls for the November month for the years ahead.

The scale parameter determined the range of the distribution to be **±35.93 mm** as maximum rainfall received in November month. This implied that environmental conservation

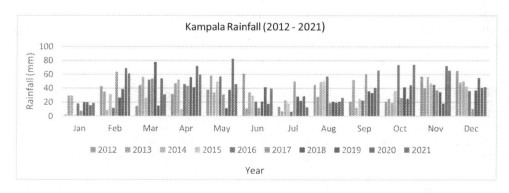

Figure 12.11 10-Year rainfall pattern for Kampala city.

Table 12.1 Kampala rainfall data (2012–2021)

Year	Jan	Feb	March	April	May	June	July	Aug	Sept	Oct	Nov	Dec
2012	2.2	42.8	14.2	31.4	37.9	60.7	13.2	45	20.8	21	56.9	65
2013	29.6	35	44.6	46.7	57.9	10.8	6.9	27.6	51.9	24.9	40.4	48.2
2014	29.6	8.9	56.2	52.1	33.4	33.7	22.8	48.8	11.9	18.6	56	49.9
2015	2.2	31.8	25.7	9.9	49.4	29.1	17.8	50.2	25.1	36.4	47	42.5
2016	18.1	11.4	52.2	45.7	57	21	6.4	57	22.5	73.9	44.9	36.1
2017	7.4	63.8	53.5	43	31	11.5	49.8	18.7	60.3	25.8	37	10.4
2018	20.1	26.1	77.9	55.7	11.2	20.9	28.2	20.4	35.5	41.4	34.3	37.2
2019	19.9	38.6	14.7	41	37.5	41.2	22	18.9	33	24.8	17.8	54.6
2020	15.3	68.9	53.9	72	82.4	17.5	28.5	20.6	40.8	44.2	72.5	40.8
2021	18.8	61.2	30.9	59.4	45.8	39.7	12.6	26.1	66	74	65.7	41.7

measures and flood protection mechanisms to mitigate the likely outcomes of the increased maximum November rainfalls must be provided. And this is where SuDS become more relevant and essential. It is the responsibility of the government to enact policies geared toward supporting SuDS so that each project development that involves a drainage component is planned in a contextual framework that views SuDS as an integral part in solving a drainage problem sustainably.

The above analysis adequately predicted that the November month rainfall would constantly increase over time. By virtue, standards, and codes of practice, the contractor was deemed to know this and must have done his independent engineering analysis. The contractor is deemed to have an appreciation of and be able to identify and manage risks to all those engaged and affected by the road construction works. The contractor, therefore, to a reasonable extent, owes a duty to ensure people's safety—those directly engaged and affected by works.

12.4.2.4 Recommendations to the contractor

1. The investigation revealed sufficient evidence that the contractor's activities, planning, and method of work significantly contributed to the flooding of the Seroma workshop. The flooding never came as a surprise because the contractor owed a duty to nearby communities in providing safeguards to mitigate the worst scenario while excavating upstream, blocking existing stormwater sewers, and filling and leveling low-lying areas. The leveling and raising of embankments changed the slope from mild to steep, consequently increasing the available energy for the runoff.
2. The November month is well-known for high rainfalls, and the two-parameter Weibull distribution indicated that the trend would continue for many years ahead. That was foreseeable, and the contractor is deemed responsible and must have incorporated it in the work plan by providing safeguards for inclement weather and also in the method statement to be compliant with standards of practice.
3. Seroma workshop suffered a loss due to uncontrolled stormwater runoff caused by the above-mentioned reasons, and the contractor was held liable to compensate.
4. It was noted that momentum change and hydrostatic pressures induced from the accumulated floods in the estate led to the collapse of the southern wing fence and the

eastern fence, which was under construction at the time. The rebuilt eastern fence was deemed to have been constructed at the contractor's cost.
5. To avoid similar events, the contractor was advised to provide adequate exit routes for stormwater runoff to enter drainage channels since the construction was getting into April, June, and May, months well-known for high rainfalls. See Figure 12.11.
6. The contractor was asked to reevaluate the hydraulic efficiency of the constructed drainage system in anticipation of developments upstream and the changing climate.

12.4.2.5 Lessons from the case for a drainage engineer or planner

As a drainage planner/engineer, below are the lessons to take from the case:

a. The future promises adequate planning for inclement weather due to climate change. Unprecedented flash floods will become increasingly common worldwide.
b. Safe systems of work must be deployed at all time and method statements must be sound enough to accommodate health and safety safeguards.
c. Section 10.7.3.2 presented the control measures for excavation work. Barriers are used to demarcate danger areas. Spoil should be removed and stored at a safe distance to prevent it from collapsing back in. The risk of blocking runoff patterns has rarely been a safety hazard. However, the future promises to look into human-induced flash floods wherever planning sites, especially those on relatively steep sites. Barriers, and heaped soils may now be considered hazards because they present a risk of blocking runoff flow patterns.
d. Construction sites on relatively high terrain areas need careful drainage evaluation because the excavations, filling, and leveling of areas rapidly change the land-use pattern of the area, and where slopes change from mild to steep, the available energy for the runoff rapidly increases.
e. On any construction site, the exit routes for stormwater runoff must be delineated properly to avoid such tragedy because it can bring heavy costs compensating for victims as a result of human-induced flash floods.
f. The hydraulic efficiency of proposed storm sewers must be evaluated holistically drawing knowledge from several stakeholders to derive a practical solution for the drainage problem.
g. The competence of staff and engineers is essential to deliver a successful project. Reworking construction slopes to avoid reoccurrence of backwaters was sufficient evidence that there was a fault that affected the hydraulic efficiency of the newly constructed storm ditches.
h. SuDS like detention ponds, infiltration systems, and RWH techniques are increasingly gaining momentum as safeguards to mitigate the impact of floods. These can be utilized as appropriate to mitigate flash floods.

12.5 EMBEDDING UN'S SDGS IN PUBLIC AUTHORITIES' MAINSTREAM DRAINAGE ACTIVITIES

The 17 sustainable development goals (SDGs) of UN are a blueprint for achieving a better and more sustainable future for all. Therefore, embedding SDGs at the core of organizations and the project strategy is one of the most critical components of the sustainability campaign. The UN's SDGs address the global challenges, such as extreme poverty and hunger, quality education, climate action, clean water and sanitation, affordable and clean energy, sustainable cities and communities, life below water, and gender equality, among others (sdgs.un.org).[6] Public authorities would prioritize the UN SDGs and task all organizations to embed the goals in their mainstream activities to achieve remarkable progress and report significant success through the nationally determined contributions.

In drainage activities, the concept of low-carbon design, climate adaptation, embracing diversity and inclusion, ecological sustainability and building robust and resilient communities must be at the forefront while planning.

The future of urban drainage promises policies to control stormwater at the source globally. This will necessitate a shift from overreliance on gray infrastructure to stormwater best management practices. Urban planning policies would call for raising structures that conform to agreed stormwater management standards that contribute to a carbon-neutral economy.

The future physical planning statutes promise to require developers to submit effective stormwater management plans to authorities before approving building plans and issuing building occupation permits. These would show collection, conveyance, treatment, and storage mechanisms convincingly as sustainable stormwater solutions. They would provide approaches on how carbon is offset in the entire estate and in the long run support public sewers from being overworked.

12.6 CONCLUSION

The future promises a number of changes in legislation trying to adopt to the changing climate. There is likely to be changes in statutory regulations as well as how the workforce is managed and trained, in order to improve the civil engineering industry's health, safety, and welfare record as far as stormwater management and drainage systems are concerned. These changes would include enforcing SuDS development as appropriate, training and equipping operation and maintenance crews with necessary knowledge and skills, widening drainage easements, and clear laws on managing and controlling human-induced flash floods. Nonconventional drainage design methods are set to gain popularity over conventional approaches because conventional approaches do not only simply translate to a conventional rainfall-runoff modeling for designing drainage system components, but also implement conventional management methods that ignore the interaction between drainage systems, the environment, and the society. Yet, sustainable planning and management of stormwater requires close interaction of system elements with society and the environment bringing out unique attributes for each catchment.

References

1. Wong, T. 'The man turning cities into giant sponges to embrace floods—BBC News.' Accessed May 15, 2022. www.bbc.com/news/world-asia-china-59115753
2. Rosamond, Hutt. (2017) 'China is about to get its first vertical forest,' World Economic Forum. May 4th 2017. www.weforum.org/agenda/2017/05/china-is-about-to-get-its-first-vertical-forest/
3. Zhdanov Mikhail Georgievich. (2016). RU2663443C2 - Polyhedral composite pipe (options) - Google Patents. https://patents.google.com/patent/RU2663443C2/en

4. Nie, L., Lindholm, O., Lindholm, G., & Syversen, E. (2009). Impacts of climate change on urban drainage systems—a case study in Fredrikstad, Norway. *Urban Water Journal*, *6*(4), 323–332.
5. Yazdanfar, Z., & Sharma, A. (2015). Urban drainage system planning and design—challenges with climate change and urbanization: A review. IWA Publishing 2015 Water Science & Technology | 72.2 | 2015.
6. UN Sustainable development goals (UN SDGs). https://sdgs.un.org/goals

Further reading

1. Construction (Design and Management) (CDM) Regulations (2015), UK.

Glossary

CHAPTER 1

Best management practices (BMPs) — Term used in North America and Canada to refer to sustainable drainage systems (SuDS).

Combined system — System used to collect, store, and convey both stormwater/surface water and foul water or wastewater.

Drainage master plan — A plan showing the network of drainage schemes, both SuDS and gray drainage infrastructure.

Drainage system — System that collects, stores, and conveys wastewater and stormwater to a safe discharge point.

Green infrastructure — An approach to stormwater management synonymous with sustainable drainage systems (SuDS) that protect, store, or mimic the natural water cycle.

Gray infrastructure — System made of mainly concrete and steel such as storm sewer pipes, drains, and related infrastructure such as roads and buildings.

Partially combined system — A system that deals with both stormwater and wastewater although the wastewater is considered almost insignificant because it is diluted by the voluminous surface water, especially when it rains.

Sustainable drainage system (SuDS) — New approach of dealing with stormwater that emphasizes dealing with stormwater at the source through collection, storage, or infiltration of stormwater into the ground in a way that mimics the natural water cycle.

Uncombined system — System usually referred to as a separate system in which the stormwater is considered separate from wastewater.

CHAPTER 2

Annual exceedance probability (AEP) — Probability of a natural hazard event such as rainfall or flooding event to occur annually.

Digital elevation models (DEM) — A 3-D representation of the bare ground (bare earth) topographical surface of the Earth excluding trees, buildings, and any other surface objects.

Embodied carbon — All carbon dioxide emitted during the production materials or piecing together materials such as constructing a building.

Operational carbon — Amount of carbon dioxide emitted during the operational or in-use phase of a system, such as a building or SuDS.

Geographic information system (GIS) — Computer system used for capturing, storing, checking, and displaying data related to positions on the Earth's surface.

Lifecycle (least cost) analysis — Tool or technique used to determine the most cost-effective option among different competing alternatives to purchase, own, operate, maintain, and dispose of an object or process, when each is equally appropriate to be implemented on technical grounds.

Low-carbon design — Design when implemented would associate with low levels of carbon dioxide throughout the lifecycle phases.

Private finance initiatives (PFI) — A procurement method where the private sector finances, builds, and operates infrastructure and provides long-term services and facilities management through long-term contractual arrangements.

Site visit — A visit in an official capacity to examine a site's adequacy and suitability for an enterprise or specific work.

Topographical surveys — Survey that locates all the surface features of a property and depicts all the natural features and elevations.

Traditional public sector (TPS) financing — Common method of contracting where the client such as government selects service providers for work and finances the project using public funds.

Utility map — Map showing the positioning and identification of buried pipes and cables beneath the ground.

CHAPTER 3

Bioretention garden — Landscaped gardens, which are usually depressions or shallow basins, used to slow and treat on-site stormwater runoff.

Bio-swale — Stormwater runoff conveyance systems that provide an alternative to storm sewers and drains.

Channel protection storage volume — Volume of runoff generated by a one-year, 24-hour duration storm event to prevent habitat degradation and erosion that may cause downstream enlargement and incision due to increased frequency of a bank-full and near-bank-full flows.

Detention pond — Large constructed depression in an urban landscape that receives and stores the stormwater runoff for a limited period of time.

Extreme flood volume — The storage volume required to control those infrequent but large storm events in which overbank flows reach or exceed the boundaries of the 100-year floodplain.

Floodplain — An area of low-lying ground adjacent to a river, formed mainly of river sediments and subject to flooding.

Greenhouse gases (GHGs)	Gases in the Earth's atmosphere that trap heat.
Infiltration system	A device or practice such as a basin, trench, rain garden, or swale designed specifically to encourage infiltration, but does not include natural infiltration in pervious surfaces such as lawns, redirecting of rooftop downspouts onto lawns or minimal infiltration from practices, such as swales or road side channels designed for conveyance and pollutant removal only.
Legal point of stormwater discharge (LPSD)	Point specified by the municipal council to which stormwater from a property must be discharged.
Low-impact development (LID)	Term used to refer to systems and practices that use or mimic natural processes that result in the infiltration, evapotranspiration, or use of stormwater in order to protect water quality and associated aquatic habitat.
Overbank flood protection volume (Q_p)	The volume controlled by structural practices to prevent an increase in the frequency of out-of-bank flooding generated by the development.
Public–private partnership (PPP)	Long-term contract between a private party and government entity, for providing a public asset or service, in which the private party bears a significant risk and management responsibility.
Recharge volume	Portion of the water quality volume used to maintain groundwater recharge rates at development site.
Retention pond	Ponds with a permanent pool of water designed with additional storage capacity to attenuate surface runoff during rainfall events.
Site-specific SuDS	Sustainable drainage systems designed for a specific location and that can withstand the environmental and weather conditions of the site.
Total suspended solids (TSS)	Dry weight of suspended particles that are not dissolved in a sample of water that can be trapped by the filter that is analyzed using a filtration apparatus.
Underdrain perforated pipe	Pipe with holes in the sides which is installed underground to collect subsurface water and transport it to a surface outlet.
Water quality volume (WQ_v)	Amount of stormwater runoff from any given storm that should be captured and treated in order to remove a majority of stormwater pollutants on an average of annual basis.

CHAPTER 4

Average recurrence interval (ARI)	The average period between the recurrence of a storm event of a given rainfall intensity.
Catchment area	Areas of land where runoff collects to a specific zone.
Conventional method/product	One that is usually used or that has been in use for a long time.

Extreme value analysis (EVA)	Statistical tool to estimate the likelihood of the occurrence of extreme values based on a few basic assumptions and observed/measured data.
Flow frequency analysis	Statistical calculation of the probability of occurrence of flood of specific magnitude in a river, the specific period is called return period and the flow can be ranked as maximum possible flow.
Hydrologic cycle	The hydrologic cycle describes the methods or processes by which water in the hydrosphere moves from surface reservoirs, into the atmosphere, and back to the surface again, over and over.
Hydrologic model	A simplification of a real-world system (e.g., surface water, soil water, wetland, groundwater, and estuary) that aids in understanding, predicting, and managing water resources.
Modified rational method	An extension of the rational method to produce simple runoff hydrographs.
Nonconventional methods	Method not conforming to convention, custom, tradition, or usual practice (unconventional).
Peak flow discharge	The maximum rate of flow during a storm, usually in reference to a specific design storm event.
Rainfall intensity	The rate at which rainfall is expressed in mm/h. It can be determined with the help of automatic rain gauges.
Rational method	Method for determining peak flow discharges that expresses rainfall intensity and area as independent variables.
Soil Conservation Service curve number (SCS-CN) method	The Soil Conservation Service curve number method is one of the most popular methods for computing the runoff volume from a rainstorm and accounts for most of the runoff producing watershed characteristics, such as soil type, land use, hydrologic condition, and antecedent moisture condition—originally developed by Soil Conservation Service, U.S. Department of Agriculture.
Transport Road Research Laboratory (TRRL) method	Developed from the time-area concept of catchment response by TRRL, UK. In designing stormwater sewerage systems for towns, city suburbs, and new developments of around 200 to 400 hectares with varied surface characteristics, a TRRL method is required which takes into account differences in storm rainfall over the catchment area.

CHAPTER 5

Bed slope	Slope of the channel bed.
Critical depth	Depth where the energy of the flow has been minimized, that is, where maximum discharge and minimum energy occur.
Critical flow	The flow at depth equal to the critical depth.
Energy equation	Mathematical formulation of the law of conservation of energy.
Friction slope	The friction slope (S_f in Manning's equation) is the slope of the energy grade line.

Froude number	A dimensionless parameter measuring the ratio of 'the inertia force on an element of fluid to the weight of the fluid element', that is, the inertial force divided by gravitational force.
Headwater depth	The depth of water above the culvert inlet bottom is known as the headwater depth.
Hydraulic jump	The rise of water level, which takes place due to the transformation of unstable shooting flow (supercritical) to the stable streaming flow (subcritical flow).
Hydraulic radius	A measure of channel flow efficiency is defined as the ratio of the cross-sectional area of fluid flow, A, to the length of the wetted perimeter, P.
Hydrostatic pressure	The force that is created by standing or resting water. The pressure exerted by a fluid at equilibrium at a given point within the fluid due to the force of gravity. Hydrostatic pressure increases in proportion to depth measured from the surface because of the increasing weight of fluid exerting downward force from above.
Incompressible fluid	A fluid that does not change the volume of the fluid due to external pressure.
Link	Drain or conduit.
Momentum equation	The momentum equation is the mathematical formulation of the law of conservation of momentum.
Node	Manhole, junction, basin, inlet, or outlet.
Normal depth	The depth of flow in a channel or culvert when the slope of the water surface and channel bottom is the same and the water depth remains constant. Normal depth occurs when gravitational force of the water is equal to the friction drag along the culvert and there is no acceleration of flow.
Specific discharge	The discharge observed at a point in the stream/channel network per unit width of the stream or channel.
Specific energy	The energy at a cross section of an open channel flow with respect to the channel bed. The concept of specific energy is very useful in defining critical water depth and in the analysis of open channel flow.
Stationary control volume	A stationary control volume is fixed in space. As a result, it will only analyze fluid pass through that volume of space.
Subcritical flow	The flow at which depth of the channel is greater than critical depth, velocity of flow is less than critical velocity, and slope of the channel is also less than the critical slope is known as subcritical flow. Flow is slow or tranquil.
Supercritical flow	The flow at which depth of the channel is less than critical depth, velocity of flow is greater than critical velocity, and slope of the channel is also greater than the critical slope is known as supercritical flow.
Wetted perimeter	Portion of the cross-sectional perimeter of a container that's in contact with water.

450 Glossary

Inlet control	Flow through the culvert is limited by culvert entrance characteristics.
Outlet control	Flow through the culvert is limited by friction between the flowing water and the culvert barrel.

CHAPTER 6

Courant number	A dimensionless value representing the time a particle stays in one cell of the mesh. It must be below 1 and should ideally be below 0.7. If the Courant number exceeds 1, the time step is too large to see the particle in one cell, it 'skips' the cell.
Diffusive wave	The diffusive wave occurs when the inertial acceleration is much smaller than all other forms of acceleration, or in other words when there is primarily subcritical flow, with low Froude values.
Dynamic wave	Solves the full Saint Venant equations.
Eigenvalues	Eigenvalues are the special set of scalar values that are associated with the set of linear equations most probably in the matrix equations for which nontrivial solutions exist.
Finite-difference method (FDM)	The finite difference method relies on discretizing a function on a grid. To use a finite difference method to approximate the solution to a problem, one must first discretize the problem's domain. This is usually done by dividing the domain into a uniform grid. In numerical analysis, finite-difference methods (FDM) are a class of numerical techniques for solving differential equations by approximating derivatives with finite differences. Both the **spatial domain** and **time interval** (if applicable) are discretized, or broken into a finite number of steps, and the value of the solution at these discrete points is approximated by solving algebraic equations containing finite differences and values from nearby points.
Kinematic wave	For the kinematic wave, the acceleration and pressure terms in the momentum equation are neglected, and hence the name kinematic referring to the study of motion exclusive of the influence of mass and force.
Flow routing	Flow routing is a procedure to determine the time and magnitude of flow (i.e., the flow hydrograph) at a point on a watercourse from known or assumed hydrographs at one or more points upstream.
Wave attenuation	Attenuation is the gradual reduction in the intensity of a signal or a beam of waves which is propagating through a material medium.
Wave propagation	The *path* taken by the wave to travel from the point in a channel to another, that is, the ways in which waves travel and spread.
Monoclinical rising wave	A uniformly progressing translatory wave of stable form.

Steady flow	If the quantity of water entering and leaving the reach does not change, then the flow is considered steady. Steady flow assumes that the **pressure at the surface is constant** and the hydraulic grade line is at the surface of the fluid. Thus, the flow depth and velocity do not change with time at a point.
Unsteady flow	Flow where the flow depth and velocity change with time at a point.
Wave celerity	Velocity with which a wave advances.

CHAPTER 7

Capillary rise	The rise in a liquid above the level of zero pressure due to a net upward force produced by the attraction of the water molecules to a solid surface (e.g., soil).
Filtration	The action or process of filtering something, that is, the process in which solid particles in a liquid or gaseous fluid are removed by the use of a filter medium that permits the fluid to pass through but retains the solid particles.
Flood defense structures	Structure built to protect communities from floods.
Gravity flow	This is a flow of water or liquid drawn under the force of gravity, and fall to the ground, for example, water from surrounding areas can be absorbed by the soil then flow by gravity to areas underneath the pavement structure—in pavement with high air voids (above 8%–9%), water can percolate down through the pavement structure itself by the force of gravity.
Infiltration	Infiltration happens when water enters the ground surface but does not come out thus increasing the moisture content of the soil.
Mohr–Coulomb theory	A mathematical model (see yield surface) describing the response of brittle materials such as concrete, or rubble piles, to shear stress as well as normal stress. Most of the classical engineering materials somehow follow this rule in at least a portion of their shear failure envelope.
Nonwoven geotextile	A felt-like fabric made by thermally bonding polypropylene or a mixture of polypropylene and polyester fibers and then finishing using needle punching, calendering, and other methods.
Percolation	The vertical movement of water beyond the root zone to the water table.
Permeability	Permeability, as the name implies (ability to permeate), is a measure of how easily a fluid can flow through a porous medium.
Pipe jacking	A technique for installing underground pipelines, ducts, and culverts without open-cuts (generally referred to in the smaller diameters as 'micro-tunneling').
Retaining wall	A wall built to hold back earth/soil or water.

Seepage	Seepage is the lateral movement of subsurface water that takes place when there is difference in water levels on the two sides of the structure such as a dam or a sheet pile.
Shear strength	A term used in soil mechanics to describe the magnitude of the shear stress that a soil can sustain. The shear resistance of soil is a result of friction and interlocking of particles and possibly cementation or bonding at particle contacts.
Subsurface	The stratum or strata below the earth's surface, that is, below the ground level.

CHAPTER 8

Angle of internal friction	A measure of the ability of a unit of rock or soil to withstand a shear stress.
Bedding class	The quality of material laid below a pipe that supports the pipe against the top and adjacent soil load on the pipe.
Boussinesq's theory	Boussinesq (1842–1929) evolved equations that can be used to determine stresses at any point P at a depth z as a result of a surface point load.
Complete case	If the plane of equal settlement is located above the embankment, then the installation is termed as complete trench condition or complete projection condition according to the direction of the shearing force.
Concrete pipe crushing strength	The greatest compressive stress that a concrete pipe can sustain without fracture.
Differential settlement	Term used in structural engineering for a condition in which a structure's support foundation settles in an uneven fashion, often leading to structural damage.
Factor of safety	The ratio of a structure's absolute strength (structural capability) to actual applied load. This is a measure of the reliability of a particular design. In general, it is a term used to describe how much stronger a system or structure is than it is required to be to fulfill its purpose under expected conditions.
Incomplete case	If the plane of equal settlement is within embankment, then the installation is incomplete trench condition or incomplete projection condition.
Marston's theory	The load on the installed storm sewer pipe is equal to the weight of soil prism, which is termed as interior prism, on the pipe minus or plus the frictional shearing force transferred to the soil over the pipe by the trench wall side or exterior soil prisms on either side of interior prism.
Negative projection cases	In the case of negative embankment condition, the pipe is placed in a trench which is narrow compared with pipe size and trench depth. In this installation condition, the top of the pipe is below the original ground surface and the fill material over the pipe exceeds the original ground level surface.

Glossary 453

Plane of equal settlement	When the embankment is sufficiently high, it is assumed that the shearing forces will terminate at some horizontal plane in the embankment fill; this plane is termed the plane of equal settlement.
Positive projection cases	This is when the pipe is installed in a wide trench or on the ground and an embankment installed over the top.
Projection ratio	Proportion of the pipe outer diameter which is above the bedding level or the natural ground level.
Rankine's equation	A stress field solution that predicts active and passive earth pressure. It assumes that the soil is cohesionless, the wall is frictionless, the soil-wall interface is vertical, the failure surface on which the soil moves is planar, and the resultant force is angled parallel to the backfill surface. It was developed by William John Macquorn Rankine in 1857.
Settlement-deflection ratio	The ratio between the settlement of the fill and the settlement of the ground underlying the pipe.
Soil unit weight	The soil unit weight is the density of the backfill material.

CHAPTER 9

Capital expenditure (CAPEX)	Funds used by an organization or individual to acquire, upgrade, and maintain physical assets such as property, plants, buildings, technology, or equipment. CAPEX is often used to undertake new projects or investments.
Carbon sink	Anything, natural or otherwise, that accumulates and stores some carbon-containing chemical compound for an indefinite period and thereby removes carbon dioxide (CO_2) from the atmosphere.
Lifecycle (least cost) cost analysis (LCCA)	Tool or technique used to determine the most cost-effective option among different competing alternatives to purchase, own, operate, maintain, and dispose of an object or process, when each is equally appropriate to be implemented on technical grounds.
Lifecycle impact assessment (LCIA)	A qualitative and quantitative assessment of the environmental impact of a product based on the resource, energy consumption data, and various emission data provided after the inventory analysis.
Maintenance cost	The cost incurred by an organization, individual, or business to keep their assets in good working condition.
Operational expenditure (OPEX)	The money a company spends on an ongoing, day-to-day basis in order to run a business or system.
Present worth	An economic concept that states an amount of money today is worth more than that same amount in the future.
Project design life	Project design life is the number of years of relatively maintenance-free performance. Design life is the forecast life expectancy of products based on their design.

Residual value	In case a material, system, or structure has a service life greater than the project design life, then it would have a residual future current money value (residual value), which should be discounted back to a present constant money value utilizing the inflation/interest factor and subtracted from the original cost.
Safe net-zero goal	Safe net-zero goal is the internationally agreed upon goal for mitigating global warming in the second half of the century, and the Intergovernmental Panel on Climate Change (IPCC) concluded the need for net-zero CO_2 by 2050 to remain consistent with 1.5°C temperature rise goal.
Service life	Service life is the number of years of service that a material, system, or structure will provide before rehabilitation or replacement is required

CHAPTER 10

Trench shoring	The process of bracing the walls of a trench to prevent collapse and cave-ins.
Environment and social impact assessment (ESIA)	Assessment carried out before project implementation to predict the environmental and social consequences that a future project/intervention might entail and proposes measures to mitigate potential negative impacts.
Environmental and social management plan (ESMP)	This is a plan drawn to present the environmental management, mitigation, monitoring, and institutional measures to be taken during project implementation and operation, to reduce adverse environmental and social effects to acceptable levels and enhance positive effects.
Environmental impact assessment (EIA)	Assessment carried out before project implementation to predict the environmental consequences that a future project/intervention might entail and proposes measures to mitigate potential negative impacts.
Environmental management system (EMS)	A framework that helps an organization achieve its environmental goals through consistent review, evaluation, and improvement of its environmental performance.
Environmental scoping report	Report used to make an early assessment of potential environmental and cultural heritage impacts and opportunities that could have a major or severe impact on the project.
Environmental screening	A mandatory review of all proposed investment subprojects undertaken with the purpose of categorizing them by expected environmental risks and impacts, filtering out proposals grossly detrimental for the environment, and determining appropriate extent and type of environmental assessment to be applied to the investment subprojects accepted for further processing.

Excavation	Excavation consists of using tools, equipment, or explosives for the purposes of moving soil, rocks, or other materials. Excavation is undertaken for a number of purposes, and different types of excavation are classified either by their specific purpose or the type of material being excavated.
Hazard	A hazard is a potential source of harm.
Risk	The effect of uncertainty on objectives or the possibility of something bad happening.

CHAPTER 11

Cast-in-place manhole	It is a method of constructing in which the walls and slabs of the manholes are cast on-site in formwork. Often known as poured-in-place or cast in situ manhole, it is a method of concreting that is carried out in situ or in the finished location of the concrete component.
Combination inlet	Combination inlets are composed of a kerb and gutter inlet acting as a single unit.
Drop manholes	A type of manhole where the vertical pipe is permitted to flow between the main sewer and branch sewer. This type of manhole is used in steep areas or where the elevation of the inlet pipe is more than the elevation of the outlet pipe.
Grease and oil trap	Grease and oil traps help prevent excessive buildup of grease and solidified oil within pipes, drains, and associated systems.
Gutter inlet	Gutter inlets (also called horizontal inlets) have horizontal openings in the gutter which are covered by gratings through which stormwater flows.
Junction box	Synonymous with an electrical junction box, it is a node where links interconnect.
Kerb inlet	Kerb inlets (also called vertical inlets) have vertical openings in the road kerbs through which stormwater enters and flows.
Lamp-hole	
Manhole	A manhole (utility hole, maintenance hole, or sewer hole) is an opening to a confined space such as a shaft, utility vault, or large vessel. Manholes are often used as an access point for an underground public utility, allowing inspection, maintenance, and system upgrades. The majority of underground services have manholes, including water, sewers, telephone, electricity, storm drains, district heating, and gas.
Outfall	An outfall is the discharge point of a waste stream into a body of water; alternatively, it may be the outlet of a river, drain, or a sewer where it discharges into the sea, a lake, or ocean.
Precast manhole	Precast concrete manhole sections are produced off-site in a controlled environment, thus exhibiting high quality and uniformity and transported for installation on the site.

Stormwater regulators	A product designed to be installed in structures to regulate or restrict the flow as it is released from a water body or site. Stormwater regulators are used when the municipal drainage network cannot accomodate large stromwater flows—serving as an alternative for slowing the flow of stormwater. The installed flow control or regulator is used primarily in stormwater and process sewage systems where other devices can be exposed to temporary overload due to irregular nature of the rainfalls. The stormwater regulator can be installed directly in the detention/retention tanks or the manholes.
Virtual reality	This typically refers to computer technologies that use software to generate realistic images, sounds, and other sensations that replicate a real environment.

CHAPTER 12

Combined sewer flow (CSO)/system	Combined sewer systems are sewers that are designed to collect stormwater runoff, domestic sewage, and industrial wastewater in the same pipe.
Drainage easement	A document that is usually attached to a property deed that gives access to part of the property to a third party, usually a municipality, for the purpose of maintaining drainage.
Flash floods	This is the rapid flooding of low-lying areas distinguished from regular floods by having a timescale of fewer than six hours between rainfall and the onset of flooding.
Flood warning systems	Systems designed for flood forecasting to save lives, protect property, and limit damage to infrastructure by monitoring rainfall and water levels and alerting when conditions approach dangerous levels.
Forest city	The prototype of a city composed by vertical forests.
Heat islands	These are urbanized areas that experience higher temperatures than outlying areas.
Sponge city	A concept developed in China that emphasizes the traditional practices of managing and conserving stormwater and stormwater runoff in a way that mimics the natural environment and that considers the climate of the area.
Trenchless technology	A form of underground construction that requires the use of few or no trenches at surface or street level.
Vertical forests	If the exterior (sides and top) and vacant parts of the building are covered with vegetation, then the building as a whole is considered as a vertical forest. During the construction of the building, balcony-like projections are created to plant trees that serve both as balcony and a small forest. Vertical forest is a model for a sustainable residential building, a project for metropolitan reforestation contributing to the regeneration of the environment and urban biodiversity without the implication of expanding the city upon the territory.

Index

Note: Page numbers in **bold** indicate tables; those in *italics* indicate figures. Endnotes are indicated by the page number followed by "n" and the note number e.g., 151n1 refers to note 1 on page 151.

AASHTO LRFD Bridge Design Specifications 288–9, 296
AASHTO LRFD design method 289
acceleration due to gravity (g) 40, 42, 45
achievable SuDS 58
advantages of SuDS 300
angle of internal friction 249–50, 278
annual exceedance probability (AEP) 37, 118, 400
artificial rills 8
ASTMC-1131 formula 315
ASTM procedure 310, 314
AutoCAD CIVIL 3D 35, 237
AutoCAD Storm, and Sanitary Analysis 35, 154
average recurrence interval (ARI) 32, 37, 49, 118, 131, 141, 168, 181, 307, 323, 324, 400

back-and-forth 32, 36–8, 49, 111–12, 169, 331
backfill 17, 21, 66, 193, 197, 239–40, 243, 245, 251–2, 261–2, 264, 267–70, 272–4, 276–7, 279, 285, 288, 296–7, 311, 341, 344–5, 389
backwater effects 43–5, 212–13, 231–5, 270
bedding class 36, 40, 252, 273, 288, 290, 292–3, 297, 307
bedding factor 290–1
bedding materials 17
bed slope (S_o) 32, 43, 153, 157–61, 163, 172, 175–6, 181, 194, 203, 208, 211–12, 215, 231
bentley StormCAD 35
best management practices (BMPs) 2–3, 16, 45, 54, 113, 299, 365, 414, 417, 419, 443, 445
biochemical oxygen demand (BOD) 100
biodiversity 23, 58, 108–9, 348, 358, 456; and resilience 303
bioretention 5–6, 9–10, 14, 23, 26, 29, 32, 34, 38–9, 56, 58–66, 75–81, 89–112, 144, 149–50, 232, 240, 242, 247, 260, 302–3, 305–7, 324, 330–2, 367, 403, 410, 446; description and purpose 74; design criteria 76; garden 32, 34; system 60, 61

bio-swale 6–7, 27, 29, 61–2, 86–8, 108, 301, 304–5, 420, 446
boundary conditions 204, 208, 220, 226–9, 231, 236
Boussinesq formula 288
Boussinesq's theory 273–4, 452
Boussinesq velocity distribution coefficient 153, 173
brickwork 34, 295, 382, 387, 393; manholes 389; sewers 311
British Standards Institute (BSI) 293

cantilever retaining wall 21, 264
capillary rise 240–1, 451
capital cost 36, 57, 74, 112, 305, 311, 314, 322, 426
capital expenditure (CAPEX) 309, 453
carbon offsets 324
carbon sink 2, 53, 89, 111, 301, 324, 332, 453
cast-in-place manhole 455
catchment area 18, 56, 94, 117, 120, 126, 183, 187, 308–9, 435, 447–8
catchment characteristics 22–3, 125
channel 2, 6, 8, 14, 22, 27, 32–3
channel protection storage volume 78, 81, 446
characteristic equations 222, 227, 229
characteristic grid 224
circular ecology 325, 332
circular pipe 33, 42, 155, 157, 163–4, 170–1, 184
coastal flood defenses 260
combined and uncombined systems 8
combined sewer overflows (CSOs) 9, 55
complete case negative projection 285
complete embankment case or wide-trench case (positive projection) 278
concrete pipe crushing strength 291, 452
concrete shrinkage 272
conduit 15, 32–3, 43, 48–9, 118, 155, 163, 168–70, 179, 212, 219, 231, 233–4, 280, 294, 338, 400, 449

457

conservation 1, 3, 5, 10, 96, 118–19, 122, 126, 148, 153, 156, 199, 206, 208, 212, 220, 299, 303–4, 312, 335, 359, 365, 424, 440, 448–9; of the ecological system 103; of energy 153, 154, 156; of momentum 153, 154, 210
constraints 3, 15, 20–1, 28–9, 31–4, 38, 40, 42, 44, 45, 47–9, 54–6, 58–62, 65, 68; and data collection 16
construction (Design and Management): regulations (2007) 366; regulations (2015) 50, 378, 429
construction materials 14, 25, 30, 274, 294, 367, 378
continuity equation 117, 204–6, 209–12, 214, 219, 233–5
contributory negligence 430–2
control volume 45, 171–2, 203, 208–11, 449
conventional tunneling 248–49
corrosion of storm sewers 272
corrugated steel pipes 294
cost implications 427
counterfort retaining walls 265, 266
critical flow 32, 41, 43–4, 161–3, 168, 191, 196, 219, 227, 231, 448
critical plane 280–2
critical slope 29, 153, 161, 163, 181, 190, 194, 196, 449
crushing strength 40, 290–1, 293, 307–8, 375, 452

deep drainage 247
deep percolation 247
detention pond 7–8, 22, 31, 33, 60–1, 63, 81–7, 110, 113, 121, 150, 154, 227, 232, 234, 254, 305–7, 409, 429–30, 442, 446; cost 85; description and purpose 82; design criteria 85
differential settlement 30, 48, 270, 295, 412, 452
diffusive wave 450
digital elevation model (DEM) 19, 47, 445
digital monitoring: of sewers 425
digital monitoring of sewers 425
disadvantages of SuDS 300, 305
downspout disconnection 7, 32, 56, 63, 100, 101, 107, 408; description and purpose 100
drainage assets 4, 11, 16, 18–19, 123, 330, 381, 401–2, 412–13, 416, 426–8
drainage construction 25, 335–6, 345–8, 366, 374, 376–7, 434, 438–9
drainage easement 4, 24, 48, 196, 200, 267, 414, 422–3, 434–5, 443, 456
drainage mats 245
drainage networks 2, 4, 16, 27, 30–4, 37, 43–4, 49, 85, 112, 116, 149, 156, 219, 399, 434
drainage planning 13, 19, 23, 27, 35, 38, 119, 134, 299, 300, 311, 367, 419
drainage prioritization 16, 37, 47
drainage schemes 2, 14–16, 20, 23–5, 36, 40, 53, 131, 179, 300, 303, 310–11, 324,

329–30, 335, 345–6, 348, 367, 369, 381–2, 416, 426–7, 445
drainage system 1–3, 8–10, 13–19, 22, 24
Driscoll equation 70–1, 80, 330, 331
dynamic wave 31, 40, 45, 49, 204–5, 211–12, 219–20, 230, 233–5, 450
dynamic wave routing 31, 45, 49, 219–20, 230, 233–4

earth overburden 249, 253, 273, 274, 297, 388, 390
earthen embankment structures 255
easement 4, 16, 20, 24, 32–4, 41–2, 48, 69, 155, 157, 196, 200, 267, 306, 345, 414, 422–3, 434–5, 443, 456
East African Transport Road Research Laboratory (EA TRRL) flood model 122
ecological sites 23, 303
eigenvalues 222, 450
embedded carbon (eCO$_2$) 14, 32, 53, 112, 169, 231–2, 293, 307, 323, 325–8, 330–2
embodied carbon 14, 25, 41, 190, 325–6, 328, 330, 332, 446
energy equations 153, 156–8, 170, 179, 199, 231
environmental and social issues 23
environmental and social management plan (ESMP) 335, 345, 350, 355, 454
environmental impact assessment (EIA) 350, 454
environmental LCA (ELCA) 309
environmental management 4, 13, 310, 335, 345, 348–50, 355, 363–5, 378, 402, 454
environmental management system (EMS) 348–9, 364, 378, 454
environmental scoping report 454
environmental screening 351, 454
environment and social impact assessment (ESIA) 345, 454
environment lifecycle cost analysis 293
EPA SWWM 35, 120
equipment 14–15, 21, 31, 35, 38, 57, 81, 112, 145, 262, 270, 306, 313, 325, 332, 336, 339, 340, 346, 347, 351, (continuous)
erosion 23, 27, 43, 45, 47, 66, 74, 81, 83, 85, 90, 96, 101, **106–8**, 110, 112, 134, 149, 154, 205, 256, 260, 265–6, 295, 304, 347, 352, 355–6, 358–9, 365, 394, 404–7, 411, 415–17, 446; of banks 112; and corrosion of pipes 27; and pipe corrosion 154; and sedimentation 304, 415; and sediment control 23, 365; and sloughing 74
excavation 25, 38, 74, 181, 186, 191, 247–9, 264–5, 311, 336–47, 351, 356, 359, 366, 369–74 (continuous); and confined spaces 366; to lay sewers 24; to level 235; of trenches 336
extreme flood volume 78, 81, 446
extreme value analysis (EVA) 115, 134, 145, 299, 448

Index 459

factor of safety (FOS) 163, 290–1, 307, 452
finite-difference approximation 213
finite difference method (FDM) 220, 450
finite element method 203, 211
flash floods 21, 22, 47, 53, 110, 111, 170, 171, 178, 191, 192, 194, 205, 253–6, 258, 267, 428–35, 438–40, 442–3
flood defense: barriers 3; control mechanism 429; control structures 22; structures 22, 110–11, 253–5, 258, 260, 451; systems 21, 22, 255; walls 22, 110, 254, 258–60
floodplain 20, 23–4, 30, 81, 108–10, 254, 301, 420–2, 446
flood warning systems 115, 421, 456
flow frequency analysis 134, 448
flow regime 42, 158
flow routing 30, 34, 40, 182, 203–6, 208–9, 212–14, 231, 233, 237, 450; techniques 30, 34, 203, 205–6, 233, 237
foundation drain 244–5, 268
French drain 66, 242–3
friction forces 171, 209
friction slope 27, 43, 153, 157, 163, 169, 181, 194, 200, 203, 211–12, 220, 231, 233, 448
Froude number 29, 40, 41, 43, 153, 154, 160, 161, 163, 169, 174, 181, 219, 232, 308, 323, 449
funding opportunities 55–6, 426

gabion mesh walls 266
geographic information system (GIS) 19, 446
gravity flow 240, 451
gravity forces 45, 171–3, 177
gravity retaining wall 263
gray infrastructure 1, 14–16, 28–31, 34, 36, 47, 49, 53, 57, 85, 112, 120
grease and oil traps 396, 455
green bonds 56
greenfield 35, 54–6, 63–4, 82, 100, 302, 346, 419–20
greenhouse gas 14, 144, 191, 301, 346, 447; (GHG) emissions 51–2; effects 1
green infrastructure 1, 39, 56, 98, 193, 302, 445
green roofs 4, 7, 58, 63, 102–5, 114, 149, 245, 403, 408
green spaces 6, 10, 47, 56, 62–3, 98, 100, 144, 301, 303–4, 407, 420
green streets 6
groundwater movement 247
groundwater recharge 58, 65–6, 74, 87, 104, **106–8**, 115–44, 151, 247, 302, 304, 324, 447
Gumbel extreme value distribution 115, 134, 135
gutter inlets 31, 384–5, 455

hazard 2–3, 19, 24, 26, 33, 59–60, 89, 93, 181, 192, 196, 205, 255, 335, 337, 353, 357, 360, 366, 367–71, 373, 402, 407, 412–13, 417, 428–9, 442, 445, 455
headwater depth 153, 203, 219, 230, 449
Health and Safety Executive's (HSE's) Approved Code of Practice (ACoP), L144 366
heat island 7, 90, 102–3, 301, 456
high-density polyethylene (HDPE) pipes 294
hydraulic jump 43–5, 176–8, 200–1, 231–2, 433, 438, 449
hydraulic model 23, 32, 156, 170, 175, 204, 207, 213, 219, 220, 232, 234–6, 300
hydraulic radius 30, 32–3, 41–3, 48–9, 121, 153, 155–7, 169–70, 200, 213, 220
hydrologic cycle 54, 116–17, 205, 448
hydrologic soil groups (HSGs) 308
hydrology 17, 24, 29, 30, 36, 38–40, 54, 64, 89, 111, 115, 117–18, 120–21
hydrostatic forces 171–3, 440
hydrostatic pressure 97, 179, 240–1, 261–2, 268, 438, 441, 449

impermeable area 78–9
impermeable membrane 21, 258–9
impervious 2, 4, 6, 16, 19, 22, 25, 47, 54, 57–8
imposed loads 273–4, 290
incomplete case negative projection 286
incomplete wide-trench case (positive projection) 279
incompressible fluid 171, 449
infiltration system 5, 6, 9, 14, 17, 21, 26, 29, 34, 39, 58
infiltration test 17, 60, 66, 150, 240, 376
inflation/interest factor 312, 313, 315–17, 323, 454
InfoDrainage 35
initial abstraction 122–3, 151
inlet 2, 4, 20, 31–3, 44, 93, 154, 175, 181–3, 190, 192, 194, 196
inlet control 182, 450
Innovyze 35, 237
inspection and maintenance checklist 401, 416
intensity-duration-frequency curves (IDFs) 142, 181
interlocking (lego) blocks 22
internal water load 44, 269, 272–4, 289–90, 297
invert elevations 154, 184

junction boxes 31, 45, 336, 374, 394, 412
junctions 4–5, 31–2, 45, 85, 175, 182, 194–5, 204–5, 231–2, 235, 272, 401

Kabale municipality 45, 48, 141–5
Kampala flyover construction 432, 433
kerb inlets 383–5, 455
kinematic wave 31, 43, 49, 204–6, 211–15, 218, 231–3, 235, 450

kinematic wave routing 31, 43, 213, 215, 218, 231, 233

lateral inflows 34, 43, 155, 175–6, 182, 194–5, 204, 219
least cost analysis (LCA) 36, 39–40, 446
legal and policy framework 376, 398, 419
legal point of stormwater discharge (LPSD) 57, 300, 400–1, 419, 426, 447; fees 57
lifecycle cost analysis (LCCA) 4, 36, 293, 309, 323, 427
lifecycle impact assessment (LCIA) 310, 453
lifecycle (least cost) cost analysis (LCCA) 453
link 31–3, 42–5, 49, 56–7, 62, 85, 118, 154, 182, 190, 196, 213, 219–20, 231, 234–5, 244, 308–9, 336, 338, 344, 449
loading conditions 250–2, 268–9, 271–2, 275, 277, 293, 375
loads that act on buried pipes 273
log-normal distribution 115, 134–5, *147*, *148*
log-pearson type III distribution 115, 134–5
low-carbon planning 13
low-impact development (LID) 1, 25, 53, 447

maintenance cost 53, 74, 86, 104, 110, 299, 304, 309, 311, 313, 315–22, 405, 407, 409–10, 426–7, 453
manholes 2, 4, 5, 9–10, 31–4; backdrop 27–8; brick 386, 388–90; deep 5, 391–2; mini- 31–3; pre-cast 25, 386–8
Manning's equation 43, 118, 157, 163, 164, 168, 170, 189–90, 195, 212, 214
Marston's equation 273, 275–8
Marston-Spangler Load Analysis Theory 274
Marston's theory 273–4, 297, 452
material service life 311–12
MATLAB 35, 220, 232
maximum drainage area 61, 66, 68
MicroDrainage 35, 237
micro-tunneling 20, 33, 38, 49, 232, 247–8, 369, 377, 424, 451
mild slope 32, 43
modified rational method 85, 119–20, 122, 150, 448
Mohr–Coulomb theory 250, 451
momentum equation 31, 40, 117–18, 153, 170, 174, 176, 178–9, 199, 205
momentum flux 173, 219, 231
momentum principle 34, 170, 172
monoclinical rising wave 206, 450

National Corrugated Steel Pipe Association (NCSPA) 294, 297
National Pollutant Discharge Elimination System (NPDES) 3, 24
native species 25, 61, 151

negative projection cases 269, 277–8, 282, 285, 452
negligence 267, 341, 378, 430–4
Newton's second law 170, 210
node 32–3, 42–3, 45, 85, 118, 154, 175, 182, 190, 196, 205, 213, 219, 220, 226–7
nonconventional design methods 129
nongovernmental organizations (NGOs) 16, 56, 304
nonwoven geotextile 242, 451
normal depth 153, 163, 181, 194, 199, 235, 449
normal distribution 115, 134–5, 147–48
normal flow 32, 44, 155, 163, 170, 181, 194, 219, 233–4

open channel flow 157, 159, 172, 176, 199, 204, 208, 223–4, 228–9, 449
open spaces 6, 7, 98
operation and maintenance 13, 19, 36–8, 40, 63, 76, 304–6, 318, 322, 330, 349–50, 352, 355, 363, 369, 378, 381–2, 398, 405
operational carbon 14, 37, 44, 112, 309, 323–5, 330, 332
operational expenditure (OPEX) 28, 57, 309, 453
outfall 1, 32–3, 126, 175, 181–3, 189, 192, 234–6, 338, 344
outlet 1–2, 4, 27, 31, 33, 44, 58, 66, 77, 81–2, 84–7, 96
outlet control 154, 219, 356, 450

partly full pipe 161
peak discharge 75, 82, 87, 95, 103, 120, 125, 129, 135, 136, 145
peak flow 10, 23–4, 27, 29, 37–8, 45, 63–4, 66, 82–3, 90
percolation 58, 150, 241, 247, 410, 451
perforated pipes 67, 242, 262, 268
permeability 18, 30, 67–8, 76, 79, 89, 93, 127, 239–40
personal protective equipment 367, 369
pipe jacking 247–48, 268, 376, 424, 451
pipe joint 270, 376, 413
pipe laying 338, 343
pipe laying and jointing 343
pipe strength 273, 290, 293, 341, 375
plane of equal settlement 279–80, 282–3, 285–6, 452
planter boxes 4, 8, **63**, 90–1, 240, 247, 302, 305, 410
PLAXIS 256
pollutants 2–3, 5–6, 14, 24, 26–7, 38, 53, 61, 65, 74–5
polyhedral pipes 424–5
porous pavers 6, 65, 92–3, 406
positive projection cases 278, 453
potential hotspots 61, 68

present worth 299, 312, 453
private finance initiatives (PFI) 56, 446
private sector financing initiatives 57
probability density function (PDF) 115, 130–1, 151
Prof. Yu Kongjian 420
project design life 36, 311–16, 323, 453
projection ratio p 280–7, 453
public–private partnership (PPP) 37, 57, 447

quality control 24, 54, 63, 66, 118, 234–6, 374, 375, 400
quantity of flow 26, 28–9, 33, 37–8, 58, 118, 231, 295

rainfall intensity 97, 120, 129, 136, 140, 150, 447
rainwater harvesting 4, 7, 26, 62–3, 94–7, 107, 144, 191
Rankine's equation 273, 453
rational method 85, 116, 118–23, 133, 136
reactive forces 171–2
recharge volume 78, 80, 447
rectangular channel 40, 153, 156, 161–3, 177, 198, 200, 221–2
regulatory framework 76, 306
rehabilitation cost 299, 311, 313, 315–17, 322, 426
reinforced concrete pipe (RCP) 30, 291, 293–4, 329, 331, 424–5
reliability analysis 306–8, 311, 322
replacement cost 299, 311, 313–15, 322, 426
residual value 36, 299, 311, 313, 315, 317, 323, 454
restoration programs 22–3, 30, 105, 110, 112, 115, 118, 134, 148, 299
retaining wall 21–2, 178, 191–3, 196–7
retention pond 8, 31–2, 63, 81, 82, 86
Riemann invariants 223, 226
risk 4, 7, 14–16, 20, 23–5, 32, 36
river Kiruruma 48–9, 145, 147–8
root intrusion 411, 414
routine maintenance 67, 325, 332, 364, 381, 403–5, 407, 414, 416
routine oversight 381–2, 398, 403–4, 410, 416
Rugumayo, A. 139, 152
runoff 1–10, 13–17, 19
runoff curve number (CN) 123, 151n1, 152
runoff hydraulic modeling and Simulation 234

sacred sites 30, 39, 62, 303, 378
safe net-zero goal 323, 454
safe systems of work 369, 431, 442
safety 20, 25, 33, 37, 39, 42, 48, 50, 54; in drainage construction 366
Saint Venant equations 117, 203–8, 210–11, 220–2, 234, 237, 450

seepage control 255, 256
segmental retaining walls 264, 265
sensitivity analysis 308
separate systems 1, 9, 11
service life 36, 295, 311, 312–16, 322–3, 388, 454
settlement-deflection ratio 280
settlement-deflection ratio for negative projection 286
settlement-deflection ratio r_{sd} 280
17 SDGs 53
shear strength 25, 193, 239, 249–52, 267–8, 270, 452
sheet or bored pile walls 265
site-specific SuDS 18, 55–6, 58, 324, 447
soakaways 8, 410
socioeconomic data 18
socket and spigot pipes 271, 294
software applications 35
Soil Conservation Service curve number (SCS-CN) 118, 122, 448
soil unit weight 453
space 61; constraints 28, 56, 61–2, 89, 110, 182, 306
specific discharge 153, 162, 178, 186–9, 194–5, 449
specific energy 42–3, 45, 153–4, 160, 162–3, 167, 173, 178
specific force 45, 173–4
specific SuDS maintenance practices 405
sponge city 296–7, 420–1, 456
stabilization trenches 244
standard pipework drainage solutions 4, 55, 303–4, 414, 419
standards 24, 30, 35–6, 38, 55, 63, 76
stationary control volume 171–2, 449
statistical models 130
steady flow: routing 40, 182, 212–13
steep slope 21, 27–8, 32, 43, 59, 67, 89, 154, 169, 340, 392, 410, 428, 429
storm sewer appurtenances 382
storm sewer pipes 5, 36, 271, 274, 288, 294, 308, 375
stormwater regulators 375, 397, 398, 456
structural design of buried pipes 36, 49, 278, 280, 289, 293, 296
subcritical flow 32, 40–2
subpavement drains 244–5
subsoil drains 244
sub-surface drainage 1–2, 21, 239, 240
SuDS hydrology 150–1
SuDS (sustainable drainage system) 1–7, 9–11, 13, 14
supercritical flow 32, 40–1, 43
surcharging 32, 45, 231
sustainability 1, 3, 4, 10, 13

sustainability goals 10, 18, 23, 33
sustainable development 1, 3, 4, 8
Swales 9, 27, 32

technical release 55 (TR-55) graphical method 122
topographical map 18–19, 46–7, 195, 434–5, 437, 439
topographical survey 18–20, 40, 45, 179–80, 195, 435, 446
total suspended solids (TSS) 26, 61, 105, 447
traditional public sector (TPS) funding 15, 37, 446
Transport Road Research Laboratory (TRRL) 117, 122, 125, 448
tree planting 5, 38, 47, 144, 301, 324
tree trench 5, 38, 56, 61, 63, 65, 75, 91–3, 107, 296, 301, 377, 405, 418
trenchless technology 247–8, 403, 424, 456
trench shoring 337, 454
tunneling 24–5, 33, 48, 247–9, 424

Uganda 20, 45, 128, 136, 137, 141, 145, 146
Uganda's Ministry of Works and transport drainage design manual 118
UK 3, 20, 30, 50, 56, 63, 114, 117–18
UK modified rational method 120
UN Brundtland report 51
underdrain perforated pipe 67, 76, 79, 447
underground drainage systems 247
underseepage 256
unit hydrograph 119, 123–6
unplasticized polyvinyl chloride (uPVC) 295; pipes 294, 300

unsteady flow 207, 219, 451
urban drainage 1, 11, 20, 32, 150, 152, 206, 235, 237, 419
urban drainage design 206
urbanization 4, 9, 14, 24, 47, 58, 119, 123, 130, 134, 152, 195
urbonas equation 70, 80
USA 3, 21, 24, 49, 56, 70, 74
U.S. Army Corps of engineers 118, 237, 312, 323
utility maps 19–20, 179, 353, 354, 371
utility service corridors 19

vegetated kerb extensions 7, 59, 101
velocity 22, 27–8, 32, 37–9; self-cleansing 154, 191
vertical forests 423–4, 456
volumetric runoff coefficient (R_v) 70–1, 80

water movement in basements 258
water quality volume (WQ_v) 65, 69–71, 77–8, 80, 86, 112–13
water table 17, 26, 38, 59–60, 66–9
waterproofing 7, 91, 102, 104, 258, 259, 262, 268
Watkins and Fiddes method 136–7
wave attenuation 129, 204, 450
wave celerity 40–1, 161, 203, 212, 214, 222–4, 226, 228, 451
wave propagation 237, 450
Weibull distribution 115–16, 133, 141–2, 307, 318, 320, 441
wetted perimeter 41, 153, 155, 209, 449

zero projection case 288